Increasing Productivity of Intensive Rice Systems Through Site-Specific Nutrient Management

Edited by A. Dobermann,
C. Witt, and D. Dawe

2004

IRRI

CIP data will be provided on request.

Copublished outside The Philippines:

SCIENCE PUBLISHERS, INC.
Post Office Box 699
Enfield, New Hampshire 03748
United States of America

Internet site: *http://www.scipub.net*

sales@scipub.net (marketing department)
editor@scipub.net (editorial department)
info@scipub.net (for all other enquiries)

ISBN 1-57808-266-8

© 2004 International Rice Research Institute

Suggested citation:
Dobermann A, Witt C, Dawe D, editors. 2004. Increasing productivity of intensive rice systems through site-specific nutrient management. Enfield, NH (USA) and Los Baños (Philippines): Science Publishers Inc., and International Rice Research Institute (IRRI). 410 p.

Layout and Design: Ariel Paelmo
Figures and Illustrations: Ariel Paelmo

Published by Science Publishers, Inc. Enfield, NH, USA
Printed in India

Contents

Foreword

Rice yield gains have slowed down in recent years, particularly in regions with early adoption of Green Revolution technologies. Although scientists are developing new germplasm to raise current yield ceilings, future yield increases are likely to occur in smaller increments than in the past. These yield increases will require more knowledge-intensive forms of soil and crop management that increase the efficiency of production inputs and, at the same time, do not harm the local and global environment. The integrated and efficient use of nutrients is one of the key issues for sustainable resource management in the world's most intensive rice systems.

This book summarizes research conducted from 1994 to 2001 to develop a new concept and the tools needed for site-specific nutrient management (SSNM) in irrigated rice systems and the tools needed for applying it in farmers' fields. As part of IRRI's Irrigated Rice Research Consortium (IRRC), the Reaching Toward Optimal Productivity (RTOP) project[1] has evolved into one of the largest and most important agronomic research projects in the world. More than 100 researchers and support staff in six countries collaborate in this unprecedented network of strategic on-farm and on-station research, representing disciplines such as soil science, agronomy, soil microbiology, pest management, and socioeconomics.

After reviewing the economics of rice production and productivity trends in Asia, most of the book presents the principles of a new SSNM concept and results of a first phase of field-testing conducted from 1997 to 2001 at numerous sites in Asia. This approach represents a "new school" of plant nutrient management in which much emphasis is given to quantifying crop nutrient needs and using the crop as an indicator of soil nutrient supply. Initial results were reviewed in September 1999, during a workshop at Thanjavur, Tamil Nadu, India. Since then, the theoretical development of new nutrient management concepts has continued, approaches have

---

[1]From 1994 to 2000, the project was called Reversing Trends of Declining Productivity in Intensive Irrigated Rice Systems (RTDP).

been simplified, and new tools such as a nutrient decision support system and a *Practical Guide for Nutrient Management* in rice have been developed.

This book demonstrates how long-term intellectual and financial support of different stakeholders for strategic, interdisciplinary on-farm research results in finding generic solutions for resource management with a high impact potential. Site-specific nutrient management has potential for improving yield and nutrient efficiency in irrigated rice to close existing yield gaps. The major challenge will be to retain the success of the approach while reducing the complexity of the technology as it is disseminated to farmers.

Ronald P. Cantrell
Director General
IRRI

Acknowledgments

Research in the RTDP project was funded by the Swiss Agency for Development and Cooperation (SDC), the International Fertilizer Association (IFA), the Potash & Phosphate Institute and the Potash & Phosphate Institute of Canada (PPI/PPIC), and the International Potash Institute (IPI). On behalf of all scientists involved in the RTDP project, we wish to thank all donors, the International Rice Research Institute (IRRI), and the national agricultural research and extension systems for supporting our work since 1994. Please see further acknowledgments at the end of individual book chapters. We are very grateful to Bill Hardy and staff of the Communication and Publications Services of IRRI for editorial assistance, design work, and printing.

List of abbreviations and acronyms

AEN	agronomic efficiency of fertilizer N (Δkg grain yield kg^{-1} fertilizer N)
ai	active ingredient (pesticides)
CLRRI	Cuu Long Rice Research Institute, Omon, Vietnam
DS	dry season
ER	early rice
FFP	farmers' fertilizer practice
GRF	gross return over fertilizer cost
HYS	high-yielding season
IEK	internal efficiency of K (kg plant K kg^{-1} grain yield)
IEN	internal efficiency of N (kg plant N kg^{-1} grain yield)
IEP	internal efficiency of P (kg plant P kg^{-1} grain yield)
IKS	indigenous K supply
INS	indigenous N supply
IPM	integrated pest management
IPS	indigenous P supply
NISF	National Institute for Soils and Fertilizer, Hanoi, Vietnam
LR	late rice
LYS	low-yielding season
PhilRice	Philippine Rice Research Institute, Philippines
PEN	physiological efficiency of N (Δkg grain yield Δkg^{-1} plant N)
PFPN	partial factor productivity of N (kg grain yield kg^{-1} fertilizer N)
PTRRC	Pathum Thani Rice Research Center, Pathum Thani, Thailand
REK	recovery efficiency of applied fertilizer K (kg plant K kg^{-1} fertilizer K)
REN	recovery efficiency of applied fertilizer N (kg plant N kg^{-1} fertilizer N)
REP	recovery efficiency of applied fertilizer P (kg plant P kg^{-1} fertilizer P)
RIR	Research Institute for Rice, Sukamandi, Indonesia
RTDP	Reversing trends of declining productivity (1997-2000)
RTOP	Reaching toward optimal productivity (2001-04)
SBRES	Suphan Buri Rice Experiment Station, Suphan Buri, Thailand

SSNM	site-specific nutrient management
SWMRI	Soil and Water Management Research Institute, Tamil Nadu, India
TNRRI	Tamil Nadu Rice Research Institute, Tamil Nadu, India
WS	wet season
Y_{max}	potential or maximum yield
ZU	Zhejiang University, China

Part 1

1| Introduction

A. Dobermann and C. Witt

Intensive irrigated rice systems evolved as a result of the Green Revolution in Asia and have become one of the most important food production systems in the world. The release of semidwarf, short-duration, high-yielding varieties such as IR8 (1966), IR20 (1969), IR36 (1976), IR64 (1985), and IR72 (1988) triggered investments in irrigation infrastructure and allowed farmers to grow two to three rice crops per year. Worldwide, about 79 million ha of rice are grown under irrigated conditions (55% of the global harvested area), accounting for about 75% (440 million t of rice per year) of the annual rice production. Irrigated double- and triple-crop mono-culture rice systems alone occupy a land area of 24 million ha in tropical and subtropical Asia (Huke and Huke 1997) or 49 million ha of rice harvested annually. These cropping systems account for more than 40% of the global rice supply and nearly 60% of all irrigated rice produced.

Under tropical conditions, stable yields of 3 to 5 t ha^{-1} in the wet season (WS) and 5 to 7 t ha^{-1} in the dry season (DS) are common in farmers' fields, but yields average only about 60% of the yield potential of the present generation of modern rice varieties. From 1967 to 1984, rice production in Asia grew 3.2% annually, mainly because of yield increases (2.5% y^{-1}). However, growth rates declined to 1.5% y^{-1} (production) and 1.2% y^{-1} (yield) from 1984 to 1996 (Dawe and Dobermann 1999). This slowdown was partly due to lower rice prices and the slowdown in demand growth because of secular trends in population and per capita consumption of rice (Dawe 1998), but concern was raised about resource degradation, yield gaps, and declining yield growth in rice-rice and rice-wheat systems of Asia (FAO 1994a,b, 1997, 2001). In line with this, concern was raised about a long-term yield decline in unfertilized plots as well as in treatments with the "best recommended" fertilizer rates in rice-rice systems at Philippine experiment stations (Cassman and Pingali 1995, Flinn et al 1982, Flinn and De Datta 1984, Ponnamperuma 1979) and in some rice-wheat long-term experiments (LTE) in South Asia (Nambiar 1994). More recent analysis of yield trends suggests, however, that yield declines in long-term experiments appear to be not widespread at current production levels (Dawe et al 2000), although they may be more common in rice-wheat systems (Duxbury 2001). Where yield declines occur, they often result from inadequate management. Both climatic factors and adjustment of crop management can contribute to reversing a yield decline (Dobermann et al 2000).

Farm surveys conducted in the 1980s and '90s provided some evidence for stagnating or declining productivity, soil fertility, and resource-use efficiency in intensive rice and wheat areas of the Philippines, China, India, Bangladesh, Nepal, and Pakistan (Ali 1996, Byerlee 1992, Byerlee and Siddiq 1994, Cassman and Pingali 1995, Huang and Rozelle 1995). Yields have stagnated since the mid-1980s in some large rice production domains where farmers were early adopters of modern irrigated rice production technologies (Cassman and Dobermann 2001). It was generally thought that declining soil fertility was a major cause of declining productivity and rice farmers in Asia often claimed that they needed to apply more N to obtain yields similar to those of 10 or 20 years ago.

In 1992, the 4th External Program and Management Review of IRRI concluded that (1) there were unexplained declines in yields of rice, especially at the highest levels of intensive cultivation, (2) this was a critical problem for the future, and (3) insufficient information to reach any firm conclusions existed (CGIAR 1992). It was recommended that "...*IRRI lead a major research effort, enlisting the best talents available in the world, to seek solutions for this complex of problems—a task that may take a decade or longer to complete.*" In 1996, a FAO expert consultation recommended "...*an urgent need to develop an FAO/IRRI/NARS joint program to identify the causes of and arrest the downward yield trends. .... UNDP and other donors are urged to provide the necessary funding in support of this priority program*" (FAO 1997).

However, the basis for drawing solid conclusions about the extent and possible causes of a yield or productivity decline was weak. Conflicting views evolved, which mainly resulted from confusion about the proper use of terminology ("yield decline," "productivity decline," "yield stagnation," "deceleration of yield growth," ...) and the paucity of data to test the hypotheses proposed (Dawe and Dobermann 1999). It was unclear how representative the observations from a few long-term experiments and farm surveys were for the irrigated rice ecosystem in Asia. Was soil quality decreasing under intensive rice cropping? Were yield or productivity declines caused by generic processes or mainly by local factors such as the wrong soil and crop management? In many long-term experiments, the lack of thoroughly measured soil and plant characteristics and the lack of archives for storing soil and plant samples have made it very difficult to study changes in soil properties and crop response over time. Most farm surveys conducted were based on secondary data (district/provincial statistics) or purely socioeconomic surveys (farmer interviews) and lacked thoroughly measured time series of socioeconomic and biophysical data. Methodological problems often caused uncertainties about the relevance of productivity measures such as total factor productivity (TFP).

Recent estimates are that rice yields in Asia must increase by about 14% from 2000 to 2010 and by 25% from 2000 to 2020 (Table 1.1). Assuming that there is a decline in rice area as shown in Table 1.1 and that rice yields grow at the same rate in irrigated and rainfed systems, average yields of irrigated rice must rise from about

Table 1.1. Projected changes in the global harvested area, yield, and production of rice.

Actual[a]	1991	1993	2000				
Area (10^6 ha)	147.5	145.4	153.8				
Yield (t ha^{-1})	3.5	3.6	3.9				
Production (10^6 t)	515	530	599				

			2000-10		2000-20	
Projected[b]	2010	2020	%	% y^{-1}	%	% y^{-1}
Area (10^6 ha)	151.8	150.0	−1.3	−0.13	−2.5	−0.12
Yield (t ha^{-1})	4.4	4.9	13.9	1.31	24.9	1.12
Demand (10^6 t)	673	729	12.4	1.18	21.8	0.99

Yield of irrigated rice (t ha^{-1})[c]	1991	2000	2010	2020
	4.9	5.4	6.1	6.7

[a]Actual production (FAO statistics, paddy). [b]Modified IMPACT model projections (paddy, assuming 1 t paddy = 0.67 t milled rice). [c]Actual (1991) and estimated average yield of irrigated rice assuming yield growth rates shown in the table above (for both irrigated and nonirrigated rice). Source: D. Dawe, IRRI, October 2001.

5.3 t ha^{-1} in 1998 to 6.5 to 7.0 t ha^{-1} in 2020 to keep inflation-adjusted prices approximately constant. For tropical areas, this represents average yields of about 7.7 t ha^{-1} in the DS and 5.7 t ha^{-1} in the WS. About 30% of all farmers must achieve yields of >8 t ha^{-1} and 15% >9 t ha^{-1} in at least one crop per year (Dobermann 2000). This goal can be achieved if (1) the yield potential of tropical lowland rice will be raised to 12 t ha^{-1} in the DS and 8 to 9 t ha^{-1} in the WS and (2) improved crop management technologies will be implemented to increase average farm yields to about 70% of the increasing yield potential. Improving nutrient and pest management through more knowledge-intensive, dynamic, and site-specific technologies will have to play a major role to achieve the increases in rice yields required.

Therefore, the project on Reversing Trends of Declining Productivity in Intensive Irrigated Rice Systems (RTDP) began in 1994 in key irrigated rice domains in Asia to gather more knowledge about the extent of a possible productivity decline at the farm level and to identify mitigation options to secure future increases in rice production. The specific project objectives were to

1. Conduct long-term biophysical and socioeconomic farm monitoring to establish trends in total factor productivity (TFP), partial factor productivities (PFP), and inherent soil nutrient-supplying capacity.
2. Develop and validate site-specific nutrient management (SSNM) technology for intensive rice systems at eight sites in six countries of South and Southeast Asia.

3. Improve understanding of biotic and abiotic processes governing soil nutrient-supplying capacity.
4. Integrate SSNM with integrated pest management (IPM) to establish site-specific crop management (SSCM) practices.

The RTDP project monitored socioeconomic and biophysical indicators of system performance in long-term experiments and 205 farmers' fields in irrigated rice domains and produced practical strategies and tools to manage, preserve, and improve the irrigated system's resource base. In phase I (1994-96) of the RTDP project, farm monitoring research began at five sites with tropical climate and a rice-rice cropping system (Nueva Ecija, Philippines; Tamil Nadu, India; Cantho Province, Vietnam; Suphan Buri, Thailand; West Java, Indonesia) to identify the major constraints to productivity. The initial focus was on biophysical and socioeconomic farm monitoring, quantification of nitrogen (N)-use efficiency, understanding of soil organic matter chemistry, and monitoring of the indigenous nutrient supply as a measure of soil quality. In phase II of the project (1997-2000), the work expanded to three new sites with subtropical climate (Zhejiang Province, China; Red River Delta, Vietnam; Uttar Pradesh, India) and broadened to include studies on soil microbial characteristics and nutrients such as phosphorus (P) and potassium (K). The results of the monitoring work led to the development of a new approach for SSNM, which was tested in 205 farmers' fields. Long-term experiments were established at all project sites and mainly used for strategic research related to objectives 3 and 4. Of particular importance were studies on soil organic matter and soil microbial biomass and its relationship with nutrient cycling (Olk et al 1996, 1998, 1999, Olk and Senesi 2000, Reichardt et al 1996, 2000, Witt et al 1998), and research on potassium and phosphorus (Dobermann et al 1996a,b,c).

The papers in this book mainly present results related to objectives 1 and 2, with a focus on evaluating the results of on-farm trials on SSNM conducted in Asia from 1997 to 2000. Chapter 2 provides a summary of the methodology for socioeconomic and agronomic on-farm research used in the project. Results of the long-term productivity monitoring (1) are presented in Chapters 3 and 4, but recent studies on yield trends in long-term experiments were published elsewhere (Dawe et al 2000, Dobermann et al 2000). Chapter 5 provides a detailed description of the general scientific approach for SSNM as it evolved over time and was implemented in the detailed on-farm studies. Chapters 6 through 12 summarize results obtained in SSNM studies in rice domains of India, Thailand, the Philippines, Indonesia, Vietnam, and China. Chapter 13 reports on recent work on improving nutrient management in West Africa, whereas Chapter 14 shows a diagnostic modeling concept for understanding yield formation in irrigated rice and as it is possibly affected by SSNM. In Chapters 15 and 16, both the agronomic and economic performances of SSNM were summarized across all on-farm sites in Asia, allowing general conclusions about the potential impact of this new technology and how it must be refined for practical use. The last two chapters provide details on the simplification of the SSNM technology and a summary of policy recommendations, research needs, and concepts for extending site-specific nutrient management. Many of the concepts and tools developed are being used to fine-tune, simplify, and expand the use of SSNM across Asia in the third phase of the project, which is now continuing under its new name—Reaching Toward Optimal Productivity (RTOP)—in 2001-04.

References

Ali M. 1996. Quantifying the socioeconomic determinants of sustainable crop production: an application to wheat cultivation in the Tarai of Nepal. Agric. Econ. 14:45-60.

Byerlee D. 1992. Technical change, productivity and sustainability in irrigated cropping systems of South Asia: emerging issues in the post-green revolution era. J. Int. Dev. 4:477-496.

Byerlee D, Siddiq A. 1994. Has the Green Revolution been sustained? The quantitative impact of the seed-fertilizer revolution in Pakistan revisited. World Dev. 22:1345-1361.

Cassman KG, Dobermann A. 2001. Evolving rice production systems to meet global demand. In: Rice research and production in the 21st century: Proceedings of a symposium honoring Robert F. Chandler, Jr. Los Baños (Philippines): International Rice Research Institute. p 79-100.

Cassman KG, Pingali PL. 1995. Extrapolating trends from long-term experiments to farmers' fields: the case of irrigated rice systems in Asia. In: Barnett V, Payne R, Steiner R, editors. Agricultural sustainability in economic, environmental, and statistical terms. London: J. Wiley & Sons. p 64-84.

CGIAR. 1992. Report of the fourth external program and management review of the International Rice Research Institute (IRRI). Rome: TAC Secretariat, FAO.

Dawe D. 1998. Reenergizing the green revolution in rice. Am. J. Agric. Econ. 80:948-953.

Dawe D, Dobermann A. 1999. Defining productivity and yield. IRRI Discussion Paper No. 33. Makati City (Philippines): International Rice Research Institute. 13 p.

Dawe D, Dobermann A, Moya P, Abdulrachman S, Bijay Singh, Lal P, Li SY, Lin B, Panaullah G, Sariam O, Singh Y, Swarup A, Tan PS, Zhen QX. 2000. How widespread are yield declines in long-term rice experiments in Asia? Field Crops Res. 66:175-193.

Dobermann A. 2000. Future intensification of irrigated rice systems. In: Sheehy JE, Mitchell PL, Hardy B, editors. Redesigning rice photosynthesis to increase yield. Makati City (Philippines), Amsterdam: International Rice Research Institute, Elsevier Science. p 229-247.

Dobermann A, Cassman KG, Sta. Cruz PC, Adviento MAA, Pampolino MF. 1996a. Fertilizer inputs, nutrient balance, and soil nutrient-supplying power in intensive, irrigated rice systems. II. Effective soil K-supplying capacity. Nutr. Cycl. Agroecosyst. 46:11-21.

Dobermann A, Cassman KG, Sta. Cruz PC, Adviento MAA, Pampolino MF. 1996b. Fertilizer inputs, nutrient balance, and soil nutrient-supplying power in intensive, irrigated rice systems. III. Phosphorus. Nutr. Cycl. Agroecosyst. 46:111-125.

Dobermann A, Dawe D, Roetter RP, Cassman KG. 2000. Reversal of rice yield decline in a long-term continuous cropping experiment. Agron. J. 92:633-643.

Dobermann A, Sta. Cruz PC, Cassman KG. 1996c. Fertilizer inputs, nutrient balance, and soil nutrient-supplying power in intensive, irrigated rice systems. I. Potassium uptake and K balance. Nutr. Cycl. Agroecosyst. 46:1-10.

Duxbury JM. 2001. Long-term yield trends in the rice-wheat cropping system: results from experiments in Northwest India. J. Crop Production 3:27-52.

FAO. 1994a. Land degradation in South Asia: its severity, causes and effects upon people. World Soil Resources Report 78. Rome: FAO.

FAO. 1994b. Sustainability of rice-wheat production systems in Asia. In: Paroda RS, editor. Proceedings of an expert consultation on the sustainability of the rice-wheat production systems on different agro-ecological settings in Asia, 1993. Bangkok: FAO Regional Office for Asia and the Pacific. p 1-209.

FAO. 1997. Trends of yield and productivity of modern rice in irrigated rice systems in Asia. Int. Rice Comm. Newsl. 46:19-25.

FAO. 2001. Yield gap and productivity decline in rice production. Proceedings of the expert consultation held in Rome, 5-7 September 2000. Rome: FAO. 470 p.

Flinn JC, De Datta SK. 1984. Trends in irrigated-rice yields under intensive cropping at Philippine research stations. Field Crops Res. 9:1-15.

Flinn JC, De Datta SK, Labadan E. 1982. An analysis of long-term rice yields in a wetland soil. Field Crops Res. 5:201-216.

Huang JK, Rozelle S. 1995. Environmental stress and grain yields in China. Am. J. Agric. Econ. 77:853-864.

Huke RE, Huke EH. 1997. Rice area by type of culture: South, Southeast, and East Asia. A revised and updated database. Manila (Philippines): International Rice Research Institute. 59 p.

Nambiar KKM. 1994. Soil fertility and crop productivity under long-term fertilizer use in India. New Delhi: ICAR.

Olk DC, Brunetti G, Senesi N. 1999. Organic matter in double-cropped lowland rice soils: chemical and spectroscopic properties. Soil Sci. 164:633-649.

Olk DC, Cassman KG, Mahieu N, Randall EW. 1998. Conserved chemical properties of young humic acid fractions in tropical lowland soil under intensive irrigated rice cropping. Eur. J. Soil Sci. 49:337-349.

Olk DC, Cassman KG, Randall EW, Kinchesh P, Sanger LJ, Anderson JM. 1996. Changes in chemical properties of soil organic matter with intensified rice cropping in tropical lowland soils. Eur. J. Soil Sci. 47:293-303.

Olk DC, Senesi N. 2000. Properties of chemically extracted soil organic matter in intensively cropped lowland rice soils. In: Kirk GJD, Olk DC, editors. Carbon and nitrogen dynamics in flooded soils. Makati City (Philippines): International Rice Research Institute. p 65-87.

Ponnamperuma FN. 1979. Soil problems in the IRRI farm. IRRI Thursday Seminar, 8 November 1979. Los Baños (Philippines): International Rice Research Institute.

Reichardt W, Inubushi K, Tiedje J. 2000. Microbial processes in C and N dynamics. In: Kirk GJD, Olk DC, editors. Carbon and nitrogen dynamics in flooded soils. Makati City (Philippines): International Rice Research Institute. p 101-146.

Reichardt W, Mascarina GB, Padre B, Doll J. 1996. Microbial communities of continuously cropped, irrigated rice fields. Appl. Environ. Microbiol. 63:233-238.

Witt C, Cassman KG, Ottow JCG, Biker U. 1998. Soil microbial biomass and nitrogen supply in an irrigated lowland rice soil as affected by crop rotation and residue management. Biol. Fertil. Soils 28:71-80.

Notes

Authors' addresses: A. Dobermann, International Rice Research Institute (IRRI), Los Baños, Philippines, and Department of Agronomy and Horticulture, University of Nebraska, P.O. Box 830915, Lincoln, NE 68583-0915, USA. E-mail: adobermann2@ unl.edu; C. Witt, IRRI.

Citation: Dobermann A, Witt C, Dawe D, editors. 2004. Increasing productivity of intensive rice systems through site-specific nutrient management. Enfield, N.H. (USA) and Los Baños (Philippines): Science Publishers, Inc., and International Rice Research Institute (IRRI). 410 p.

2 | Methodology for socioeconomic and agronomic on-farm research in the RTDP project

A. Dobermann, G.C. Simbahan, P.F. Moya, M.A.A. Adviento, M. Tiongco, C. Witt, and D. Dawe

2.1 Specific hypotheses and general research approach

Three major hypotheses are being tested in long-term on-farm trials at nine domains in Asia:

1. *Total factor productivity (TFP) is declining on intensive irrigated rice farms in Asia.* Measurements are made on a wide range of rice farms of all biophysical and socioeconomic variables required to estimate TFP by (1) production functions and (2) the sum of partial factor productivities of all production inputs. Socioeconomic data collected are also being used to analyze the economics of rice production in all domains sampled, including variation among domains and among farms within each domain. These measurements started in 1994 (119 farms in five domains) or 1997 (86 farms in four domains) and continue for an indefinite period to obtain data on medium- and long-term trends of productivity.

2. *Rice yields per unit fertilizer N addition are declining as a result of a decline in the supply of N from indigenous (nonfertilizer) sources.* In farmers' fields, plant N accumulation in small plots receiving no N fertilizer (0-N plots) is used as an index of the indigenous N supply (INS) from all sources other than fertilizer over a growing season. These 0-N plots are embedded within a field that is otherwise under the farmers' management. The 0-N plots are moved to a different location within the field in each crop grown to avoid residual effects. Measurements in 0-N plots and sampling plots of the farmers' fertilizer practice (FFP) include yields and yield components, plant N accumulation, and various soil properties. These measurements are used to (1) estimate changes in the use efficiency of applied fertilizer N under on-farm conditions under which most production factors are not held constant, (2) relate changes in the productivity measured in FFP plots to changes in the INS, and (3) compare changes in the INS and N-use efficiency in farmers' fields with those measured in long-term trials. The design allows quantification of the changes in N response functions across farms, but not within the same farm over time. These measurements started in 1994 (119 farms in

11

five domains) or 1997 (86 farms in four domains) and continue for an indefinite period to obtain data on medium- and long-term trends of INS and N-use efficiency.

3. *Rice yields, profit, plant NPK uptake, and the efficiencies with which N, P, and K fertilizer are used can be increased by applying NPK fertilizers on a field- and cropping-season–specific basis.* Nutrient omission plots are embedded within farmers' fields (0-N, 0-P, and 0-K plots) to estimate the indigenous supply. The indigenous N supply (INS) is estimated as total plant N accumulation at maturity in the 0-N plot; the indigenous P supply (IPS) is estimated as total plant P accumulation at maturity in the 0-P plot (+NK plot); the indigenous K supply (IKS) is estimated as total plant K accumulation at maturity in the 0-K plot (+NP plot). Field-specific values of INS, IPS, and IKS are used to work out field- and season-specific fertilizer recommendations for each farm (see Chapter 5). This recommendation is tested in a large site-specific nutrient management (SSNM) plot. In this plot, the same data are collected as in the FFP field to compare yields, profit, plant NPK uptake, N-use efficiencies, NPK input-output balances, and effects on soil fertility over time. Measurements of INS, IPS, and IKS were conducted on all 205 farms at nine sites for four consecutive rice crops grown during 1997 to 1998. The SSNM plot was established in 1997 (119 farms at five domains) or 1998 (86 farms at four domains) and continues for an indefinite period to obtain data on medium- and long-term performance. It remains at the same location.

2.2 Experimental approach

At each project site, on-farm and on-station long-term trials were conducted using standard protocols for design and data collection. A general description of the procedures relevant for this book is given below. More detailed information and results not described in this book can be found elsewhere (Cassman et al 1994, Olk et al 1999, Olk and Moya 1998, RTDP 1997, 1998, 1999, Simbahan 1999). Procedures deviating from those described will be explained in the respective chapters for each experimental site.

Domains and cropping systems
The RTDP project includes nine sampling domains in six countries. Six domains (Maligaya, Suphan Buri, Omon, Sukamandi, Aduthurai, and Thanjavur) represent rice monoculture systems of the humid or subhumid tropics. Three domains (Hanoi, Jinhua, Pantnagar) are located in subtropical regions and include rice-upland crop systems (Table 2.1). At each site, the principal study area is a sampling domain located within a very large ($\geq$100,000 ha) irrigated rice production area. On-farm monitoring in each domain is conducted on 18 to 26 farms, usually located within a radius of 15–20 km around a research station and clustered into several villages. The

Table 2.1. Experimental locations, crop calendar, and institutions participating in the project on "Reversing Trends of Declining Productivity in Intensive Irrigated Rice Systems."

Domain	Region, country	Project start	No. of farms	Cropping system	Crop calendar
Maligaya **PhilRice**	Central Luzon Philippines	1994	26	Rice-rice	Dry-season rice / Wet-season rice
Suphan Buri **PTRRC**	Central Plain Thailand	1994	24	Rice-rice	Dry-season rice / Wet-season rice
Omon **CLRRI**	Mekong Delta Vietnam	1994	24	Rice-rice-rice	Dry-season rice / Wet-season rice / Wet-season rice
Sukamandi **RIR**	West Java Indonesia	1994	20	Rice-rice	-season rice / Dry-season rice / Wet
Aduthurai **TNRRI**	Tamil Nadu India	1994	25	Rice-rice	Pulses / Kuruvai-rice / Thaladi-rice
Thanjavur **SWMRI**	Tamil Nadu India	1997	18	Rice-rice	Pulses / Kuruvai-rice / Thaladi-rice
Hanoi **NISF**	Red River Delta Vietnam	1997	24	Rice-rice-maize	Spring-rice / Summer-rice / Maize
Jinhua **ZU**	Zhejiang China	1997	21	Rice-rice (hybrid)	Winter crop / Early-rice / Late-rice
Pantnagar **GBPUAT**	Uttar Pradesh India	1997	23	Rice-wheat	Rabi wheat / Kharif rice

Month: Jan Feb Mar Apr May Jun Jul Aug Sep Oct Nov Dec

sample farmers at each site were not chosen at random. The major criteria for selecting farmers were
- represent the most common soil types in the region,
- represent the most typical cropping systems and farm management practices in the region,
- represent a range of socioeconomic conditions (small to large farms, poor to rich farmers),
- reasonable accessibility to allow frequent field visits, and
- farmer interest in participating in the project over a longer term.

In some domains, the number of sample farms was larger (usually 25–30 farms) during the first years of the project. For various reasons, work at some farms was discontinued, mainly to reduce the workload involved, but also in some cases because of poor cooperation by the farmer. All data presented in this book refer to a total of 205 farms (Table 2.1) at which research continues. Note that this number differs from that in earlier project publications.

Experimental design
From 1997 to 1999, five treatments were used in all on-farm trials (Fig. 2.1):
- **–F plots—no fertilizer applied; 0 N—no N fertilizer applied**. The –F plots served three purposes: (1) as a general index of the soil nutrient-supplying capacity and its changes with time (measured since 1994), (2) as a reference plot for estimating N-use efficiencies (see below), and (3) to estimate the INS as a requirement for working out a field-specific fertilizer recommendation. The –F treatments were implemented at TNRRI, SWMRI, PhilRice, RIR, PTRRC, and CLRRI. The 0-N plots received 30 kg P fertilizer ha^{-1} and 50 kg fertilizer K ha^{-1} (+PK plots), but no N to ensure that nutrients other than N did not limit plant N uptake from indigenous sources. The 0-N plots were implemented at ZU and NISF in 1997 to 1999, and used only for objectives 2 and 3.[1]
- **0 P—no P fertilizer applied**. At all domains, the 0-P plots received N and K fertilizer at high rates (+NK plots) to ensure that nutrients other than P did not limit plant P uptake from indigenous sources. Depending on the site and climatic season, the N rates varied from 120 to 180 kg N ha^{-1} and K rates varied from 100 to 150 kg K ha^{-1}. The 0-P plots were used to estimate the IPS as a requirement for working out a field-specific fertilizer recommendation. This treatment was discontinued after four successive crops had been grown from 1997 to 1998.

---

[1]From 1994 to 1996, the on-farm trials at TNRRI, RIR, PTRRC, CLRRI, and PhilRice included –F and +PK plots. Grain yield and INS measured in +PK plots tended to be slightly larger than in –F plots, but the differences were often not statistically significant. It was concluded that using either –F or +PK plots is appropriate for estimating INS and N-use efficiencies, at least in environments with no severe P or K deficiency.

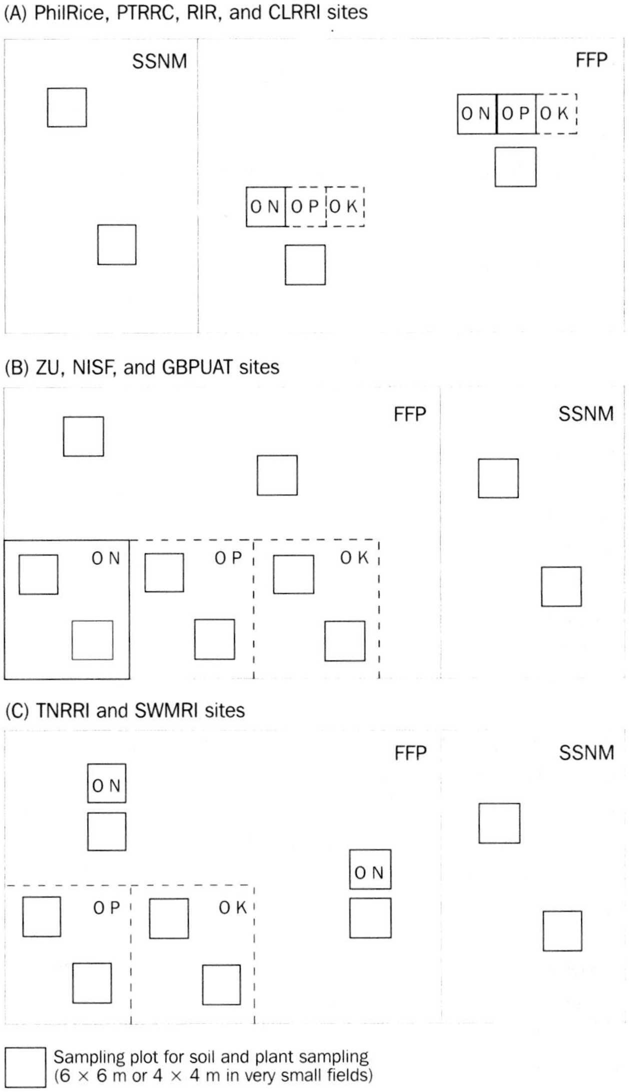
(A) PhilRice, PTRRC, RIR, and CLRRI sites
SSNM
FFP
O N
O P
O K
O N
O P
O K
(B) ZU, NISF, and GBPUAT sites
FFP
SSNM
O N
O P
O K
(C) TNRRI and SWMRI sites
FFP
SSNM
O N
O N
O P
O K
Sampling plot for soil and plant sampling
(6 × 6 m or 4 × 4 m in very small fields)

Fig. 2.1. Experimental design used for comparing the site-specific nutrient management (SSNM) with the farmers' fertilizer practice (FFP).

- **0 K—no K fertilizer applied**. At all domains, the 0-K plots received N and P fertilizer at high rates (+NP plots) to ensure that nutrients other than K did not limit plant K uptake from indigenous sources. Depending on the site and climatic season, the N rates varied from 120 to 180 kg N ha^{-1} and P rates varied from 25 to 40 kg P ha^{-1}. The 0-K plots were used to estimate the IKS as a requirement for working out a field-specific fertilizer recommendation. This treatment was discontinued after four successive crops had been grown from 1997 to 1998.
- **SSNM—site-specific nutrient management**. A single large plot (200 to 1,000 m^2) in which nutrient applications were prescribed on a field- and crop-specific basis. Chapter 5 gives a detailed description of the SSNM approach used and the management of this plot. This treatment started in 1997 (CLRRI, TNRRI, PhilRice, PTRRC, RIR) or 1998 (NISF, ZU, SMWRI) and continues.
- **FFP—farmers' fertilizer practice**. A single large plot (the remaining farmer's field) in which all fertilizer management is done by the farmer, with no interference by the researcher.

Where applicable, blanket doses of other nutrients were applied to all treatments to prevent deficiencies other than N, P, or K. This included application of Zn at TNRRI and SMWRI. Varieties grown were usually the same in all treatments and chosen by the farmer. All other crop management operations (land preparation, crop establishment, irrigation, weed control, insect and disease control) were done by the farmer and were the same in all treatments. In some cases, researchers had to take special measures such as preventive pest control or providing high-quality seed to the farmer. This will be described in Chapters 6 to 12.

Three different experimental designs were used, depending on the local preferences (Fig. 2.1). Example A in Figure 2.1 represents a strip-plot design with omission plots embedded into the FFP field, whereas, in example B, the nutrient omission treatments occupied only one part of the field, which allowed easier plot management. Each treatment contained two replicate sampling plots per farm.[2] The 0-N, 0-P, and 0-K treatments were separated from the surrounding field by bunds and were moved to a different location after each crop grown to avoid residual effects caused by nutrient depletion, whereas the SSNM plot remained at the same location. Sampling plots within the SSNM and FFP also rotated from crop to crop.

---

[2]From 1994 to 1996, the on-farm trials at TNRRI, RIR, PTRRC, CLRRI, and PhilRice had three replicate sampling plots per treatment. However, statistical analysis showed that two sampling plots were sufficient to estimate the field mean.

Socioeconomic data collection

All socioeconomic data were collected at the whole-farm level, that is, including the field used for the agronomic research (Fig. 2.1) as well as other fields belonging to the same farmer. Information collected for each farm included

- General farm data such as demographic and other general farm characteristics (collected once),
- Specific farm data for each crop cycle, including income from different sources, credit information, rice area planted, grain yield (farmers' estimate, not adjusted to a standard moisture content), and labor and material inputs for all crop management operations (land preparation, crop establishment, fertilizer/manure application, pest control, irrigation, harvest, postharvest operations).

Socioeconomic data were collected on at least two trips to each farmer per crop cycle. A supplementary interview about farmers' fertilizer and crop residue management was conducted in 1997 (RTDP 1998). Secondary data such as prices or production and input-use statistics were collected at the appropriate administrative level, for example, district or provincial level.

Agronomic data collection

All soil and agronomic data were collected at the single field/single treatment level, that is, only for the field used for the agronomic research. Two 6 m × 6-m plots were sampled for each treatment and the samples were processed separately (Fig. 2.1). In domains with very small fields and treatment parcels (NISF, ZU), the size of the sampling plots was reduced to 4 m × 4 m. Figure 2.2 shows the principal sampling design within a single sampling plot. Note that not all of the data collected will be presented in this book. Soil and plant measurements included

- Soil sampling to determine general soil properties (0–15-cm depth). At NISF and ZU, this was done as initial soil sampling before the start of the on-farm trials by collecting 10 to 15 soil cores from the entire field. In all other domains, soil samples collected from 0-N plots in 1995 or 1996 (three replicate 0-N plots with four soil cores per plot) were used for this analysis. Analytical methods followed standard guidelines (van Reeuwijk 1992).
- Soil sampling to determine available soil nutrients at the tillering stage. Samples were collected from 20 to 30 days after transplanting or sowing (DAT or DAS). In each sampling plot, three or four soil cores from 0–15-cm depth were collected using a standard sampling tube (Fig. 2.2). The soil from these cores was mixed into one composite sample per sampling plot. Two standard soil samples were included in each batch of samples analyzed. Standard determinations on dried soil involved
  - samples from 0-N plots: total Kjeldahl-N (Bremner 1996), total organic C (Walkley 1947), hot and cold 2 M KCL-extractable NH_4-N (Gianello and Bremner 1986),

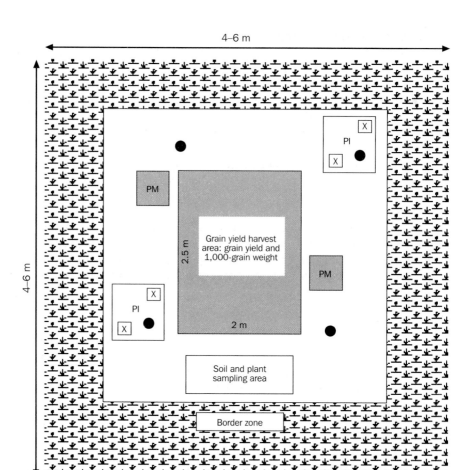

4–6 m

4–6 m

2.5 m

2 m

X

PI

X

PM

Grain yield harvest area: grain yield and 1,000-grain weight

PM

PI

X

X

Soil and plant sampling area

Border zone

● Soil cores collected 20–30 DAT (DAS) (only 0-N, 0-P, 0-K plots)
X Resin capsules for *in situ* nutrient extraction (only 0-N, 0-P, 0-K plots)
PI 0.5-m^2 quadrat for plant sampling at panicle initiation (occasionally)
PM Plant sample collected at physiological maturity (0.25-m^2 quadrat or 12 hills) to determine
— grain and straw yield (harvest index)
— plant nutrient concentrations in grain and straw
— yield components

Fig. 2.2. Principal design of soil and plant sample collection in the sampling plots.

— samples from 0-P plots: 0.5 M $NaHCO_3$-extractable Olsen-P (Olsen et al 1954) and 0.03 M NH_4F + 0.025 M HCl-extractable Bray-1 P (Bray and Kurtz 1945), and

— samples from 0-K plots: 1 N NH_4-acetate-extractable K.

● Plant sampling to determine total dry matter and nutrient accumulation at the panicle initiation stage (Fig. 2.2). Plants were cut from two 0.50-m^2 quadrats in the sampling zone surrounding the harvest area, combined into one composite sample per sampling plot, dried at 70 °C, and ground. Analysis of N, P, and K content followed standard procedures (Walinga et al 1995).

● Plant sampling to determine grain yield, straw yield, total nutrient uptake, and yield components at maturity. Grain yields were obtained from a central 5-m^2 harvest area at harvestable maturity (Fig. 2.2). The 1,000-grain weight was determined using a subsample of the oven-dry grain yield from the 5-m^2 harvest area. A 12-hill plant sample (or two 0.25-m^2-quadrat samples in direct-seeded rice) was collected at physiological maturity when 90% to 95% of all grains had lost their green color. This sample was taken before harvestable maturity to determine the grain to straw ratio, yield components, and nutrient concentrations in plant tissue because it avoids leaf loss from wind, rain, and decomposition in this humid tropical climate. Grain and straw subsamples from the 12-hill sample were dried to constant weight at 70 °C. Straw yields were estimated from the oven-dry grain yield of the 5-m^2 harvest area and the grain to straw ratio of the 12-hill sample. All yields (t ha^{-1}) are reported at a standard moisture content of 0.14 g H_2O g^{-1} fresh weight. Nutrient concentrations in grain and straw were measured by digestion with H_2SO_4-salicylic acid-H_2O_2, distillation, and titration (N), colorimetry (P), or flame emission spectroscopy (K) (Walinga et al 1995). Two standard plant samples were included in each batch of samples analyzed.

● Scores of crop management quality for each SSNM plot. Five categories— (a) land preparation, (b) water supply and management, (c) occurrence of weeds, rats, or snails, (d) occurrence of insect pests or diseases, and (e) other problems (lodging, seed quality, typhoons)—were scored on a scale of 0 (very good, no problem), 1 (moderate, some problems), and 2 (poor, severe problems).

Other measurements not reported here included (1) chlorophyll meter (SPAD) readings of the uppermost fully expanded leaf at 7- to 10-d intervals in the SSNM and FFP treatments, (2) chemical analysis of nutrient concentrations in irrigation water, (3) *in situ* adsorption of nutrients on ion exchange resin capsules placed into 0-N, 0-P, and 0-K plots, and (4) the amount of crop residues left in the field after harvest (RTDP 1997). From 1998 to 1999, a detailed assessment of pest incidence and injury levels (rodents, snails, insects, diseases, weeds) was made in the SSNM and FFP plots following a standard procedure in all domains and for three successive rice crops grown (IRRC 1998, Savary et al 1996).

All the researchers used uniform templates for field and laboratory data collection, data management, and statistical analysis. After various iterations of quality

Table 2.2. Hypothetical example showing the effect of site-specific nutrient management on agronomic characteristics over a period of four rice crops grown.

Parameter	Levels[a]	Treatment[b] SSNM	Treatment[b] FFP	Δ[c]	P>|T|[c]	Effects[d]	P>|F|[d]
Grain yield (GY)	All	6.45	5.96	0.49	0.003	Village	0.580
(t ha^{-1})	Year 1	6.19	5.97	0.22	0.424	Year	0.000
	Year 2	6.72	5.94	0.78	0.000	Season	0.813
	DS	6.97	6.46	0.51	0.035	Year × season	0.879
	WS	5.95	5.47	0.48	0.012	Village × crop	0.468

[a]All = all four crops grown from 1997 DS to 1999 WS; Year 1 = 1997 DS and 1998 WS; Year 2 = 1998 DS and 1999 WS; DS = 1997 DS and 1998 DS; WS = 1998 WS and 1999 WS. [b]FFP = farmers' fertilizer practice; SSNM = site-specific nutrient management. [c]Δ = SSNM − FFP. P>\|T\| = probability of a significant mean difference between SSNM and FFP. [d]Source of variation of analysis of variance of the difference between SSNM and FFP by farm; Prob>\|F\| = probability of a significant F-value.

control, all data were stored in a central project database, which was released at regular intervals on a CD-ROM (Simbahan 1999). In all chapters of this book, the evaluation of the on-farm performance of SSNM will be presented in a uniform manner for four consecutive rice crops grown during 1997 to 1999 at each site (Table 2.2). Common methods of data analysis and display are described below.

2.3 Data analysis

Definitions of N-use efficiency

Definitions of N-use efficiencies followed the framework described by Cassman et al (1998):

$$PFPN = GY_N/FN \tag{2.1}$$
$$AEN = (GY_N - GY_0)/FN \tag{2.2}$$
$$REN = (UN_N - UN_0)/FN \tag{2.3}$$
$$PEN = (GY_N - GY_0)/(UN_N - UN_0) \tag{2.4}$$
$$IEN = GY_N/UN_N \tag{2.5}$$

where PFPN = partial factor productivity of applied N (kg grain yield per kg N applied), AEN = agronomic efficiency of applied N (kg grain yield increase per kg N applied), REN = apparent recovery efficiency of applied N (kg N taken up per kg N applied), PEN = physiological efficiency of applied N (kg grain yield increase per kg fertilizer N taken up), IEN = internal efficiency of N (kg grain per kg N taken up), GY_N is the grain yield in a treatment with N application (kg ha^{-1}, FFP or SSNM), FN is the amount of fertilizer N applied (kg ha^{-1}), GY_0 is the grain yield in the 0-N plot without N application, UN_N is the total plant N accumulation measured in aboveground biomass at physiological maturity (kg ha^{-1}) in plots that received N, and UN_0 is the total N accumulation in the 0-N plot.

With proper nutrient and crop management, PFPN should surpass 50 kg grain kg^{-1} N applied. Note that AEN = PEN × REN. With proper nutrient and crop management, AEN should be ≥20 kg grain yield increase kg^{-1} N applied. REN largely depends on the congruence between plant N demand and the quantity of N released from applied N. With good crop management and plant-based N management strategies, REN of >0.5 kg kg^{-1} can be achieved. PEN represents the ability of a plant to transform a given amount of acquired fertilizer nutrient into economic yield (grain). It depends on genotypic characteristics, climate, plant density, water supply, supply of all nutrients, and the level of pest incidence. In a healthy rice crop with no significant constraints to growth, PEN should be close to 50 kg grain kg^{-1} N taken up from fertilizer. IEN includes N taken up from indigenous and fertilizer sources and is an aggregate measure of nutritional status. Under conditions of optimal nutrition and few other constraints to growth, IEN should be close to 68 kg grain kg^{-1}plant N (Witt et al 1999).

Definition of economic parameters
Chapter 3 presents a complete economic analysis of farmers' practices. Net return from rice (US$ ha^{-1} crop^{-1}) is defined as the net return over paid-out cost, not including the value of family labor. Here, we define other terms used to assess the performance of SSNM in comparison with the FFP in chapters 6 to 12. Note that all calculations were made using US$ as the standard currency and fixed average prices of fertilizers and rice at each site.

$$GR = pR \times GY/1,000 \qquad (2.6)$$
$$TFC = pN \times FN + pP \times FP + pK \times FK \qquad (2.7)$$
$$GRF = GR - TFC \qquad (2.8)$$
$$\Delta GY = GY_{SSNM} - GY_{FFP} \qquad (2.9)$$
$$\Delta TFC = TFC_{SSNM} - TFC_{FFP} \qquad (2.10)$$
$$\Delta Profit = GRF_{SSNM} - GRF_{FFP} \qquad (2.11)$$

where GR = gross return (US$ ha^{-1}), TFC = total fertilizer cost (US$ ha^{-1}), GRF = gross return above fertilizer cost (US$ ha^{-1}), ΔGY = difference in grain yield (tha^{-1}), ΔTFC = difference in TFC (US$ ha^{-1}), ΔProfit = profit increase or decrease by SSNM over FFP (US$ ha^{-1}), pR = price of rice (US$ kg^{-1}), GY = rice yield (t ha^{-1}), pN = price of N fertilizer (US$ kg^{-1}), pP = price of P fertilizer (US$ kg^{-1}), and pK = price of K fertilizer (US$ kg^{-1}).

Analysis of variance (ANOVA)
For each domain, ANOVA was performed on the differences (Δ) between SSNM and FFP measured on each farm for four consecutive rice crops grown (Δ = SSNM – FFP). Two crops were grown in year 1 (usually one DS and one WS crop) and two crops in year 2. The fixed-effects model (below, with the degrees of freedom, DF, for one domain, PhilRice, as an example) was used to analyze the data. All effects, except Village, were tested against the residual. The Village effect was tested against

Farm within Village as an error term. For variables with missing observations, the denominator mean square was adjusted using the Satterthwaite approximation (Satterthwaite 1946). Crop was partitioned into three orthogonal components (i.e., Year, Season, and Year × Season interaction), which allows for the specific testing of Year or Season main effects and the Year × Season interaction effect. PROC GLM of SAS was used to analyze the data (SAS Institute Inc. 1988).

Village (V)	DF	2
Farm within Village (F)	DF	24
Crop (C)		
Year 1 vs year 2	DF	1
DS vs WS	DF	1
Year × Season	DF	1
Village × Crop (V × C)	DF	6
Residual	DF	72

Because sampling domains were not selected randomly, only fixed effects were used, which limits the assumption that the sample villages and farms selected within the domain are truly representative for the whole domain. Note that the size of the V × C interaction term is important for the interpretation. If this effect is large and significant, results have less extrapolation potential to other villages, that is, they may be quite village- and season-specific.

Table 2.3 shows an example of how all ANOVA results were displayed in chapters 6 to 12. In this case, data would be interpreted as follows:

1. The average yield increase of SSNM over FFP across all four crop cycles was 0.49 t ha^{-1} and significant ($P = 0.003$).
2. The average yield difference between SSNM and FFP was not significant in the first year (0.22 t ha^{-1}, $P = 0.424$), but highly significant in the second year (0.78 t ha^{-1}, $P = 0.000$).
3. The average yield difference between SSNM and FFP was significant in the DS (0.51 t ha^{-1}, $P = 0.035$) and in the WS (0.48 t ha^{-1}, $P = 0.012$).
4. The yield increase (Δ) in year 2 (0.78 t ha^{-1}) was significantly larger than the yield increase achieved in year 1 (0.22 t ha^{-1}; see crop-year interaction in the last column, $P = 0.000$). In other words, it appears that the performance of SSNM has improved over time.
5. The average yield increase (Δ) achieved in both dry seasons (0.51 t ha^{-1}) was not significantly different from the average yield increase achieved in the two wet-season crops (0.48 t ha^{-1}; see crop-season interaction in the last column, $P = 0.813$). In other words, it appears that SSNM performed equally well in the DS and WS crops, although, in absolute terms, DS crops yielded about 1 t ha^{-1} more than WS crops.
6. The V × C interaction is not significant, that is, this level of performance of SSNM could probably also be achieved on other farms within the same domain.

Table 2.3. Typical crop residue management practices in all monitoring domains, the estimated fraction of the total amount of straw remaining in the field (fRes), and the estimated fraction of nutrients lost from crop residues by burning or leaching (fLoss).

Domain	Crop residue management	fRes	fLoss N	P	K
Maligaya, PhilRice	Hand-harvest and threshing on-site. Moderate straw removal. Remaining residues are incorporated (standing stubble) or burned (heaps of loose straw).	0.5	0.2	0.0	0.1
Suphan Buri, PTRRC	Combine-harvest. Small straw removal. Remaining residues are burned.	0.9	0.9	0.1	0.2
Omon, CLRRI	Hand-harvest with some straw removal. Remaining medium-long stubble is burned.	0.6	0.9	0.1	0.2
Sukamandi, RIR	Hand-harvest and threshing on-site. Moderate straw removal. Remaining residues are incorporated (standing stubble) or burned (heaps of loose straw).	0.5	0.2	0.0	0.1
Aduthurai, TNRRI Thanjavur, SWMRI	Hand-harvest. Almost complete straw removal. Remaining short stubble is incorporated.	0.2	0.1	0.0	0.1
Hanoi, NISF	Hand-harvest. Complete straw removal. Remaining short stubble is incorporated.	0.1	0.1	0.0	0.1
Jinhua, ZU	Hand- or combine-harvest. Complete or no straw removal. Early rice (ER): crop residues are incorporated or removed. Late rice (LR): all residues are left in the field over winter.	ER 0.5 LR 0.9	0.1 0.2	0.0 0.1	0.1 0.2
Pantnagar, GBPUAT	Combine-harvest. Small straw removal. Remaining residues are burned.	0.9	0.9	0.1	0.2

Other methods

Bar graphs were used to show the effect of site-specific nutrient management on selected agronomic and economic parameters as well as fertilizer use over a period of four crops. Fig. 2.3 depicts a hypothetical example for grain yield means in FFP and SSNM treatments with (1) error bars representing the standard deviation of the treatment means to assess the spatial variability of yield and (2) the standard error to compare treatment means. The latter was chosen for the across-site analysis (Chapter 15), in which the spatial variability of parameters was of less concern. Means and standard deviation were presented in chapters 6–12 to capture the spatial variability of parameters at individual sites. In addition, P values were included for each season on top of the treatment bars to indicate the probability of treatment means being significantly different. Note therefore that the P values correspond with the standard error bars in the figures presented in chapter 15 (and in Fig. 2.3B), but not with the bars representing the standard deviation in the figures of chapters 6–12 (and in Fig. 2.3A).

Boxplots were used to display the variability and trends over time of variables such as INS, IPS, and IKS. The solid horizontal line in each box represents the me-

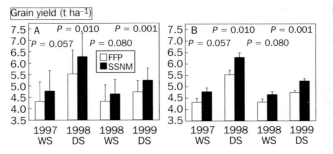

Fig. 2.3. Hypothetical example showing the effect of site-specific nutrient management (SSNM) on grain yield over a period of four crops with error bars representing (A) the typical standard deviation (SD) and (B) standard error (SE) of the treatment means. FFP = farmers' fertilizer practice.

dian, the box represents 50% of all measured data (= the range from the 25th to the 75th percentile), the fences moving upward and downward from the box represent the 10th and 90th percentiles, and the dots represent extreme values (outliers).

To assess the effect of crop management quality on the performance of SSNM, all data sets were split into farms with no or only minor crop management problems (SSNM+) and farms with one or more severe constraints to growth other than NPK (SSNM–). Using the field scores given for the different crop management factors (see above), an overall score was assigned to each farm × crop data set. The average score (ScAve) was computed as the average of the five individual scores for categories a to e and compared with the maximum score of any individual factor (ScMax). The overall score (ScTotal) used for classifying data sets into SSNM+ (ScTotal 0 or 1) and SSNM– (ScTotal 2) was then derived as

$$\text{ScTotal (no problems)} \quad = 0 \quad \text{if ScAve} < 0.5 \text{ and ScMax} < 2$$
$$\text{ScTotal (slight problems)} \quad = 1 \quad \text{if } 0.5 \leq \text{ScAve} < 1.0 \text{ and ScMax} < 2$$
$$\text{ScTotal (severe problems)} \quad = 2 \quad \text{if ScAve} \geq 1.0 \text{ or ScMax} = 2$$

Analysis of variance was performed to compare the two groups of farms (SSNM+ vs SSNM–) for each crop grown in each domain.

Estimating NPK input-output balances

A general formula for estimating the nutrient budget (B) for a rice field is

$$B = M + A + W + N_2 - C - PS - G \tag{2.12}$$

where (all components measured in kg elemental nutrient ha^{-1}) M = added inorganic and organic nutrient sources, A = atmospheric deposition (rainfall and dust), W = irrigation, floodwater, and sediments, N_2 = biological N_2 fixation (N only), C = net crop removal with grain and straw (total uptake – nutrients in crop residues returned to the soil), PS = percolation and seepage losses, and G = gaseous losses (denitrification, NH_3 volatilization).

We used the following assumptions in applying equation 2.12 to our data sets:

1. Measured values for fertilizer nutrient input (M). Where applicable, input from manure was estimated from the total amount of manure applied. This mainly refers to all 24 farms in North Vietnam (NISF domain), where all farmers applied farmyard manure at average rates of about 10 t ha^{-1} crop^{-1}. With few exceptions, most farmers in all other domains did not apply significant amounts of manure and did not grow green-manure crops.

2. Inputs from atmosphere (A) or water (W) were unknown, but we assumed that they were equivalent to losses by leaching (PS) so that these three terms were not considered in our calculations. Most farmers in the NISF, CLRRI, PhilRice, TNRRI, PTTRC, ZU, and RIR domains used surface (canal) water with perhaps low nutrient concentrations for irrigation.

3. An average N input from BNF of 50 kg N ha^{-1} crop^{-1} was used for all farms. This is roughly equivalent to the average INS measured across all domains and seasons. Because all fields have been under intensive rice cultivation for at least 20 years, we assume that the INS has reached a quasi-equilibrium stage equivalent to the net input from BNF.

4. Crop removal with grain and straw was measured for each crop cycle. The net amount of NPK returned to the field from crop residues was deducted from the total removal assuming (1) a typical percentage of straw remaining in the field and (2) a typical percentage of nutrients lost from the crop residues because of burning or leaching (Table 2.3). For example, at TNRRI and SWMRI, most straw is removed from the field so that only about 20% of the crop residues remain in the field. Burning is not common at these sites so that losses of N, P, and K from stubble were estimated to be only 0–10% (Table 2.3). Some of this occurs because of grazing by cattle and goats.

5. Crop residue management assumed a typical practice in each domain, which was based on our surveys, but there was more variation in this among farms within each domain.

Notes about data interpretation

In several domains, farmers tried to partially copy the SSNM practice in their own FFP plots. This refers mainly to attempts to follow a similar pattern of N applications, but in some cases we also noticed that farmers changed their fertilizer rates. This "copycat" effect cannot be quantified, but the statistical comparison of SSNM with FFP is probably somewhat biased and tends to underestimate the gains achieved by the SSNM as compared with the FFP.

Another way to assess SSNM performance is to compare the results to the baseline data set collected before any interference occurred by establishing an SSNM plot in the farmer's field. This is reasonable because yearly data sets for each domain include the same farmers and two consecutive crops (climatic seasons) in each year.

References

Bray RH, Kurtz LT. 1945. Determination of total organic and available phosphorus in soils. Soil Sci. 59:39-45.

Bremner JM. 1996. Nitrogen—total. In: Sparks DL, Page AL, Helmke PA, Loeppert RH, Soltanpour PN, Tabatabai MA, Johnston CT, Sumner ME, editors. Methods of soil analysis. Part 3. Chemical methods. SSSA Book Series No. 5. Madison, Wis. (USA): Soil Science Society of America and American Society of Agronomy. p 1085-1121.

Cassman KG, Aragon EL, Matheny EL, Raab RT, Dobermann A. 1994. Soil and plant sampling and measurements. Part 1: Soil sampling and measurements. Part 2: Plant sampling and measurements. Video and supplement. Los Baños (Philippines): International Rice Research Institute.

Cassman KG, Peng S, Olk DC, Ladha JK, Reichardt W, Dobermann A, Singh U. 1998. Opportunities for increased nitrogen use efficiency from improved resource management in irrigated rice systems. Field Crops Res. 56:7-38.

Gianello C, Bremner JM. 1986. A simple chemical method of assessing potentially available organic nitrogen in soil. Commun. Soil Sci. Plant Anal. 17:195-214.

IRRC (Irrigated Rice Research Consortium). 1998. Pest impact assessment and crop residue management research of the Irrigated Rice Research Consortium. Workplan 1998-2000. Makati City (Philippines): International Rice Research Institute. 85 p.

Olk DC, Cassman KG, Simbahan GC, Sta. Cruz PC, Abdulrachman S, Nagarajan R, Tan PS, Satawathananont S. 1999. Interpreting fertilizer-use efficiency in relation to soil nutrient-supplying capacity, factor productivity, and agronomic efficiency. Nutr. Cycl. Agroecosyst. 53:35-41.

Olk DC, Moya PF. 1998. On-farm management of applied inputs and native soil fertility. IRRI Discussion Paper Series No. 23. Makati City (Philippines): International Rice Research Institute. 212 p.

Olsen SR, Cole CV, Watanabe FS, Dean LA. 1954. Estimation of available phosphorus in soils by extraction with sodium bicarbonate. USDA Circ. 939. Washington, D.C. (USA): U.S. Government Printing Office.

RTDP. 1997. Reversing trends of declining productivity in intensive irrigated rice systems, phase II (1997-2000): manual for research on site-specific nutrient management. Manila (Philippines): International Rice Research Institute.

RTDP. 1998. Reversing trends of declining productivity in intensive irrigated rice systems. Progress report 1997. Makati City (Philippines): International Rice Research Institute. 131 p.

RTDP. 1999. Reversing trends of declining productivity in intensive irrigated rice systems. Progress report 1998. Makati City (Philippines): International Rice Research Institute. 244 p.

SAS Institute Inc. 1988. SAS/STAT user's guide, Release 6.03 edition. Cary, N.C. (USA): SAS Institute Inc. 1,028 p.

Satterthwaite FE. 1946. An approximate distribution of estimates of variance components. Biometrics Bull. 2:110-114.

Savary S, Elazegui FA, Teng PS. 1996. A survey portfolio for the characterization of rice pest constraints. IRRI Discussion Paper Series No. 18. Makati City (Philippines): International Rice Research Institute.

Simbahan GC. 1999. RTDP databases user's manual. Makati City (Philippines): International Rice Research Institute. 58 p.

van Reeuwijk LP. 1992. Procedures for soil analysis. 3rd ed. Wageningen (Netherlands): ISRIC.

Walinga I, van der Lee JJ, Houba VJG, van Vark W, Novozamsky I. 1995. Plant analysis manual. Dordrecht (Netherlands): Kluwer Academic Publishers.

Walkley A. 1947. A critical examination of a rapid method for determining organic carbon in soils: effect of variations in digestion conditions and of inorganic soil constituents. Soil Sci. 63:251-263.

Witt C, Dobermann A, Abdulrachman S, Gines HC, Wang GH, Nagarajan R, Satawathananont S, Son TT, Tan PS, Tiem LV, Simbahan GC, Olk DC. 1999. Internal nutrient efficiencies of irrigated lowland rice in tropical and subtropical Asia. Field Crops Res. 63:113-138.

Notes

Authors' addresses: A. Dobermann, Department of Agronomy and Horticulture, University of Nebraska, P.O. Box 830915, Lincoln, NE 68583-0915, USA, e-mail: adobermann2@unl.edu; G.C.Simbahan, P.F. Moya, M.A.A. Adviento, M. Tiongco, C. Witt, and D. Dawe, International Rice Research Institute, Los Baños, Philippines.

Citation: Dobermann A, Witt C, Dawe D, editors. 2004. Increasing productivity of intensive rice systems through site-specific nutrient management. Enfield, N.H. (USA) and Los Baños (Philippines): Science Publishers, Inc., and International Rice Research Institute (IRRI). 410 p.

3 | The economics of intensively irrigated rice in Asia

P. F. Moya, D. Dawe, D. Pabale, M. Tiongco, N.V. Chien, S. Devarajan, A. Djatiharti, N.X. Lai, L. Niyomvit, H.X. Ping, G. Redondo, and P. Wardana

Almost 90% of the world's rice is produced in Asia, with about 130 million hectares of cropped area devoted to rice production annually. China is the world's leading producer, growing nearly two-fifths of Asia's total on 32 million ha. India produces nearly a quarter on about 44 million ha. The remaining one-third of Asian production comes primarily from Indonesia, Vietnam, Bangladesh, Thailand, Myanmar, Japan, and the Philippines (Table 3.1).

Nearly 60% of the rice area in Asia is irrigated, where rice is often grown in monoculture with two or even three crops a year depending upon water availability (Huke and Huke 1997). The irrigated ecosystem provides about three-fourths of Asia's rice, but there are concerns about the sustainability of production in this system (Cassman and Pingali 1995). The sustainability of this system may be threatened either because of resource degradation leading to declining productivity or because of a lack of economically exploitable new technologies that will allow farm production and profitability to meet future increases in demand at prices acceptable to con-

Table 3.1. Average annual rice production and area harvested in selected Asian countries, 1997-99 (FAO 2001).

Country or area	Production $(10^6$ t)	Percent of world	Area harvested $(10^6$ ha)	Percent of world
China	201.2	34	31.8	21
India	128.9	22	44.2	29
Indonesia	49.8	8	11.6	8
Vietnam	29.4	5	7.4	5
Bangladesh	28.8	5	10.2	7
Thailand	23.2	4	10.0	7
Myanmar	17.9	3	5.6	4
Japan	11.7	2	1.8	1
Philippines	10.5	2	3.7	2
Other Asian countries	34.2	6	10.9	7
Asia	535.6	91	137.2	90
World	586.3	100	153.0	100

sumers. These two concerns coincide with the two major socioeconomic research objectives of the "Reversing Trends of Declining Productivity" (RTDP) project.

Earlier research using farm-level data (Cassman and Pingali 1995) reported declining factor productivity in intensive rice areas of the Philippines and India. However, the geographic extent and nature of this declining productivity were not known. Thus, the first socioeconomic research objective of the RTDP project was to assess trends in productivity at the farm level in a large number of intensive irrigated farming systems. Because total factor productivity (TFP) is a measure superior to partial factor productivities (PFPs) of individual factors of production (Dawe and Dobermann 1998), the project has collected large amounts of data on outputs and input use to establish a long-term database with which to measure TFP. Furthermore, because no single indicator is a good measure of sustainability (Byerlee and Murgai 2000), the RTDP project has also collected extensive biophysical data on soil quality to relate trends in TFP to trends in indicators of resource quality (see Chapter 2). Trends in TFP at the RTDP sites are discussed briefly in Chapter 4, but it is still too early to make definitive conclusions regarding such trends after only 5 years of data collection.

Another important research objective is to understand socioeconomic and technological differences in intensive irrigated rice production systems across Asia in order to provide information on the context within which technology adoption (e.g., site-specific nutrient management) takes place. Context is crucially important because a given technology might be appropriate for some production systems but not for others when there are differences across systems in fertilizer management, labor costs, structure of the labor force, and other factors. The purpose of this chapter is to describe and quantify the nature of on-farm production in some key representative intensive irrigated rice-based farming systems of Asia.

3.1 Irrigated rice production systems in selected countries of Asia

Before the Green Revolution, paid-out rice production costs per unit of land were low because farmers did not use large quantities of purchased inputs such as fertilizer and pesticide and labor was often supplied by the family or provided in a reciprocal manner by friends and neighbors. However, the new seeds of the Green Revolution, coupled with the expansion of irrigation, have made rice production more of a commercial enterprise in many areas. As a result, cultural practices for rice production today sometimes differ substantially across sites, even within the class of intensive irrigated farming systems. For example, rice farmers use different quantities and types (e.g., family, hired) of labor and hire that labor using a variety of institutional arrangements. Many of these differences are due to varying levels of economic development. For example, in countries such as Thailand, where rapid economic development has occurred, farmers have moved away from a traditional labor-intensive rice production system to a highly mechanized system. In India and northern Vietnam, current systems are still relatively labor-intensive. Even within sites, however, yields and input use can vary markedly from year to year. For example, quantities and tim-

Table 3.2. List of years included in the analysis (high-yielding season = HYS, low-yielding season = LYS).

Site	Years included		Total number of crops
	HYS	LYS	
Central Luzon, Philippines	1995-99	1994-99	11
Central Plain, Thailand	1995-96, 1998-99	1994-95, 1997-99	9
Mekong Delta, Vietnam	1995-96, 1998-99	1994-95, 1997-99	9
West Java, Indonesia	1995-96, 1998-99	1995, 1997-99	8
Tamil Nadu, India	1995-99	1994-95, 1997-98	9
Red River Delta, Vietnam	1997-99	1997-99	6
Zhejiang, China	1998-99	1998-99	4

ing of inorganic fertilizer use vary considerably from year to year, as will be shown later. Some of this may be due to changing socioeconomic circumstances, but adaptation to changing climatic conditions and biophysical circumstances also probably plays a large role.

To quantitatively assess these differences in production practices, detailed data on farm inputs, outputs, and cultural practices were collected from farms at seven of the RTDP sites described in Chapter 2.[1] The villages where these farm-level data were collected were selected to be representative of the local irrigated rice ecosystems in terms of cropping pattern and soil type. For example, if soil type varied within the region, then villages were selected to represent each major type. Within a village, farmers were chosen not at random but based on their willingness to cooperate for a relatively long period of time. However, an effort was made to choose farms that cover a range of sizes. Farmers were also usually clustered in some way so as to minimize the travel time and costs involved in data collection.

Unless stated otherwise, the data presented in this chapter are averages of several years, depending on the availability and completeness of the data set of the project. Table 3.2 lists the specific years included for each site. The sites span most of the important low-income rice-producing countries in Asia, with Bangladesh and Myanmar being the two exceptions. At least two crops of rice are grown annually at each site, with some farmers in the Mekong Delta of Vietnam and the Central Plain of Thailand occasionally growing three crops in a year. Each of the two seasons at each site is classified as either a high-yielding season (HYS) or low-yielding season (LYS),

---

[1]Socioeconomic data from Thanjavur are not included in this chapter because its socioeconomic characteristics are similar to those of Aduthurai and because insufficient data are currently available for detailed analysis.

depending upon which season generally gives the highest yields at each particular site. At the tropical sites, the HYS is usually the dry season (DS) and the LYS is the wet season (WS), although the site in Sukamandi is an exception. The relatively high WS yields in Sukamandi may be due to a long aerobic fallow period after the DS (2–3 mo) that increases the availability of mineral N and P for the succeeding rice crop. For the two subtropical sites, the HYS is the late rice crop in Zhejiang and the early (spring) rice crop in northern Vietnam.

Sample farms and basic farm characteristics

Table 3.3 summarizes the basic characteristics of the sample farms at each of the sites. Farms at the RTDP sites (and in Asia more generally) are much smaller than farms in developed countries with land surpluses such as the United States and Australia. Farm sizes vary across sites, with mean farm size ranging from about 0.3 ha in northern Vietnam to 4 ha or more in the Central Plain of Thailand and Tamil Nadu, India.[2] Furthermore, farms at most sites are split into multiple parcels. This tendency is most pronounced in China and Vietnam, communist countries where land reforms intentionally gave each family multiple plots of varying quality. At the RTDP site in Zhejiang, each farmer has on average about four parcels, with an average parcel size of just 0.3 ha. In northern Vietnam, the situation is even more extreme, with each farmer having on average seven parcels that average just 0.04 ha in size! Recent relaxation of previous land market restrictions is now allowing some consolidation of these small parcels.

Most of the farmers in our samples at the tropical sites own the land they till or are leaseholders making payments that will lead to eventual ownership. Leaseholding is particularly common in Central Luzon, Philippines (more than half our sample), and in the Central Plain of Thailand (about one-third of our sample). There was only a small amount of share tenancy at the sites in the Philippines and Indonesia. At the sites in northern Vietnam and China, the government is the owner of all our sample farms, but the farmers still have full control over all production decisions.

With regard to age, a majority of the farmers in the sample are presently in their late forties or early fifties and have many years of farming experience that have produced a set of farming practices based on local conditions. The average level of education of the farmers is generally around seven years of school. Nearly all are literate and can read, write, and communicate with agricultural extension workers and researchers on farm-related matters.

Farm-level yields

The mean area planted to rice in the parcel monitored for the socioeconomic study is generally small, ranging from 0.05 ha in northern Vietnam to about 3 ha in Tamil

---

[2]One very large farm of nearly 32 ha skews the mean farm size for Tamil Nadu. The median farm size for the sample farmers in Tamil Nadu is 2.8 ha.

Nadu, India (Table 3.3). Farmers often have more than one parcel, however, and more than one of these parcels may be planted to rice. Thus, the data on mean area planted are not indicative of the importance of rice, but merely indicate the size of the area in which our data were collected.

The yield data in Table 3.4 are calculated using farmers' estimates of production and area for the whole parcel at field moisture content (MC) levels shortly after harvest (estimated to be about 20–24% MC). These calculations of yield are generally similar to the data from crop-cuts presented elsewhere in this book (after adjustments for moisture content).

As is well known, yields can vary substantially at the same site from year to year because of climatic fluctuations. Thus, to give a good indication of relative yield across sites, we need to present data that are averages of several years (Table 3.4). In both the HYS and LYS, the highest yields were achieved in Zhejiang, where yields averaged 6.8 t ha^{-1} in the HYS and 5.7 t ha^{-1} in the LYS, for a total annual average rice production of about 12.5 t ha^{-1}. The high yields at Jinhua result mainly from the use of hybrid rice in the HYS, better growing conditions relative to the other sites, relatively balanced fertilizer use, and fertile soils with a large indigenous nutrient supply. Relatively high two-crop yields of 11.4 t ha^{-1} were also achieved at the other subtropical site (northern Vietnam). This figure excludes the grain yield obtained from the maize crop that is planted by most farmers in this area after the summer rice crop. At the tropical sites, average two-crop yields varied in a relatively narrow range from 9.7 t ha^{-1} at Sukamandi to 10.9 t ha^{-1} in Tamil Nadu. If yields were calculated on an annual basis for the tropical sites, farmers in the Mekong Delta would have the highest annual yields. Nearly all of our sample farmers in that area plant three crops of rice per year and data for the lowest yielding third crop are not included in Table 3.4. Some of our sample farmers in the Central Plain of Thailand also planted three crops of rice in a year, but this is much less common than in the Mekong Delta and occurred only when water availability was unusually favorable.

Farm-level yields vary considerably among farms within each site (Table 3.5). The difference between maximum and minimum yields for a particular site and season was often in the range of 3 to 4 t ha^{-1} in the HYS and even larger in the LYS (4 to 7 tha^{-1}) when climatic conditions are typically more variable. At first glance, this might suggest that there are large exploitable yield gaps between average (or poor) farmers and progressive farmers. To explore this issue quantitatively, pair-wise regressions were estimated with farmer yields in one cropping season as the dependent variable and farmer yields for the same group of farmers in another cropping season as the independent variable (biophysical crop-cut yield data were used for these regressions; similar results hold when using whole-parcel yield data). The R^2 values of these regressions measure the percentage of variation in farmer yields in one season that can be explained by farmer yields in the other season. If the same farmers consistently obtain the highest yields in their area, then the R^2 of this regression will be relatively high. This is true even if average yields in one year are high because of good growing conditions in the area but are low the next season because of bad weather. If the progressive farmers are able to obtain higher than average yields

Table 3.3. Basic socioeconomic characteristics of sample farms, RTDP sites 1999.

Site	No. of farms	Farm size (ha)			Area of parcel monitored (ha)			No. of parcels	Age of farmer	Education (y)	Family size
		Mean	Minimum	Maximum	Mean	Minimum	Maximum				
Central Luzon, Philippines	26	2.18	0.50	6.28	2.12	0.50	6.28	1.38	54	7.6	6
Central Plain, Thailand	23	4.00	0.96	7.96	2.13	0.96	4.00	2.48	49	4.6	5
Mekong Delta, Vietnam	20	0.97	0.26	2.20	0.73	0.23	1.80	2.00	49	7.4	6
West Java, Indonesia	20	0.69	0.16	2.84	0.69	0.20	2.84	1.25	45	6.8	4
Tamil Nadu, India	24	4.30	0.37	31.73	3.13	0.29	24.93	2.00	47	11.5	6
Red River Delta, Vietnam	24	0.31	0.14	0.55	0.05	0.02	0.10	7.38	50	7.0	6
Zhejiang, China	21	1.18	0.15	5.87	1.21	0.13	6.67	3.81	44	7.1	4

Table 3.4. Average yield on RTDP farms, 1994-99 (high-yielding season = HYS, low-yielding season = LYS).

Site	Rice yield (t ha [1])		
	HYS	LYS	Sum
Central Luzon, Philippines	6.38	4.13	10.51
Central Plain, Thailand	5.17	5.08	10.25
Mekong Delta, Vietnam	5.96	3.89	9.85
West Java, Indonesia	5.58	4.12	9.70
Tamil Nadu, India	5.79	5.13	10.92
Red River Delta, Vietnam	5.96	5.44	11.40
Zhejiang, China	6.79	5.71	12.50

Table 3.5. Range in rice yields of farms in low-yielding and high-yielding seasons, 1999.

Site	Rice yield (t ha [1])			
	Mean	Minimum	Maximum	CV(%)
Low-yielding season				
Central Luzon, Philippines	3.6	0.4	7.2	42
Central Plain, Thailand	5.3	1.4	6.8	20
Mekong Delta, Vietnam	3.6	1.6	5.6	31
West Java, Indonesia	4.4	1.9	7.0	28
Tamil Nadu, India	5.3	4.4	6.9	13
Red River Delta, Vietnam	6.2	4.5	7.6	16
Zhejiang, China	5.7	3.0	9.6	29
High-yielding season				
Central Luzon, Philippines	5.6	3.9	7.7	17
Central Plain, Thailand	5.5	4.0	7.0	13
Mekong Delta, Vietnam	5.6	3.2	6.9	18
West Java, Indonesia	5.3	3.0	9.7	34
Tamil Nadu, India	5.9	4.7	6.8	10
Red River Delta, Vietnam	6.2	4.4	7.9	16
Zhejiang, China	6.7	3.7	7.7	15

under most conditions (good and bad weather), then the R^2 of the regression will be high even if there is wide variation between average yields from one year to the next.

A total of 157 such regressions were performed, covering the farms at all seven sites discussed in this chapter. The average R^2 of the regressions was just 0.11 (Table 3.6 shows results for each of the seven sites separately). In 15 of the 157 cases, the correlation was positive and statistically significant at the 1% level, and it was positive and significant at the 5% level in another 13 cases. Since the regressions explain just 11% of the variance in yield, it does not appear that there is a strong relationship between farmers who obtain a high yield in one year and those who obtain a high

Table 3.6. R^2 of regressions between yield in year s and yield in year t (see text for explanation), "measured" and "true" interfarmer yield gaps.

Site	Av R^2	Number of crops	Number of farms	Measured yield gap (kg ha^{-1})	True yield gap (kg ha^{-1})	Difference (absolute)	Difference (%)
Central Luzon, Philippines	0.07	9	25	1,002	442	560	56
Central Plain, Thailand	0.07	7	24	1,209	931	278	23
Mekong Delta, Vietnam	0.06	7	23	940	725	215	23
West Java, Indonesia	0.20	7	20	1,008	529	478	47
Tamil Nadu, India	0.11	8	20	1,114	979	136	12
Red River Delta, Vietnam	0.24	6	24	1,355	1,109	246	18
Zhejiang, China	0.12	6	21	1,004	536	468	47
Average	0.11			1,090	750	340	31

yield the next year. In fact, in 26 cases (nearly one-sixth of the time), the coefficient on yield in one year (as an explanatory variable for yield in a different year) was negative, implying an inverse relationship. In such cases, farmers with a higher than average yield one year typically obtained a lower than average yield in the other year. The set of regressions was also estimated for yields in –F treatments and the results were largely similar (i.e., very low R^2).

This analysis implies that the magnitude of the yield gap between the best farmer and average farmer is usually overstated if data from only one crop are used in the calculation. To quantify the magnitude of this problem, the interfarmer yield gap between progressive farmers and below-average farmers (defined as third-quartile yield minus first-quartile yield) was calculated for each individual crop at each site. The average of these yield gaps for each site was calculated and termed the "measured" interfarmer yield gap, as shown in Table 3.6. A second interfarmer yield gap (again, third-quartile yield minus first-quartile yield) was also calculated using average yields across all crops for each farmer and this was termed the "true" interfarmer yield gap. On average across sites, the magnitude of the true yield gap is about 0.75 t ha^{-1} compared with a measured gap of 1.09 t ha^{-1}. The interfarmer yield gap is thus overstated by 0.34 t ha^{-1} (31%) on average. Furthermore, the true yield gap as defined here still includes differences caused by soil quality, and it will not be possible to close all of these gaps. It would be an interesting research question to explore how much of this yield gap can be eliminated in theory, but it is not clear how to design an appropriate experiment. One could have researchers take over the management of plots owned by good and average farmers for several years and see if the difference in yields remains. But since researchers are not likely to manage the plots in a profit-maximizing fashion (as farmers do to the best of their ability), it is not clear what the results of such an experiment would mean. In other words, the researchers might be able to raise yields more for the farmers with the lowest yields, but only by using input quantities that are not profitable for farmers.

Table 3.7. Mean fertilizer N, P, and K use (kg ha⁻¹) by RTDP farmers, 1994-99 (high-yielding season = HYS, low-yielding season = LYS).

Site	N		P		K	
	HYS	LYS	HYS	LYS	HYS	LYS
Central Luzon, Philippines	130	88	15	13	22	18
Central Plain, Thailand	112	99	21	21	1	1
Mekong Delta, Vietnam	90	95	14	14	10	13
West Java, Indonesia	109	113	9	8	10	7
Tamil Nadu, India	115	103	23	23	35	31
Red River Delta, Vietnam	103	94	20	13	60	56
Zhejiang, China	174	171	17	19	63	60
Mean	119	109	17	16	29	26

Some of the variations in yields across farmers from year to year may be explained by variations in input use other than fertilizer (fertilizer use is not responsible because the same pattern occurs in –F plots as noted earlier). For example, if farmer A achieved a higher yield than farmer B in year 1, but realized a lower yield in year 2, it might be because farmer A used more fertilizer or labor than farmer B in year 1 but less in year 2. To test this hypothesis, several production functions were estimated for the farmers at Maligaya and Tamil Nadu, but the coefficients on the input variables were typically not statistically significant and not always of the expected sign (results not shown).

The difficulty in achieving reliable statistical estimates of the production function from farm-level data may seem surprising at first glance, but it is consistent with some reasonable assumptions regarding economic behavior. When estimating a production function for a group of farmers, one implicitly assumes that all farmers face the same set of biophysical constraints. If true, and if all farmers in a given area face the same set of input and output prices (a reasonable assumption within our study sites where marketing systems are well developed), then all farmers should use the same quantities of inputs if they maximize profits. Yet the data show wide variations in input use across farmers within a given site for a given crop, both for fertilizer and for other inputs (e.g., see Table 3.7). Different farmers facing the same production function and the same set of input and output prices might apply different amounts of fertilizer if some farmers did not understand the benefits of fertilizer. A similar outcome might result if some farmers faced credit constraints and were unable to purchase fertilizer at the proper time. Yet, at the RTDP sites, where farmers have been cultivating modern varieties for many years, farmers clearly understand that more nitrogen fertilizer means higher yields (up to some maximum). Fertilizer also represents a small portion of the value of production and credit constraints are not important for most farmers at the RTDP sites (Moya 1998). This may be because many rice farmers receive income from many sources other than farming and they can use this income to purchase fertilizer. Thus, knowledge and credit constraints are not adequate to explain the wide variations in input use across farms within a given site.

The fact that different farmers use different quantities of the same inputs thus suggests that they face different production functions. If this is the case, then it is not reasonable to estimate a production function within a site, and there is no a priori reason to expect that higher yields will be correlated with higher levels of input use. But if the tremendous variation in yields across farmers from year to year cannot be explained by differential input use, it is not clear what factors do explain it. Part of the variation may be due to climatic variation, even within a site. A more likely possibility is that timing of input use is a critical factor and that farmers are not able to consistently make the proper adjustments for every crop in response to uncertain and changing weather conditions. Researchers face similar problems on experiment stations. In any event, the lack of consistency in farm yields from one year to the next is an important topic for future research.

Fertilizer use

Inorganic fertilizer management practices vary considerably both across and within sites. Farmers in Zhejiang, China, applied the highest amount of N fertilizer across all the sites, an average of more than 170 kg N ha^{-1} (Table 3.7). These high rates of N application were not the cause of the relatively high yields achieved at this site, however. In fact, researchers in the RTDP project were able to reduce N applications at this site by more than 60 kg N ha^{-1} in the two crops in 1999 and still raise yields by more than 400 kg ha^{-1} (see Chapter 12). N rates at the other sites were much lower, usually in the range of 90–115 kg N ha^{-1}.

Average prices paid by farmers for N fertilizer from 1994 to 1999 varied substantially across sites, ranging from US$0.23 kg^{-1} N in Tamil Nadu, India, to $0.50 kg^{-1} N in Maligaya, Philippines. Besides Tamil Nadu, prices were also relatively low in Sukamandi, Indonesia. Urea prices in India and Indonesia were low primarily because of government pricing policies that resulted in subsidies for farmers. These subsidies remain in place in India today, but they have recently been removed in Indonesia. In addition, N prices were $0.40 kg^{-1} N in Zhejiang and from $0.47 to $0.50 kg^{-1} N at the four sites in Vietnam, Thailand, and the Philippines.

Farmers usually applied more N in the HYS than in the LYS, although the differences were generally not large. This is consistent with applying more N during periods when solar radiation is higher, as is generally true in the HYS. The largest differential application was in the Philippines, where farmers applied just 88 kg N ha^{-1} in the WS, but 130 kg N ha^{-1} in the DS, perhaps because this is also the site with the largest estimated difference in climatic yield potential between the two seasons (see Table 5.1). Farmers in Indonesia applied more N in the LYS, but at this site the LYS is the DS, despite greater solar radiation during this period. The WS gives higher yields at this site, possibly because of a long fallow before the WS that results in high N mineralization (Dobermann et al 2000). Thus, farmers' behavior at this site is still consistent with applying more N when solar radiation is relatively high.

Farmers in southern Vietnam may be an exception to this general rule. In southern Vietnam, there are almost 3 months of flooding immediately before the DS (HYS). This flooding provides a rich source of nutrients for the soil and farmers may perceive less need for fertilizer as a result. However, although the flooding may provide a good source of P and K for the succeeding crop, any N contained in the floodwaters will be prone to losses, and this is probably not a good rationale for adjusting N rates between seasons.

Phosphorus (P) fertilizer was used in moderate amounts at all sites. Across sites, with the exception of Sukamandi, the average amount applied ranged from 13 to 23 kg P ha^{-1} crop^{-1}. Rates of P application did not vary substantially between the HYS and the LYS. Phosphorus use in Indonesia was low from 1997 to 1999 primarily because of the financial crisis that hit Southeast Asia beginning in 1997. This caused a sharp depreciation in the value of the Indonesian currency (the rupiah) and led to a substantial increase in domestic fertilizer prices. In 1995-96, P rates at Sukamandi were about 13 kg P ha^{-1}, but declined to just 6 kg P ha^{-1} from 1997 to 1999. Average prices for P paid by farmers also varied somewhat across sites, ranging from an average over the past five years of $0.97 to $1.32 kg^{-1} P. As with N, prices for P were relatively low in India and Indonesia and most expensive in the Philippines and Thailand.

Compared with phosphorus use, potassium use was more variable across sites. Farmers from northern Vietnam and Zhejiang, China, applied significantly higher amounts of K fertilizer (56–63 kg ha^{-1} crop^{-1}) than farmers at the other sites. This may be the result of government information campaigns promoting balanced fertilization that were effective within a system of central planning. In contrast, farmers from Thailand essentially apply no K fertilizer. Potassium may be less necessary at this site because nearly all straw (which is high in K) is left in the ground after harvest by combines. In Indonesia, K use declined sharply because of the economic crisis, just as with phosphorus. In 1995-96, K use at Sukamandi was about 17 kg K ha^{-1}, but this fell to just 3 kg K ha^{-1} from 1997 to 1999. K use was slightly higher in the HYS than in the LYS at most sites, but the difference was not substantial. Prices paid by farmers for K varied in a narrow range of $0.33 to $0.36 kg^{-1} K at all sites except Maligaya and Aduthurai. At Maligaya, the imputed price of K was relatively high during the past five years ($1.06 kg^{-1} K) because most farmers who used K obtained it from compound fertilizers, which have a high price per unit of nutrient. This appears to be slowly changing as muriate of potash (MOP) is now more easily available at this site. Nevertheless, even in the most recent seasons, about 40% of the K used by farmers at Maligaya came from compound sources. At Aduthurai, the prices of K were lower, at about $0.25 kg^{-1} K, consistent with the relatively low prices for other nutrients at this site.

The response of Indonesian farmers to the financial crisis that began in the second half of 1997 was different for P and K fertilizer compared with N. The prices of all three fertilizers rose substantially, from 50% to 75% in nominal terms (i.e., not adjusted for inflation). But N use declined just 13%, whereas P and K use declined 56% and 80%, respectively. This differential response is consistent with the different influence of these fertilizers on production. For N, most of the applied fertilizer is used by the current crop or is lost to the surrounding water and atmosphere. Little is carried over to future crops. Thus, N fertilizer is essentially a raw material input in the production process. P and K fertilizers, however, can be stored in the soil from one crop to another and the impact of increased applications in one year may not result in increased production immediately. Thus, P and K fertilizers are more akin to investment goods than raw material inputs. In the event of price increases for P and K that are perceived by farmers to be possibly temporary, it is rational to postpone purchases of these fertilizers, and that is how the farmers at Sukamandi behaved. For N, it is not possible to reduce use substantially without a serious immediate effect on production, so farmers reduced the use of urea only slightly. An important question is how long P and K use can remain low before yields are adversely affected.

Because the farmers' use of inorganic N, P, and K varied considerably within each site for a given crop (Table 3.8), we collected additional data on their fertilizer management practices using a separate questionnaire administered at the start of the second phase of the project (Moya 1998). Analysis of these survey results indicates that several considerations influence farmers' decisions on the amount of fertilizer to apply, including both biophysical and socioeconomic factors. Farmers cited biophysical factors, such as soil fertility and variety planted, as important in 55% of the responses. Thus, farmers are using their knowledge and an assessment of local conditions in determining rates of fertilizer application. Surprisingly, only a handful of farmers (7%) consider the availability of cash or credit as an important factor determining the quantity of fertilizer use. This is consistent, however, with data from other studies conducted in irrigated areas well served by infrastructure, where farmers are often net lenders to traders (Hayami et al 1999).

Although fertilizer use varies across farmers within a site for a given crop, these differences are not necessarily consistent from year to year. Just as the same farmers don't always consistently achieve the highest yields at a given site, the same farmers don't always use the highest amounts of fertilizer. Interviews with farmers indicate that they vary N rates in response to constantly changing growing conditions. Furthermore, Dawe and Moya (1999) use long-term experimental data to show that the optimal rate of N application varies substantially from year to year. Thus, it appears that farmers are trying to achieve some degree of optimality to the best of their knowledge and ability. Farmers also vary P and K rates substantially from year to year. Unlike N, these fertilizers can be stored in the soil from year to year. Thus, farmers may decide to apply P and K in some years, but not in others, a strategy that is not viable for N management. (The minimum rates of P and K are often zero in Table 3.8, although this never occurs for N.)

Farmers also differ in their frequency and timing of fertilizer application. The results in Table 3.9 show a wide variability in the frequency of fertilizer application across sites. In India and northern Vietnam, where labor is abundant and wage rates are low, farmers tend to split their application into three or four doses. In some locations where wage rates are higher (the Philippines and China), farmers limit the number of applications to once or twice, perhaps to minimize labor use. Yet, this correla-

Table 3.8. Amount of applied fertilizer nutrients on RTDP farms, 1999.

Item	Central Luzon, Philippines	Central Plain, Thailand	Mekong Delta, Vietnam	West Java, Indonesia (kg nutrient ha^{-1})	Tamil Nadu, India	Red River Delta, Vietnam	Zhejiang, China
High-yielding season							
N fertilizer							
Mean	123	112	85	104	113	109	170
Minimum	39	46	34	49	60	89	67
Maximum	176	166	211	190	205	141	300
CV (%)	28	28	45	34	28	14	33
P fertilizer							
Mean	15	19	13	3	20	26	17
Minimum	0	9	5	0	12	13	3
Maximum	43	29	32	22	31	38	29
CV (%)	69	37	49	206	32	24	35
K fertilizer							
Mean	22	0	18	5	30	69	62
Minimum	0	0	2	0	0	22	0
Maximum	91	0	46	42	62	110	187
CV (%)	92	–	64	252	54	35	59

continued on next page

Table 3.8 continued.

Item	Central Luzon, Philippines	Central Plain, Thailand	Mekong Delta, Vietnam	West Java, Indonesia (kg nutrient ha^{-1})	Tamil Nadu, India	Red River Delta, Vietnam	Zhejiang, China
Low-yielding season							
N fertilizer							
Mean	89	100	77	99	92	103	187
Minimum	29	44	26	29	54	64	86
Maximum	146	131	204	177	163	151	329
CV (%)	30	22	60	40	26	22	34
P fertilizer							
Mean	14	19	12	6	24	9	18
Minimum	0	5	3	0	9	0	9
Maximum	42	32	31	22	50	17	29
CV (%)	59	35	56	111	41	52	30
K fertilizer							
Mean	19	0	23	4	24	64	65
Minimum	0	0	0	0	0	20	0
Maximum	47	0	112	21	68	123	213
CV (%)	67	–	111	194	70	36	62

Table 3.9. Frequency of fertilizer application (no. of applications per season), RTDP farms, 1994-99.

Site	High-yielding season					Low-yielding season				
	1	2	3	>4	Av no. of applications	1	2	3	>4	Av no. of applications
	(% of total sample)					(% of total sample)				
Central Luzon, Philippines	1	56	40	3	2.5	20	70	9	1	1.9
Central Plain, Thailand	0	44	54	1	2.6	2	57	42	0	2.4
Mekong Delta, Vietnam	1	18	65	17	3.0	0	15	74	11	3.0
West Java, Indonesia	60	40	0	0	1.4	15	75	10	0	1.8
Tamil Nadu, India	0	5	62	33	3.3	1	12	66	22	3.1
Red River Delta, Vietnam	0	1	33	65	3.7	0	17	40	43	3.3
Zhejiang, China	12	81	7	0	2.0	52	48	0	0	1.5

tion between wage rates and number of fertilizer applications is far from perfect. In Thailand, which has the highest wage rates and the best off-farm employment opportunities of any RTDP site, farmers apply fertilizer in two or three splits, whereas in Indonesia (where labor use is high) farmers apply the fewest splits of any site. Thus, although there is some rough correlation between the number of fertilizer applications and wage rates, other factors also appear to be at work. More research would be helpful to illuminate the reasons behind the cross-site variability in the number of fertilizer applications because improved N management may require a higher frequency of N applications at some sites, particularly Indonesia.

All farmers at these sites apply the bulk of their fertilizer by broadcasting. Even in Indonesia, where the government strongly encouraged farmers to use urea tablets during part of this period, farmers typically broadcast the urea tablets into the field instead of placing them below the surface of the soil. Some farmers use foliar fertilizers, and these fertilizers are applied by spraying them directly on the leaves. Foliar fertilizers were most common at the site in southern Vietnam, where about one-sixth of our sample farmers used them. At other locations, the use of such fertilizers was extremely rare.

Preplant application of fertilizer is popular among the farmers of India, China, and northern Vietnam, but much less so in other locations (Table 3.10). To some extent, this may be due to a difference in methods of crop establishment, as preplant fertilizer application is low at all sites where direct seeding is widespread (Thailand, southern Vietnam, and the Philippines). In such areas, farmers typically broadcast N fertilizer within 10–20 days after sowing (DAS) to promote fast crop establishment. However, preplant N application was also low in Sukamandi, Indonesia, where transplanting is the method of crop establishment. Among the sites where preplant application is common, the percentage of N applied at this time was relatively large in Tamil Nadu and in China, usually about one-third of total N. The SSNM treatments generally reduced the amount of N applied preplant, since crop demand for N was judged to be quite low at this point in the crop cycle. Farmers typically applied most P preplant, or, in direct-seeded rice, broadcast P within 20 DAS. Potassium, however, was generally applied more evenly across the crop cycle.

Inorganic fertilizers account for the bulk of nutrients applied by farmers to the rice crop, but organic manure was also used at two sites. All of our sample farmers in northern Vietnam apply farmyard manure (FYM) to their field for each rice crop in addition to chemical fertilizers. This manure comes primarily from their own farms. Most farmers in Tamil Nadu also apply FYM, but it is primarily used in the HYS crop (*kuruvai* season). None of the farmers at the other sites applied FYM, presumably because such fertilizers are either difficult to obtain or expensive (in terms of labor) to apply. The high rates of FYM applied in northern Vietnam (12–13 t ha^{-1}) require nearly 20 person-days ha^{-1} of labor for application. This labor is all family labor that appears to have little opportunity cost. In Tamil Nadu, the quantity of FYM applied is similar to that in northern Vietnam, but it generally takes only half as long on a per-hectare basis. This labor clearly has an opportunity cost since it is predominantly hired labor, but wages in Tamil Nadu are among the lowest of the RTDP sites.

Table 3.10. Preplant application of fertilizer nutrients at RTDP farms, 1994-99.

Site	% of farmers who applied preplant	Preplant fertilizer application							
		Nitrogen		Phosphorus		Potassium	
		kg ha^{-1}	% of total	kg ha^{-1}	% of total	kg ha^{-1}	% of total
Central Luzon, Philippines	11	49	34	9	51	12	49
Central Plain, Thailand	1	20	19	11	51	0	0
Mekong Delta, Vietnam	3	17	18	2	16	1	5
West Java, Indonesia	4	79	72	8	96	10	100
Tamil Nadu, India	92	55	32	30	86	15	29
Red River Delta, Vietnam	76	13	14	18	100	4	7
Zhejiang, China	46	68	39	15	83	10	16

Furthermore, organic manure is available on-farm because of integrated crop-live-stock farming systems at this site.

Pesticide use

Farmers at the RTDP sites vary considerably in the quantity of pesticides applied to rice (Table 3.11). As with N use, total pesticide use by farmers in Zhejiang, China, was far in excess of the amounts used at other sites. This was due to the extraordinarily high use of insecticides, which was nearly four times the rate used at the site with the second highest use (Thailand). This is consistent with another recent study that also shows a very high use of insecticides in Zhejiang (Widawsky et al 1998). After China, the next largest users of pesticides were farmers in Thailand, Indonesia, and northern Vietnam. Farmers at these sites used relatively high amounts of all types of pesticide: insecticides, herbicides, and others (which include primarily fungicides and molluscicides). Farmers at the remaining sites (the Philippines, southern India, and southern Vietnam) used relatively small quantities of pesticides. Integrated pest management (IPM) extension programs that encourage less reliance on methods of chemical control have been especially active near the RTDP sites in the Philippines and Tamil Nadu, and these efforts may be partially responsible for the lower use at these sites.

Differences in pesticide use between high- and low-yielding seasons were generally small, so these data are not presented. The one exception is China, where insecticide use (active ingredient, ai) in the high-yielding season (late rice) was much greater (6.47 kg ai ha^{-1}) than use in the low-yielding season (1.46 kg ai ha^{-1}). This may be due to lower temperatures and a shorter growing season in the early rice crop, which result in less pest pressure. Insecticide use in the early rice crop in China is still in the upper range for RTDP sites, but use in the late rice crop is far above that at other sites.

To examine the effect of prices on pesticide use, prices for insecticide were calculated by dividing total insecticide costs by the quantity of active ingredient. The same was done separately for herbicides. This procedure has problems because one unit of active ingredient for one type of insecticide is not the same as a unit of active ingredient of another insecticide. On the other hand, the wide range of pesticides used, most of which are not used at all sites, makes it difficult to compare prices for specific pesticides. Based on these calculations, China has by far the lowest insecticide prices, with prices roughly one-third of the prices in Thailand, Vietnam, and Tamil Nadu. Insecticide prices at these latter sites are in turn roughly half of the prices in Indonesia and the Philippines. The high prices in the Philippines, coupled with IPM, probably account for the very low level of insecticide use at the Maligaya site. In Indonesia, however, the high prices have not deterred heavy use. Herbicide prices were similar across sites.

Pesticide use also appears to be affected to some extent by nitrogen use. These nutrient by pest interactions were quantified by estimating regressions with either insecticide or herbicide use (kg ai ha^{-1}) as the dependent variable and nitrogen use (kg N ha^{-1}) as the independent variable. Separate regressions were estimated for

Table 3.11. Pesticide use among RTDP farms, 1994-99.

Site	Insecticide	Herbicide	Other pesticide	Total	Handweeding
	(kg active ingredient ha $^{-1}$ crop $^{-1}$)				(person-days ha $^{-1}$)
Central Luzon, Philippines	0.18	0.34	0.18	0.70	0.7
Central Plain, Thailand	0.97	0.89	0.25	2.10	0.9
Mekong Delta, Vietnam	0.51	0.49	0.10	1.10	8.3
West Java, Indonesia	0.62	0.69	0.54	1.85	25.4
Tamil Nadu, India	0.29	0.11	0.01	0.41	24.7
Red River Delta, Vietnam	0.61	0.65	0.34	1.60	18.8
Zhejiang, China	3.96	0.09	0.17	4.23	0.0

Table 3.12. Effects of N use on pesticide use, RTDP sample farms. The table shows the coefficient on N in a fixed or random effects regression of pesticide use per hectare on N use per hectare for each season at each site.

Site	Insecticide use[ab]		Herbicide use	
	LYS	HYS	LYS	HYS
Tamil Nadu, India	0.0006	−0.0019	0.0025*	0.0009
Mekong Delta, Vietnam	0.0049*	0.0116*	0.0059*	0.0038*
Red River Delta, Vietnam	−0.0098	0.0092*	0.0042	−0.0080*
Zhejiang, China	0.0098*	−0.0102	0.0000	0.0001*
Central Luzon, Philippines	0.0017	−0.0006	0.0009	0.0004
Central Plain, Thailand	0.0036	−0.0015	−0.0010	0.0022
West Java, Indonesia	0.0053*	−0.0040	0.0016	−0.0007
Average	0.0023	0.0004	0.0020	−0.0002

[a]LYS = low-yielding season, HYS = high-yielding season. [b]*indicates statistically significant at 5% level or less.

each season (HYS or LYS) at each site. Farm-specific effects were controlled for using either a fixed effects or random effects model, the choice of which was based on the value of the Hausman test statistic (Hausman 1978). For insecticides, nutrient by pest interactions were found to be statistically significant at the 5% level in five of the 14 instances (seven sites by two seasons), and they were positive in each of those five cases (Table 3.12). For these five site-seasons, the magnitude of the coefficient ranged from about 0.005 to 0.01. This means that an increase in N use of 20 kg N ha^{-1} would be associated with an increase in insecticide use of about 0.1 to 0.2 kg ai ha^{-1}. In general, across all sites and seasons, the average magnitude of the effect was substantially higher in the LYS (average coefficient equal to 0.0023) than in the HYS (average coefficient equal to 0.0004).

Herbicide use also appeared to be affected by nitrogen use. Again, these interactions were statistically significant in five of the 14 cases and four of these five significant effects were positive (Table 3.12). For the four cases in which the interaction effect was positive and statistically significant, the average magnitude of the coefficient indicates that an increase in N use of 20 kg N ha^{-1} is associated with an increased herbicide use of 0.06 kg ai ha^{-1}. This is smaller than the insecticide effect. As with insecticide use, these interaction effects were typically larger in the LYS.

One might expect that herbicide use would be correlated with crop establishment method, but in general this is not the case at our sites. Among the group of sites with the lowest overall pesticide use (Tamil Nadu, the Philippines, and southern Vietnam), herbicides are relatively more important at the latter two sites, where direct seeding is common. This correlation is weak, however, when all sites are considered because Indonesia and northern Vietnam use large quantities of herbicide despite transplanting being the method of crop establishment and large quantities of labor being used for hand weeding. In general (with the exception of China), herbicide use seems to be correlated quite strongly with insecticide use, suggesting that herbicide use might be determined primarily by receptivity to and familiarity with pesticides in general.

Labor use and contractual arrangements

Labor is the most important factor of production at all sites, although the intensity of labor use varies markedly across the sites. The primary determinant of the intensity of labor use is the level of prevailing wages in the area, which are in turn determined by general conditions in the macroeconomy and the labor market. Although a discussion of the general economic conditions prevalent in each of the RTDP project countries is well beyond the scope of this chapter, the implications of these conditions for rice farmers are profound.

Wages in Suphan Buri, Thailand, are by far the highest of any of the RTDP project sites. In the past two years (1998-99), wages at this site were on average about $7.40 per 8-h day. As a result, farmers in Suphan Buri use the least amount of labor, about 14 person-days ha^{-1} crop^{-1} (Fig. 3.1A). Wages are high because of the widespread availability of jobs in the industrial and service sectors in nearby Bangkok and its surroundings. These high-paying jobs were created by the rapid development of the Thai economy during the past 30 years. High wage rates have provided incentives for farmers to reduce labor use and these incentives have led to two major developments. First, crop establishment in Thailand is by direct seeding instead of the more labor-intensive system of transplanting. Second, land preparation, harvesting, and threshing are fully mechanized. All the sample Thai farmers use a combine-harvester-thresher that can finish 1 ha of rice in 4 h with only four accompanying operators. Thus, farmers in Suphan Buri use an average of just 5 person-days ha^{-1} to carry out all harvest and postharvest operations versus more than 80 person-days ha^{-1} in northern Vietnam.

In contrast, farmers in India, Indonesia, and northern Vietnam have the most labor-intensive rice production systems, with labor use in northern Vietnam averaging about 260 person-days ha^{-1} crop^{-1} (Fig. 3.1A). At these sites, most land preparation activities are still done by human and animal power and harvesting and threshing activities are mostly done manually with the aid of simple implements such as a sickle for harvesting and a pedal thresher for threshing. Crop care activities such as

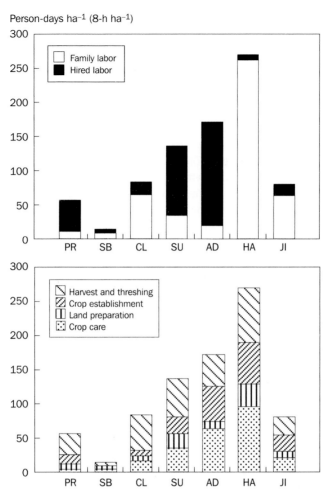

Person-days ha^{-1} (8-h ha^{-1})

Fig. 3.1. Labor distribution by activities (A) and labor use by source (B), RTDP sites, 1994-99. PR = PhilRice, SB = Suphan Buri, CL = Cuu Long, Mekong Delta, SU = Sukamandi, AD = Aduthurai, HA = Hanoi, JI = Jinhua.

hand weeding and irrigation also use considerable quantities of labor. These crop management practices are a direct result of the low wages at these sites. In Tamil Nadu, farm wages during the past two years were about $1.50 per day and in Sukamandi, Indonesia, they were just $1.25 per day. (It is difficult to gather reliable wage data in northern Vietnam because nearly all labor is unpaid family labor; see the discussion below.)

Labor use in Sukamandi increased substantially because of the economic crisis that began in late 1997. Before the crisis, average labor use was about 112 person-days ha^{-1} crop^{-1}, but this increased to 161 person-days ha^{-1} crop^{-1} postcrisis. All of the increased labor use was from additional hired labor for harvesting and threshing and additional family labor for crop care. Much of the former may have been due to the migration of laborers from urban areas, which were the hardest hit by the crisis. Thus, labor use at Sukamandi in 1998-99 has been much closer to that in Aduthurai than what is shown in Figure 3.1, which presents average data for 1994-99.

Intermediate in the level of wages and the quantity of labor use are the sites in southern Vietnam, the Philippines, and China. At Maligaya, wages are about $4.00 per day, comparable with those in southern Vietnam and China (although the prevalence of family labor at these two sites makes it more difficult to measure them reliably). At these sites, threshing is done with the use of mechanical threshers that can finish a 1-ha rice crop in 1 day with about four to eight accompanying laborers. Almost all land preparation activities, from first plowing to leveling, are done with tractors instead of the traditional bullock power. Filipino and southern Vietnamese farmers also save labor by direct seeding their crop instead of using conventional hand transplanting. In China, our sample farmers established their crop by transplanting up to and including 1999, but direct seeding started to become common in 2000. Other farmers in Zhejiang establish the crop by throwing seedlings, which also saves labor.

The labor used in rice production consists of both family labor and hired labor (all labor from outside the family is considered hired labor, whether paid in cash or in kind). The relative importance of family and hired labor varies considerably across sites (Fig. 3.1B). Labor in China and in both northern and southern Vietnam is primarily family labor because of a relatively equal distribution of land in these countries and correspondingly low numbers of rural landless. At the other extreme, there are large numbers of rural landless in the Philippines, Indonesia, and India, and these are the workers who supply the bulk of the labor requirements at these sites. The small amount of labor used in Suphan Buri is split about equally between family and hired labor.

In Tamil Nadu, hired laborers are both male and female, but daily wage rates for males are typically higher than for females by about 15–20% (for the same activities). A similar phenomenon exists at some locations in Indonesia, but this does not appear to be the practice in Sukamandi.

Because of difficulties in monitoring, it is not possible to pay laborers based on their output (e.g., weeds pulled, seedlings transplanted). Thus, hired labor is usually paid on a daily basis. However, more complex contractual arrangements exist at the RTDP sites in the Philippines and Indonesia. Transplanting, harvesting, and threshing are the three major activities affected by these more complex labor arrangements.

For harvesting, the *hunusan* system is the most popular practice among rice farmers not only at the RTDP site but also at many other places in the Philippines. It is a crop-share arrangement in which the group of harvesters is given a percentage share of the gross harvest in kind, ranging from 7% to 10% depending upon the village and the season. Percentage shares are higher in the wet season when yields are lower, harvesting is physically more difficult, it is more critical to remove the crop from the field quickly to avoid spoilage, and there is a greater area that needs to be harvested. Under the *hunusan* system, harvesting is still done manually using a sickle to cut several stalks at one time. The cut crop is then placed in small stacks for threshing. Threshing is done with portable threshers, usually operated by four to eight persons. These individuals are usually different people and not the harvesters. The payment system for threshing is also on a contractual basis according to crop share, the most common of which is 6% to 7% of the gross harvest. A small portable thresher can finish on average 5 t of paddy in 1 day. For transplanting, it is a common practice among the RTDP sample farms in Central Luzon to pay a fixed amount per hectare.

Another interesting labor arrangement in Central Luzon is the practice of hiring a so-called "permanent laborer (PL)" to manage the overall daily operations of the rice farm. This is usually practiced by farmers who own several parcels of land or who have other work outside the farm. The PL is expected to supervise all farm activities (e.g., transplanting, application of fertilizer and pesticides) and is also sometimes asked to participate directly in those activities. In return for his labor, he is given a 10–11% share of the rice crop immediately after harvest. The time spent by the PL in supervision is not included in our estimates of labor use in order to be consistent with other sites where there are no PLs and supervisory labor is not included.

As in Central Luzon, transplanting in Sukamandi is often done on a contractual basis at a fixed cost per hectare. The cost is typically lower in Sukamandi than in Central Luzon. Arrangements for harvesting and threshing in Sukamandi are also similar to those in Central Luzon, with contracting of organized groups of laborers on a share-based payment system (known as the *bawon* system in West Java). Both harvesting and threshing are done manually using sickles to cut the rice stalks (Wardana et al 1998) and thus the same group of individuals is responsible for harvesting and threshing a particular field. Payments in Sukamandi range from 10% to 25% of the harvest (depending on the village), covering both harvesting and threshing.

In Sukamandi, a limited number of farmers are practicing a system of selling the rice crop while it is still standing in the field but just before it is ready to be harvested. The buyer/trader estimates the value of the crop to be harvested and offers a fixed price to the farmer. If an agreement is reached, the buyer/trader pays the farmer in advance and then takes care of all harvesting and threshing activities, with the farmer not necessarily even knowing the quantity of total production. Under this arrangement (known as the *tebasan* system), the buyer/trader typically hires the harvesters and threshers on a daily wage basis paid in cash and does not use the *bawon* system.

3.2 Profitability of irrigated rice production systems

Farmers incur costs and receive returns in local currency, but presentation of the data in this form makes cross-country comparisons more difficult. Thus, data are presented here in $ ha^{-1} after converting with contemporaneous exchange rates. In contrast to many previous tables, we use only data from the two most recent crops to calculate costs and returns. This is because of the large exchange rate adjustments that have occurred in many Asian countries since the socioeconomic data collection began. Finally, we aggregate data across the HYS and LYS to obtain estimates of costs and returns on a two-crop basis. This simplifies the presentation of the data without losing much information because the factors underlying the differences in costs and returns across sites do not vary between seasons.

Costs of rice production

Total costs consist of all farm-related expenses that are paid either in cash or in kind from the day the farm is plowed up to the day the produce is sold or stored. Information on land rents was not collected, so these costs are excluded from the analysis. For family labor, it is necessary to impute some cost to this time. In Tables 3.13 and 3.14, we used the standard procedure in the economics literature of valuing family labor at the mean wage rate of hired labor within the study area. This assumes that there is an opportunity cost to working on the family farm and, had the family member not spent time working on the farm, he/she could have found work off the farm at the local wage rate for hired labor. This procedure will result in an overestimation of costs if the opportunity cost for family labor is lower than the market wage rate. For example, if the family member has no possibilities whatsoever for finding any gainful employment off the farm, then the true opportunity cost of labor might be close to zero (ignoring the benefits of additional leisure or time spent in taking care of the household). Although this extreme situation is unlikely at our sites, valuing family labor at the prevailing wage rate probably overstates (perhaps quite substantially) the labor costs that should be attributed to the farming enterprise. This is particularly important for the sites in China and Vietnam, and some allowance for this consideration is made in the discussion below.

Among the different costs of production, labor (hired and family) constitutes the biggest share of total costs at all sites (Table 3.13). Labor costs in labor-intensive rice systems like those in India, Indonesia, and northern Vietnam are about 70% of total costs, whereas, in Thailand, where mechanization is quite advanced, labor constitutes 33% of total costs. At the sites with intermediate levels of labor use (Philippines, southern Vietnam, China), labor accounts for a little bit more than half of total costs.

The proportion of costs spent on fertilizer is much lower than for labor. It ranges from a low of 11–13% in India and Indonesia, where fertilizer prices are low, to a high of 28% in China, where fertilizer use is very heavy. Considering only paid-out costs, however (i.e., excluding the imputed value of family labor), fertilizer often costs more than hired labor. This is true at all four sites, where family labor constitutes more than half of total labor use (China, northern and southern Vietnam, and Thailand). The proportion of costs spent on pesticides is usually much lower than that spent on fertilizer. Most other cost items are also low as a share of production costs, with the exception of Suphan Buri, where machine rental accounts for nearly one-quarter of total costs.

Table 3.13. Comparative annual costs of rice production among RTDP farms, US$ ha^{-1}, 1999.

Item	Central Luzon, Philippines	Central Plain, Thailand	Mekong Delta, Vietnam	West Java, Indonesia	Tamil Nadu, India	Red River Delta, Vietnam	Zhejiang, China
Labor	501	207	435	472	490	764	404
Hired	415	95	60	328	430	30	99
Family (imputed)	86	112	375	144	60	735	305
Fertilizer	139	125	95	73	90	145	203
Machine rental and fuel cost	109	147	40	44	70	3	32
Pesticides	47	91	44	65	7	45	52
Seeds	63	61	56	9	30	33	40
Other costs	29	4	12	7	11	78	0
Total costs per hectare	888	636	683	670	698	1068	731
Total costs per ton of paddy ($ t^{-1})	(96)	(59)	(74)	(69)	(62)	(86)	(58)
(% of total costs ha^{-1})							
Labor	56	33	64	70	70	72	55
Fertilizer	16	20	14	11	13	14	28
Machine rental and fuel cost	12	23	6	7	10	0	4
Pesticides	5	14	6	10	1	4	7
Seeds	7	10	8	1	4	3	5
Other costs	3	1	2	1	2	7	0

Table 3.14. Comparative annual profitability of rice production across RTDP farms, US$ ha^{-1}, 1999.

Site	Total value of production	Total paid-out costs[a]	Total costs[b]	Returns over paid-out costs[c]	Net return[d]
Central Luzon, Philippines	2,083	802	888	1,282	1,196
Central Plain, Thailand	1,302	524	636	778	666
Mekong Delta, Vietnam	1,160	308	683	852	477
West Java, Indonesia	1,490	526	670	964	820
Tamil Nadu, India	1,375	638	698	737	677
Red River Delta, Vietnam	1,834	334	1,068	1,500	766
Zhejiang, China	1,718	426	731	1,292	987

[a]Costs paid out by the farmer plus imputed cost of seed. [b]Includes total paid-out costs plus imputed costs of family labor. [c]Total value of production – total paid-out costs. [d]Total value of production – total costs.

Farmers in northern Vietnam incurred the highest production costs of $1,068 ha^{-1} in 1999 (Table 3.13). Most of these costs are imputed costs of family labor, however, with only one-third being paid-out costs. These relatively high total costs are substantially overstated because the opportunity cost of family labor in Vietnam is much less than market wages in the area.

The next highest level of costs was in the Philippines. The main factor responsible for these high costs appears to be relatively high labor costs (the high prices for fertilizers and pesticides play a much smaller role). With the exception of northern Vietnam, labor costs in the Philippines were the highest of any site although the quantity of labor use in the Philippines is the second lowest of all sites. However, the high labor costs in the Philippines are due in strong part to abnormally high farm-level prices for paddy (see the discussion in the next section) and are thus somewhat misleading. Because Philippine laborers participating in harvesting and threshing are paid directly in paddy, the value of this paddy is calculated using its market price. The high price for paddy also contributes to higher wages in other farm activities not paid directly in paddy by raising the value of labor services and thus the demand for labor. Thus, the high payments made to laborers in the Philippines do not necessarily translate into high living standards for these laborers because rice is an important part of their expenditures and they must pay high prices for this rice. These high costs also do not imply low net returns to rice cultivation. This is because the high costs are caused by high wages that are in turn caused by high paddy prices, which serve to raise gross revenues (see the discussion in the next section) at the same time that they raise production costs.

With the exception of northern Vietnam and the Philippines, the average annual costs of rice production were remarkably similar across sites, ranging from $635 ha^{-1} in Suphan Buri to $731 ha^{-1} in Zhejiang. This range would be even narrower if the opportunity cost of family labor in Zhejiang were valued at less than the market wage, which would lower production costs at that site.

The cost estimates in Table 3.13 do not reflect which countries have a comparative advantage for producing rice. First, these data pertain to only one particular region in each country and competitiveness with other countries depends on a comparison of relative costs at sites throughout the country. Second, our data do not consider the cost of land and the production costs of alternative crops that compete with rice for agricultural land use. Third, our calculations may overadjust for the cost of family labor, especially in China and at the two sites in Vietnam. For all of these reasons, the data presented here do not provide even an approximate guide to the comparative advantage of different countries.

Returns to rice production

The gross value of rice production per hectare varied substantially across sites (Table 3.14). Total revenue was highest in the Philippines, northern Vietnam, and China. In the Philippines, gross returns per hectare were high primarily because of relatively high prices (as opposed to high yields). The high prices received by Philippine farmers (more than 50% higher than those received by farmers in Suphan Buri and in the Mekong Delta) are due primarily to the restrictive import policy of the government that prevents the entry of cheaper rice from exporters such as Vietnam and Thailand. In China and northern Vietnam, high gross returns are partially caused by high prices, but relatively high yields also play an important role.

The net returns reported in Table 3.14 can be interpreted as returns to land and management skills. These net returns were by far the highest in the Philippines at nearly $1,200 ha^{-1} y^{-1} primarily because of the high farm-level paddy prices noted above. Net returns at the other sites in 1999 ranged from $477 ha^{-1} in the Mekong Delta to $987 ha^{-1} in Zhejiang.

These calculations of net returns assume that all rice produced on the farm is sold into the market, that is, an imputed value is assigned to the paddy rice that is retained within the household for eventual consumption. The imputed value of a unit of rice retained for home consumption is equal to the price the farmer received for the rice that was sold into the market. Not surprisingly, the share of production that is consumed at home varies substantially across sites. In Thailand, where farms are large, family incomes are high (so that per capita rice consumption is less), and farmers have an export orientation, less than 1% of production is retained for home consumption. In Tamil Nadu and Central Luzon, farm sizes are also relatively large, so that less than 15% of production is retained for home consumption. At the other extreme is northern Vietnam, where farm sizes are very small and family size relatively large. However, despite a farm size of just one-third hectare, farmers at this site sell more than one-third of their production to the market.

Net returns vary not only across sites but also within sites. Since the prices of inputs and outputs do not vary substantially within a site, intrasite variations in profitability must be due primarily to differential yields and differential input use. The most important factor is differential yields, which was discussed in detail in an earlier section. Thus, there is no further discussion here, except to note that profitability will vary widely from one farmer to another at the same site in any given season.

3.3 Summary and conclusions

There is tremendous diversity in the way rice is grown in the intensive irrigated rice production systems of Asia. Large differences occur across sites in crop management practices, driven primarily by wages and the general state of the larger economy in which farmers operate. Within a particular site, yields and fertilizer use exhibit tremendous variability: across space at a given site in a particular year and even across time for a particular farmer. Thus, although many Asian farmers are thoroughly familiar with modern rice varieties and have accumulated substantial knowledge about crop management, they are not passive managers who simply do the same thing year after year. Instead, they react dynamically to changing climatic conditions, water availability, and market prices.

Labor is the most important production cost on rice farms and many of the differences in crop management among sites are due to different levels of wages. These differences in wages affect the quantity of labor use, mechanization, the method of crop establishment, and technologies used in crop care. Labor-intensive production methods tend to predominate in India, northern Vietnam, and Indonesia, where labor is abundant and wage rates are low. In such situations, the crop is established by transplanting, more human effort goes into crop care, and harvesting methods are relatively simple. Farmers in Thailand, southern Vietnam, China, and the Philippines generally have a lower availability of labor relative to land. In these areas, the crop is direct-seeded and less human effort goes into crop care. In the extreme case of Thailand, all harvesting and threshing activities are mechanized.

For fertilizer use, several interesting patterns emerge from the analysis. The use of fertilizer and pesticides is greatest in China, which has both positive and negative aspects. Farmers there use large quantities of K, thereby avoiding mining of soil resources. On the other hand, the use of N fertilizer and insecticides is clearly excessive and contributes to negative environmental effects. Farmers at all sites treat phosphorus and potassium fertilizers differently from nitrogen. N is viewed as a raw material input that is essential for production each and every year, whereas P and K are viewed as investment goods. These latter inputs contribute to production over the long term but do not need to be applied every year. Contrary to much belief, credit does not appear to be a serious constraint to fertilizer use in these irrigated production systems. Farmers do adjust fertilizer inputs in response to changing prices, especially for P and K, but they also take many other factors into consideration, making it difficult to isolate the effect of prices. The potential effect of prices is clear from the Indonesian experience, where fertilizer use declined substantially after the Southeast Asian financial crisis caused Indonesian domestic fertilizer prices to rise.

The profitability of rice production systems depends on yields and input quantities, but it also depends in an important way on government policies, which determine the prices of those inputs and outputs. The output price policy is more important than the input price policy because the cost of material inputs is just a small share of the gross value of production. Thus, net returns are highest in the Philippines, where government trade policies result in high prices for farmers (and for consumers). Fertilizer price subsidies also affect profitability, but only to a small extent. Farmers in India are the main remaining beneficiaries of these subsidies, but international trade agreements will probably erode these distortions over time.

Because of the large role played by rice in the Asian economy, especially for poor farmers and consumers, income growth and poverty alleviation depend in an important way on productivity in the rice sector. There are two main channels by which productivity can be increased (producing more rice with the same or fewer amounts of inputs). One way is to focus on the least productive farmers and raise them to the level of the best farmers. This chapter, however, suggested that the yield gaps between these groups of farmers are probably smaller than is commonly believed. Although there are undoubtedly yield gaps between good and average farmers, it may be difficult to close them appreciably in a cost-effective manner because of differences in soil endowments across farms. Furthermore, some people always have more talent for certain endeavors than others and farming is probably not an exception.

The other method of increasing productivity is to generate new technologies or management strategies that raise productivity for all farmers. These innovations can be embodied in new seeds (e.g., hybrid rice, the new plant type, or C_4 rice) or they can improve productivity using existing germplasm. Improved methods of site-specific nutrient management are an example of the latter approach. The generation of new technologies is especially important for irrigated rice, which supplies three-quarters of Asia's rice consumption. The generation of new technologies depends to a great extent on the funding provided to agricultural research, and usually the most important policy that can be implemented in the Asian rice sector is to support the agricultural research system. Other policies, such as ones that raise or lower input prices, are generally much less effective at increasing productivity.

References

Byerlee D, Murgai R. 2000. Sense and sustainability revisited: the limits of total factor productivity. Washington, D.C. (USA): World Bank. 26 p.

Cassman KG, Pingali PL. 1995. Extrapolating trends from long-term experiments to farmers' fields: the case of irrigated rice systems in Asia. In: Barnett V, Payne R, Steiner R, editors. Agricultural sustainability: economic, environmental and statistical considerations. Chichester, West Sussex (England): John Wiley & Sons. p 63-84.

Dawe D, Dobermann A. 1998. Defining productivity and yield. IRRI Discussion Paper Series No. 33. Makati City (Philippines): International Rice Research Institute. 13 p.

Dawe D, Moya P. 1999. Variability of optimal N applications for rice in the Philippines. In: RTDP: reversing trends of declining productivity in intensive irrigated rice systems. Progress report 1998. Makati City (Philippines): International Rice Research Institute. p 83-87.

Dobermann A, Dawe D, Roetter R, Cassman KG. 2000. Reversal of rice yield decline in a long-term continuous cropping experiment. Agron. J. 92(4):633-643.

FAO (Food and Agriculture Organization of the United Nations). 2001. FAO on-line electronic database. www.fao.org.

Hausman JA. 1978. Specification tests in econometrics. Econometrica 46:1251-1271.

Hayami Y, Kikuchi M, Marciano EB. 1999. Middlemen and peasants in rice marketing in the Philippines. Agric. Econ. 20:79-93.

Huke RE, Huke EH. 1997. Rice area by type of culture: South, Southeast, and East Asia. A revised and updated database. Makati City (Philippines): International Rice Research Institute. 57 p.

Moya P. 1998. Fertilizer management practices at the farm level. In: RTDP: reversing trends of declining productivity in intensive irrigated rice systems. Progress report 1997. Makati City (Philippines): International Rice Research Institute. p 101-103.

Wardana IP, Farid A, Huelgas ZM. 1998. Baseline profile of on-farm soil fertility management in the Jatiluhur irrigation system, West Java, Indonesia. In: Olk DC, Moya PF, editors. On-farm management of applied inputs and native soil fertility. IRRI Discussion Paper Series No. 23. Makati City (Philippines): International Rice Research Institute. p 133-146.

Widawsky D, Rozelle S, Jin S, Huang J. 1998. Pesticide productivity, host-plant resistance and productivity in China. Agric. Econ. 19:203-217.

Notes

Authors' addresses: P. F. Moya, D. Dawe, D. Pabale, M. Tiongco, International Rice Research Institute, Los Baños, Philippines; N.V. Chien, National Institute for Soils and Fertilizers, Hanoi, Vietnam; S. Devarajan, Tamil Nadu Rice Research Institute, Aduthurai, India; A. Djatiharti, Research Institute for Rice, Sukamandi, Indonesia; N.X. Lai, Cuu Long Delta Rice Research Institute, Cantho, Vietnam; L. Niyomvit, Suphan Buri Rice Experiment Station, Suphan Buri, Thailand; H.X. Ping, Jinhua Agricultural School, Jinhua, Zhejiang Province, China; G. Redondo, Philippine Rice Research Institute, Maligaya, Nueva Ecija, Philippines; and P. Wardana, Research Institute for Rice, Sukamandi, Indonesia. p.moya@cgiar.org.

Acknowledgments: We would like to acknowledge all the farmer cooperators who participated in this project. Without their cooperation, this project would not be possible.

Citation: Dobermann A, Witt C, Dawe D, editors. 2004. Increasing productivity of intensive rice systems through site-specific nutrient management. Enfield, N.H. (USA) and Los Baños (Philippines): Science Publishers, Inc., and International Rice Research Institute (IRRI). 410 p.

4 | Trends in sustainability and farm-level productivity in intensive Asian rice-based cropping systems

D. Dawe

Measuring sustainability in a long-term experiment is a relatively easy task. Typically, an experiment is said to be sustainable if yield is maintained at some given level, and yield is relatively easy to measure. If yields are declining, then questions are raised about sustainability, but, if yields are increasing or are at least constant, then the system is said to be sustainable. Understanding the causes of any trend is of course more difficult, but at least the measurement of sustainability is straightforward. This procedure is entirely appropriate if the use of all inputs in the experiment is constant, as is usually the case under experimental conditions.

Under farm conditions, however, the task of measuring sustainability is considerably more complex. Farmers change input use even in the short term from year to year, often considerably, and this problem is exacerbated over the longer term with the emergence of new technologies. Thus, for example, farm yields might still be increasing, but only because the use of inputs is increasing even more rapidly. If continued rapid growth in input use is not sustainable, growth in yields might not be sustainable either. Because input use is continually changing in farmers' fields (more of some inputs, less of others), economists try to measure total factor productivity (TFP) whenever possible. This is not always possible, however, so it is helpful to have some understanding of the advantages and disadvantages of the different measures of productivity that scientists commonly use.

4.1 Measures of sustainability

Partial factor productivities (PFPs): potential for misleading results

Several commonly used measures of productivity belong to the class of partial factor productivities (PFPs). Partial factor productivity refers to the productivity of a single input. For rice, it is calculated as the quantity of grain divided by the quantity of the specific input. For example, the PFP of nitrogen fertilizer would be measured in units of kg grain kg^{-1} N. Some studies have used long-term trends in PFP-N as indicators of sustainability, but this can be misleading. Although the PFP-N in rice cultivation has declined considerably in the past 40 years, most of this decline is due to movement along a nitrogen response function as opposed to a downward shift of the response function itself. The former is not cause for concern as it merely reflects the fact that farmers took some time to learn about optimal levels of fertilizer usage. For example, survey data for a group of farmers in Central Luzon in the Philippines show that it took 10 to 15 years after the introduction of modern varieties for average N use in the wet season to increase from 10 to 60 kg ha^{-1} (loop survey, IRRI). The spread of higher levels of fertilizer use from one area to another has also taken time, requiring the transmission of knowledge and the construction of irrigation systems. Thus, as modern varieties spread and farmers learned about fertilizer, fertilizer use increased sharply. Since nitrogen response functions are highly concave, this large increase in fertilizer use has led to a sharp decline in the PFP-N. But this decline in the PFP-N is of no concern and does not imply a lack of sustainability in the system.

As an illustrative example, let us assume a fixed nitrogen response function (wet season) with parameters based on those estimated for Central Luzon farmers in Cassman and Pingali (1995). Assuming a fixed response function over time is plausible as the nitrogen responsiveness and yield potential of modern varieties have changed little since the introduction of IR8. According to data from the loop survey of IRRI's Social Sciences Division, farmers in Central Luzon increased their use of N from just 9 kg ha^{-1} in 1966 to 94 kg ha^{-1} in 1994. Using these figures of actual N use and assuming the fixed response function above, the PFP-N would decline from 340 kg rice kg^{-1} N in 1966 to just 44 kg rice kg^{-1} N in 1994, a fall of nearly 90%! This is a large decline, but it means little, as it is clear that nearly all of the increased fertilizer use reflects the learning process of farmers in approaching optimal levels of N application.

Some studies do not appear to appreciate the full magnitude of the decline in PFP-N that is engendered merely by the diminishing returns that set in because of movement along a fixed response function. For example, Woodhead et al (1994) state that "decreases (in PFP-N) may be explained in part (but not in whole) by the diminishing incremental response as fertilizer rate increases." Yet the decreases cited in their accompanying table are typically in the range of 80% to 90%, similar to those in the figure from the illustrative calculation above. Thus, it is plausible that the whole of the decline is due to naturally diminishing returns. Cassman and Pingali (1995) state that "the magnitude of the decrease in N fertilizer factor productivity, from 75 to 28 kg of grain output per kg applied N, raises the question of whether degradation of the paddy resource base due to the imposition of continuous, irrigated rice monocropping has also contributed to declining N output/input efficiency." Yet again, the decline cited is consistent with the decline from the above calculation.

It is possible to question the assumption of a fixed N response function in farmers' fields over time. If the response function has shifted inward, as appears to have happened in some long-term experiments (Dobermann et al 2000), this inward shift would contribute to a decline in PFP-N. However, the above example is merely designed to show that an inward shift of the response function is not necessary to generate a large decline in PFP-N. Thus, it is not correct to assert that a decline in PFP-N is evidence of an inward shift of the response function.

Table 4.1. Growth rates of rice yields in Asia.[a]

Country	Period	
	1967-90	1990-99
	(%)	
Bangladesh	1.8	1.2
Cambodia	0.7	2.5
China	2.7	1.1
India	2.3	1.4
Indonesia	4.0	0.0
Japan	0.4	-0.1
Korea (South)	1.9	0.6
Lao PDR	4.4	2.0
Malaysia	1.6	0.5
Myanmar	2.5	0.9
Nepal	1.2	0.2
Pakistan	1.7	2.4
Philippines	3.4	-0.4
Sri Lanka	1.6	0.6
Thailand	0.5	1.5
Vietnam	2.2	2.1
Asia	2.3	0.9

[a]Source of raw data: FAOStat on-line database.

Although PFP-N has declined over the past 30 years in much of Asia, this decline will not necessarily continue into the indefinite future. First, since most Asian farmers have been aware of the benefits of N for some time, most of the movement along the N response curve may have already occurred. Second, it is possible that farmers currently overuse N because of inefficient application. In Japan, for example, the PFP-N has increased in recent years because of a decline in N application rates (from about 110 to 95 kg N ha^{-1}) without any sacrifice in yield. The PFP-N in Japan today is still below its level in the late 1950s, however, when N applications were about 60 kg N ha^{-1}.

It is also common to calculate the PFP of land. This is measured in tons of grain per hectare of land planted, also known as yield. All increases in output must by definition come from increases in either area planted or yield per unit land. Since land is in relatively short supply in Asia, without much prospect for future expansion of area planted to rice, nearly all production gains will have to come from increased yields. Thus, yields are sometimes used as implicit indicators of sustainability.

Yields are still increasing in most countries in Asia, although the rate of growth in the past ten years was generally slower than in the early years of the Green Revolution (Table 4.1). This conclusion holds whether growth rates are calculated in percentage or absolute terms. In some countries (e.g., Indonesia, the Philippines), yields stagnated during the 1990s and it is not clear whether future yield growth will be sufficiently rapid to meet future increases in demand (Dawe 1998). Yet even this slower growth in yields may be overly optimistic to the extent that these increases are being generated by the additional use of irrigation, fertilizer, and other conventional inputs, since it may not be possible to profitably continue increasing the use of these materials. To overcome this problem, it is preferable to calculate measures of total factor productivity (TFP) as discussed in the next section.

Total factor productivity (TFP): advantages and pitfalls

Because of these problems in the interpretation of PFP measures, economists typically use other productivity indicators whenever possible. The most common of these is known as total factor productivity (TFP), which attempts to measure the productivity of all inputs taken together instead of just one input as in the case of PFP (Dawe and Dobermann 1998). Growth in TFP is essentially growth in output minus growth in input, that is, the part of output growth that cannot be accounted for by increased use of inputs. Production functions and other variants such as cost and profit functions are also used to measure trends in productivity. The interpretation of these results is similar to that of TFP: the residual growth in output after accounting for the increased use of inputs. Unfortunately, the data requirements for measuring TFP or estimating production functions are much greater than for calculating yield or the PFP-N. Data must be collected for all outputs from and inputs used in the production process, not just grain yield and fertilizer use. For example, data on total labor use in all activities are also required and these data are often not collected in farm surveys, especially over the long term.

Although calculations of TFP are much preferred to calculations of PFP whenever possible, TFP itself is by no means free of problems. It should be noted that all of these problems are shared by measures of PFP, however, so the existence of these problems is not an excuse for calculating PFP instead of TFP. Rather, they are simply issues that need to be considered in the interpretation of trends in TFP.

First, since TFP is a ratio between outputs and inputs (or the difference if one is using logarithms), small changes in either one can lead to large changes in TFP. This implies that small errors in the measurement of outputs and inputs can have large effects on the measurement of productivity. The exclusion of one input or output can have the same effect. Thus, it is important to include as many inputs and outputs as possible and to measure them as best as one can. This can never be done perfectly, but it is important to do as good a job as possible and to be aware of the limitations of any particular data set.

A second problem is posed by the inherent variability of agricultural production. Year-to-year fluctuations in the weather cause fluctuations in output that are not due to changes in input use. In such cases, there will be substantial variations in TFP that are not indicative of any underlying trends. Thus, it is best if there are many years of data, preferably consecutive. Such data sets allow a more meaningful interpretation of changes in TFP.

If possible, it is best if TFP is calculated on a cropping systems basis because calculations made for single crops that are part of larger cropping (or economic) systems can be misleading. For example, TFP for rice might be declining because of a change in transplanting date that lowers rice yields. If this change in transplanting is done to accommodate another crop in the rotation (e.g., wheat), then it may be optimal for farmers to accept lower rice yields in exchange for higher profits with the wheat crop. Thus, in this example, it is better to calculate TFP for the rice-wheat system, not for rice and wheat separately.

Changes in the quality of inputs and outputs over time also create problems for TFP calculations. For example, suppose farmers switch to rice varieties that command higher prices on the market but have lower yields. Such a switch may be optimal from the point of view of both farmers and consumers. At the same time, since TFP calculations hold input and output prices constant, this varietal switch will result in negative TFP growth. It would be wrong to be concerned about the decline in TFP caused by this switch. Similar problems can arise from changes in the quality of inputs, such as fertilizer, seeds, or labor.

Lynam and Herdt (1989) suggest that TFP is a good indicator of cropping system sustainability. They argue that it is important to continually increase productivity because without such increases the cropping system may not be sustainable. Without such gains, farm profits are likely to deteriorate and, as a result, farmland may go out of production. Such declines in production could adversely affect food prices, thus endangering the food security of the poor. Thus, declining productivity might be a leading indicator of the need for future improvements in technology (e.g., improved varieties, changes in the cropping system) or measures to halt resource degradation (e.g., salinity, erosion).

It is important to realize, however, that changes in TFP give no information as to *why* the changes occurred. Productivity can be affected both by changes in the environment and by advances in technology, and these effects are additive. It is not always easy to disentangle the separate effects of such influences. For example, rapid advances in technology might mask a deterioration of the environment if the former are greater than the latter. Thus, continual increases in productivity do not imply that the environment is not deteriorating. Rising productivity does imply, however, that any effects of environmental degradation have been more than compensated for by other means, for example, by the release of improved crop varieties or the development of improved technologies.

Alternatively, productivity stagnation might be due to insufficient funding for agricultural research. For example, Rosegrant and Evenson (1992) show that agricultural TFP growth in India is positively associated with lagged spending on agricultural research. If such funding is inadequate, there will be a productivity slowdown that is completely unrelated to environmental degradation. In general, changes in TFP do not necessarily correlate with changes in the natural resource base. This means that, if one wants to use TFP as a measure of sustainability, sustainability must be interpreted in a very broad sense to include more than just the quality of the natural resource base. Ideally, it is best if there are data that directly measure any resource degradation. These data can then be correlated with changes in TFP to construct estimates of the effects of resource degradation. Several of the studies below have incorporated such effects. The bottom line is that no single measure is likely to serve as an adequate measure of sustainability. It is important to examine several indicators, some socioeconomic and some biophysical (Byerlee and Murgai 2001).

4.2 Trends in TFP in intensive rice-based cropping systems

Despite the pitfalls just described, TFP is perhaps the best socioeconomic indicator available for monitoring long-term changes in farmers' fields. Although the data requirements are large, many studies have measured TFP trends spanning relatively long periods of time of at least ten years or more. This section will survey the results of these studies in an attempt to provide an update on the general state of knowledge regarding productivity trends in the intensive rice-based cropping systems of Asia. This discussion will be organized by geographical area: South Asia, China, and Southeast Asia.

South Asia

Rosegrant and Evenson (1992) calculated annual growth rates in total factor productivity for India, Bangladesh, and Pakistan for the crops sector in aggregate for the period 1957-85 as well as for subperiods during this time. Unfortunately, this confounds productivity growth in rice-based cropping systems with growth in cropping systems that do not include rice. Rice is the major crop in these countries, however, so the results are worthy of description. For India, they calculated annual TFP growth of 1.01% per year for the whole period. This growth was about two-thirds of the rate cited by Jorgenson and Gallop (1992) for U.S. agriculture since the 1950s. Indian TFP growth has slowed down recently, however. During the early years of the Green Revolution from 1965 to 1975, TFP growth was 1.22% per year, but this slowed to 0.98% per year in the succeeding decade. Rosegrant and Evenson (1992) were unable to determine, however, the extent to which this slowdown of growth was due to environmental degradation or simply due to an exhaustion of the potential for spreading modern varieties to new areas. In any event, TFP growth in the latter period still exceeded the growth rate in the years prior to the Green Revolution.

In Bangladesh, TFP growth was 0.84% per annum in the 1975-85 period, after negative growth in the preceding decade because of the disruptions caused by the civil war. Growth in the latter period, however, was still below the rate of TFP growth in the 1957-65 period of 1.81% per annum. In Pakistan, the situation appears to be most serious. Although growth in TFP during the entire 1957-85 period was 1.07% per year, TFP growth was –0.36% per year from 1975 to 1985.

Jha and Kumar (1998) extended the work of Kumar and Rosegrant (1994) and measured TFP growth in rice from 1971 to 1991 for various states in India that collectively account for more than 90% of India's rice production. Most states had positive TFP growth for the 20-year period, with Madhya Pradesh and Bihar being important exceptions. Of the states with positive TFP growth, roughly half had more rapid growth in the second half of this period, whereas the other half experienced a slowdown. Kumar et al (1999) measured TFP growth for the rice-wheat cropping system in Punjab, Haryana, and Uttar Pradesh for the period 1976-92. They found positive TFP growth in all three states, but this growth appeared to be slowing considerably. In Uttar Pradesh, TFP growth from 1985 to 1992 was negative. Neither of these latter two studies incorporated any variables measuring soil or water quality, so it is not possible to relate the slowdown in TFP growth (where it occurs) to environmental factors and/or exhaustion of the initial gains from the Green Revolution.

Using district-level data from the Indian Punjab, Murgai (1999) found a pattern of relatively slow productivity growth in the early years of the Green Revolution and more rapid growth once adoption of the new seeds and fertilizer was largely complete. This conclusion is counterintuitive, since the adoption of Green Revolution technologies should have led to rapid productivity gains. This puzzle leads her to point out that TFP is a biased measure of productivity when technical change is not Hicks-neutral. Hicks-neutral technical change occurs when new technologies favor the increased use of all inputs equally. But since Green Revolution technology favored the increased use of fertilizer more than the increased use of other inputs such as labor, TFP is a biased measure of technical progress under these circumstances, and the bias can be quite large. She concludes that, after making certain adjustments, TFP growth was indeed positive in the early years of the Green Revolution. In more recent periods, she found that TFP growth from 1985 to 1993 was greater than 1.5% per annum in eight of nine districts in Punjab and Haryana, even before making adjustments to remove the downward bias in TFP (although this bias is relatively smaller in later periods than in the early Green Revolution years). The only exception was in Ferozepur, where wheat-cotton is the dominant cropping system. Murgai concluded that the evidence in India's Punjab "suggest(s) that fears about unchecked reductions in productivity growth are exaggerated."

Ali and Byerlee (2002) calculated TFP growth rates on a cropping systems basis in Pakistan's Punjab from 1966 to 1994. They found positive TFP growth of 1.26% per annum for the entire period for all systems considered together. Growth was positive in the wheat-cotton, wheat-mungbean, and wheat-mix cropping systems. In the wheat-rice system, however, TFP growth was negative, especially in the early years of the Green Revolution (1966-74). This is similar to the pattern found by Murgai, and with suitable adjustments it is likely that TFP growth during this early period was in fact positive. During the most recent period (1985-94), TFP growth in the wheat-rice system was +0.88% per year.

Of the studies that focus on South Asia, Ali and Byerlee's (2002) was the only one to use extensive data on soil and water quality to measure the importance of changes in the environment for growth in TFP. Soil quality variables included organic matter content, available phosphorus, soil pH, and total soluble salts, whereas water quality variables included residual carbonate and electroconductivity. These authors found substantial deterioration of soil and water quality in all cropping systems in Pakistan's Punjab. It was most severe in the wheat-rice system, where it reduced TFP growth by 0.41% per annum during the period 1971-94. This deterioration occurred in spite of large private and government expenditures to control waterlogging and salinity. In the absence of these environmental data, TFP growth of +0.88%

per annum from 1985 to 1994 might suggest that there is no crisis of sustainability in this system. However, the environmental deterioration in this system is sobering and draws attention to the fact that TFP alone is an inadequate measure of sustainability (Byerlee and Murgai 2001).

China

Several authors have studied productivity trends in Chinese agriculture during the past 30 years. The major event affecting agricultural productivity in China during this period was the economic reforms that began in 1978. Kalirajan et al (1996) showed that growth in TFP was negative in the eight years before the reforms (1970-78), but they did not attribute this to environmental problems. After the reforms, TFP growth was positive, and many studies have focused on the contributions of various aspects of the reforms to the surge in productivity in the late 1970s and early 1980s. Lin (1992) found that the key change was the shift from the production team system to the household responsibility system, along with increased prices for farm output. Zhang and Carter (1997) argued, however, that an important component of the increase in productivity from 1980 to 1985 was due to favorable weather conditions. Thus, while these studies have investigated productivity issues, the focus has not been on sustainability or the environment, but rather on the economic reforms.

Two studies by Huang and Rozelle (1995, 1996) explicitly incorporate environmental stress variables (salinization, area prone to flood and drought, erosion) in their studies of agricultural productivity in China. These stresses have increased over time, and these authors find negative effects on productivity due to these factors. The negative effects on productivity are more than offset, however, by other factors such as new technologies (adoption of hybrid rice) and the shift from double cropping of rice to single cropping.

Southeast Asia

All of the above studies used secondary data sources, that is, they examined productivity trends using aggregate district, provincial, or national-level data. Cassman and Pingali (1995) measured productivity trends using primary farm-level survey data collected in seven or eight different years spanning a period of nearly 25 years (1966-90) in two key rice-growing areas of the Philippines, Central Luzon and Laguna. They estimated a Cobb-Douglas production function with dummy variables representing two different periods: (1) the late 1970s and early 1980s and (2) the late 1980s and early 1990s. Based on a decline in the dummy variable coefficients between the two periods, they found evidence for a cumulative productivity decline (over approximately 10 years) of about 400 kg paddy ha^{-1} in Central Luzon and 260 kg paddy ha^{-1} in Laguna. They were unable to directly correlate this decline with measurable changes in the environment because no such data were collected during the surveys. Based on other evidence, however, they suggested that the decline might be due to a degradation of the soil resource base.

Tiongco and Dawe (2000) used these same survey data, but in addition examined time series secondary data on provincial yields for these areas. They noted that the particular years in which the surveys were conducted were unrepresentative of long-term trends and unduly influenced by exogenous yield shocks. Furthermore, the survey years with unusually high yields were in the late 1970s and early 1980s (e.g., 1982 in Central Luzon), whereas the years with unusually low yields were in the late 1980s and early 1990s (e.g., 1987 and 1995 in Laguna). This coincidence led to an overestimation of trend productivity decline. After adjusting for these effects, Tiongco and Dawe (2000) found that productivity had remained essentially stagnant after the initial adoption of modern varieties spawned by the Green Revolution.

Swastika (1995) measured changes in total factor productivity for a sample of 70 farms in West Java, Indonesia, based on survey data from 1980, 1988, and 1992 using several different methods (Tornqvist-Theil index, production frontier, cost function). He found rapid growth in TFP from 1980 to 1988 but negative growth from 1988 to 1992. Whether the positive growth in the first period exceeded the negative growth in the second period depended upon the method used. The low productivity in 1992 was due to a serious pest infestation, however, so it is not clear how well these data capture the true nature of long-term trends.

4.3 The RTDP project: improving the database for analysis of productivity trends

There are at least two key dimensions along which all of the above studies can be classified. One is whether or not the studies use explicit variables that measure environmental stresses. The other is whether the studies use farm-level data or secondary data sources. The only studies that use explicit measures of soil and/or water quality are Ali and Byerlee (2002) and the two studies by Huang and Rozelle (1995, 1996). It is clearly preferable to use such data if they are available, but all too often they are not. Unless such data are available, however, any attempt at attributing changes in productivity to changes in the natural resource base is highly speculative at best. In addition, the use of variables that quantitatively capture the extent of environmental degradation allows one to begin to examine the question of whether it is more cost-effective to increase production through further technological change or by alleviating environmental stresses. This may not always be the most important question to ask, but it is certainly an important one.

The choice of primary farm-level data or secondary data is somewhat problematic, and it is not clear which is to be preferred. In general, it would seem best to use farm-level data because they are likely to be of higher quality. With secondary data sources, it is not always clear how the primary data underlying them were actually collected and how comparable the data are across different sites. Furthermore, it is of course better to avoid the loss of information entailed in aggregation of primary data to secondary data. On the other hand, a disadvantage of farm-level data is that they are typically collected less frequently, and the choice of survey years can have a profound effect on the results of the analysis if idiosyncratic weather and/or environmental conditions characterized the survey years. This appeared to be a problem in the studies by Cassman and Pingali (1995) and Swastika (1995). Tiongco and Dawe

(2000) suggested that one possible remedy was the use of selected information from secondary data sources.

The Reversing Trends in Declining Productivity (RTDP) project attempts to resolve the above problems by collecting both biophysical (environmental) and socioeconomic data for two crops every year. This approach potentially addresses all of the problems mentioned in the previous two paragraphs, and this data set will indeed be very rich. The disadvantage with such an approach is the large cost associated with collecting so many data, especially if these costs prevent the buildup of these data over the long term.

Thus far, the RTDP project has collected such data for four to five years at several sites. Although such a short period is inadequate for assessing sustainability, some of the data are presented here to give a flavor of the type of analysis that can be conducted several years from now. The analysis is conducted on a cropping systems basis, that is, high-yielding season (HYS) and low-yielding season (LYS) data are combined for a double-crop rice-rice system. The data are combined in case farmer practices that increase productivity in one season decrease productivity in another season later in the crop year (Dawe and Dobermann, 1998).

At the RTDP farms in Central Luzon, TFP declined slightly from 1994-95 to 1997-98 (see Fig. 4.1A). This was due to both reduced output per hectare and increased input use per hectare. In terms of output, annual yield declined by a little less than 3%, from 10.55 t ha^{-1} in the first crop year to 10.26 t ha^{-1} in the most recent crop year. More important, aggregate input use increased by nearly 7%. Because labor is the most important input in rice production in the Philippines, much of the increased input use was due to an increase in labor use from 119 person-days ha^{-1} in 1994-95 to 137 person-days ha^{-1} in 1997-98. Nitrogen use was identical in the first and last crop years, and thus had no effect on the change in TFP.

Because yield and input use vary from year to year because of changing growing conditions and economic circumstances, TFP can be highly variable, and the data from Central Luzon provide an interesting example of this sensitivity. If TFP is calculated on a calendar-year basis (the choice of crop year or calendar year is essentially arbitrary), then TFP increases from 1995 to 1998 (Fig. 4.1B). The higher level of TFP in 1998 relative to 1995 is due to a small increase in yield from 9.8 to 10.0 t ha^{-1}, but more importantly to a decline in labor use from 145 to 128 person-days ha^{-1}. Fertilizer use was essentially constant from 1995 to 1998 and was not responsible for the change in TFP. The main point to note, however, is the sensitivity of the calculations. Using a similar data set (just replacing data from the 1994 wet season (WS) in the crop-year calculation with data from the 1998 WS in the calendar-year calculation) changes the results quite substantially, from a 9% cumulative decline in TFP to a 16% cumulative increase in TFP. This sensitivity highlights the importance of using a relatively long time series for assessing sustainability because this allows any distinct trends to stand out amid the year-to-year fluctuations.

For a second example, Fig 4.1C shows data from Suphan Buri in the Central Plains of Thailand. The data are presented only on a crop-year basis because there are complete socioeconomic data for only three calendar years. (Data collection started

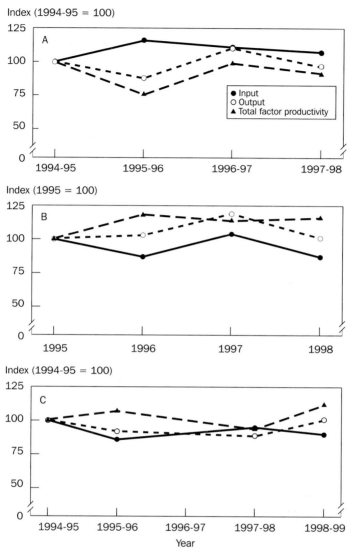

Fig. 4.1. Input, output, and total factor productivity (TFP) indices at the Reversing Trends in Declining Productivity project sites in the Philippines, 1994-95 to 1997-98 (A) and 1995 to 1998 (B), and in Thailand, 1994-95–1998-99 (C).

in late 1994 and temporarily stopped for two crops in late 1996 and early 1997. Thus, complete calendar-year data exist only for 1995, 1998, and 1999.) Here, TFP increased from 1994-95 to 1998-99 by a cumulative total of 12%, but again the variability from year to year is readily apparent. The increase in TFP is due primarily to a reduction in input use, including a decline in labor use from 35 to 27 person-days ha^{-1} and a reduction in N use from 216 to 202 kg N ha^{-1}. The use of P also declined during this period. Along with a slight increase in yields, all these changes more than offset the increased seeding rates and increased use of pesticides (especially herbicides). With only four data points, however, it is premature to speculate on what these calculations mean for the long-term sustainability of the system. Over the longer term, more substantial conclusions will be possible. Furthermore, because of the collection of biophysical data, any substantial changes in productivity can be correlated with changes in soil quality.

4.4 Conclusions

In general, most studies have found that TFP growth was relatively healthy in most intensive rice-based cropping systems in Asia in the 1970s, 1980s, and early 1990s. It is important to remember, however, that positive TFP growth can co-exist with environmental degradation, as was found by both Ali and Byerlee (2002) and Huang and Rozelle (1996). Although technological progress and improved infrastructure have more than compensated for the environmental degradation that is occurring in many locations, positive TFP growth alone does not necessarily imply that these systems are sustainable. Environmental deterioration in intensive rice-cropping systems is an important issue, especially in certain areas such as Pakistan's Punjab.

It is also important to realize that the environment is not uniformly deteriorating in all locations at all times. Data presented in Ali and Byerlee (2002) indicate a large improvement in soil salinity and a slight improvement in organic matter content since 1987 in Pakistan's Punjab. Data in Murgai (1999) similarly show large improvements in soil salinity in India's Punjab between 1972 and 1984, although more recent data were not available. These improvements in soil quality may have come at a high cost, however, so these data certainly do not indicate that environmental degradation is a problem that can safely be ignored. What they do indicate is that the issues are complex, and it is important to base future policy decisions as much as possible on data and not on preconceived biases of one sort or another. In the future, it will be important not only to understand where environmental deterioration is occurring but also to understand how much it will cost to increase production and profitability through improved environmental management. These costs are important not only in an absolute sense but also relative to the costs of achieving similar increases in production and profitability through alternative means, such as improved seeds and input management practices.

Whether or not TFP growth will continue to be healthy in the future is of course impossible to predict. However, it is risky to be overly optimistic regarding the results of the studies cited above. The calculation of TFP requires large quantities of data, and the most recent TFP estimates are not very current. Nearly all of the studies cited above stop in 1994 or sooner, and they are thus unable to fully consider the general slowdown (or, in some cases, stagnation) in yield growth during the 1990s in many parts of Asia (see Table 4.1). Yields are no higher today than they were in 1990 in Indonesia, the Philippines, or the Indian Punjab, all locations where the Green

Revolution was widely adopted many years ago (Cassman and Dobermann 2000). It will be difficult to sustain TFP growth in the future in the absence of yield growth. This argues for the continued importance of agricultural research to increase yield and productivity to reduce poverty rates as rapidly as possible (Dawe 2000).

References

Ali M, Byerlee D. 2002. Productivity growth and resource degradation in Pakistan's Punjab: a decomposition analysis. Econ. Dev. Cult. Change 50(4):839-864.

Byerlee D, Murgai R. 2001. Sense and sustainability revisited: the limits of total factor productivity measures of sustainable agricultural systems. Agric. Econ. 26(3):227-236.

Cassman KG, Dobermann A. 2000. Evolving rice systems to meet global demand. Paper presented at the Symposium on Rice Research and Production in the 21st Century (honoring Robert F. Chandler), Cornell University, Ithaca, New York, 15-17 June 2000.

Cassman KG, Pingali PL. 1995. Extrapolating trends from long-term experiments to farmers' fields: the case of irrigated rice systems in Asia. In: Barnett V, Payne R, Steiner R, editors. Agricultural sustainability: economic, environmental and statistical considerations. West Sussex (England): John Wiley & Sons. p 63-84.

Dawe D. 2000. The role of rice research in poverty alleviation. In: Sheehy JE, Mitchell PL, Hardy B, editors. Redesigning rice photosynthesis to increase yield. Proceedings of the Workshop on The Quest to Reduce Hunger: Redesigning Rice Photosynthesis, 30 Nov.-3 Dec. 1999, Los Baños, Philippines. Makati City (Philippines): International Rice Research Institute and Amsterdam (The Netherlands): Elsevier Science B.V. p 3-12.

Dawe D. 1998. Reenergizing the green revolution in rice. Am. J. Agric. Econ. 80:948-953.

Dawe D, Dobermann A. 1998. Defining productivity and yield. IRRI Discussion Paper Series No. 33. Makati City (Philippines): International Rice Research Institute. 13 p.

Dobermann A, Dawe D, Roetter R, Cassman KG. 2000. Reversal of rice yield decline in a long-term continuous cropping experiment. Agron. J. 92:633-643.

Huang J, Rozelle S. 1995. Environmental stress and grain yields in China. Am. J. Agric. Econ. 77:853-864.

Huang J, Rozelle S. 1996. Technological change: rediscovering the engine of productivity growth in China's rural economy. J. Dev. Econ. 49:337-369.

Jha D, Kumar P. 1998. Rice production and impact of rice research in India. In: Pingali PL, Hossain M, editors. Impact of rice research. Bangkok (Thailand): Development Research Institute, and Manila (Philippines): International Rice Research Institute. p 279-291.

Jorgenson DW, Gollop FM. 1992. Productivity growth in U.S. agriculture: a posterior perspective. Am. J. Agric. Econ. 74:745-750.

Kalirajan KP, Obwona MB, Zhao S. 1996. A decomposition of total factor productivity growth: the case of Chinese agricultural growth before and after reforms. Am. J. Agric. Econ. 78:331-338.

Kumar P, Rosegrant MW. 1994. Productivity and sources of growth for rice in India. Econ. Polit. Weekly, Dec. 31.

Kumar P, Joshi PK, Johansen C, Asokan M. 1999. Sustainability of rice-wheat based cropping systems in India: socio-economic and policy issues. In Pingali P, editor. Sustaining rice-wheat production systems: socio-economic and policy issues. Rice-Wheat Consortium Paper Series 5. New Delhi (India): Rice-Wheat Consortium for the Indo-Gangetic Plains. 99 p.

Lin JY. 1992. Rural reforms and agricultural growth in China. Am. Econ. Rev. 82:34-51.

Lynam JK, Herdt RW. 1989. Sense and sustainability: sustainability as an objective in international agricultural research. Agric. Econ. 3:381-398.

Murgai R. 1999. The green revolution and the productivity paradox: evidence from the Indian Punjab. World Bank Policy Research Working Paper 2234. Washington, D.C. (USA): World Bank.

Rosegrant MW, Evenson RE. 1992. Agricultural productivity and sources of growth in South Asia. Am. J. Agric. Econ. 74:757-761.

Swastika DKS. 1995. The decomposition of total factor productivity growth: the case of irrigated rice farming in West Java, Indonesia. Ph.D. dissertation. University of the Philippines Los Baños. 197 p.

Tiongco M, Dawe D. 2000. Long-term productivity trends in a sample of Philippine rice farms. Poster presented at the XXIVth International Conference of Agricultural Economists, Berlin, Germany, 13-18 Aug. 2000.

Woodhead T, Huke R, Huke E. 1994. Areas, locations, and ongoing collaborative research for the rice-wheat systems of Asia. In: Paroda RS, Woodhead T, Singh RB, editors. Sustainability of rice-wheat production systems in Asia. Bangkok (Thailand): Regional Office for Asia and the Pacific (RAPA) and Food and Agriculture Organization of the United Nations. p 68-96.

Zhang B, Carter CA. 1997. Reforms, the weather, and productivity growth in China's grain sector. Am. J. Agric. Econ. 79:1266-1277.

Notes

Acknowledgments: The author gratefully acknowledges the assistance of Marites Tiongco in performing the TFP calculations using data from the Reversing Trends in Declining Productivity project.

Author's address: International Rice Research Institute, Social Sciences Division, DAPO Box 7777, Metro Manila, Philippines. E-mail: d.dawe@cgiar.org.

Citation: Dobermann A, Witt C, Dawe D, editors. 2004. Increasing productivity of intensive rice systems through site-specific nutrient management. Enfield, N.H. (USA) and Los Baños (Philippines): Science Publishers, Inc., and International Rice Research Institute (IRRI). 410 p.

Part 2

The evolution of site-specific nutrient management in irrigated rice systems of Asia

A. Dobermann and C. Witt

5.1 The need for site-specific nutrient management

Crop management over the past four decades in Asia was driven by the increasing use of external inputs and blanket recommendations for fertilizer use over wide areas. However, future gains in productivity and input-use efficiency will require soil and crop management technologies that are more knowledge-intensive and tailored to the specific characteristics of individual farms and fields (Dobermann and White 1999, Pingali et al 1998). Recent on-farm research has demonstrated that large variability exists in soil nutrient supply, nutrient-use efficiency, and crop response to nutrients among rice farms or single rice fields (Adhikari et al 1999, Angus et al 1990, Cassman et al 1996b, Dobermann et al 1996a, Donovan et al 1999, Oberthür et al 1996, Olk et al 1999, Wopereis et al 1999). Managing this variability has become a principal challenge for further increasing the productivity of these intensive rice areas.

Terms such as site-specific nutrient management (SSNM) have often been defined as managing within-field variability in relatively large fields using georeferenced variable rate technology (Pierce and Nowak 1999). However, we consider SSNM to be a general concept for optimizing the supply and demand of nutrients according to their variation in time and space. In this sense, "site-specific" is defined as the dynamic field-specific management of nutrients in a particular cropping season. A single rice field of usually 0.2–1.0 ha size is probably the smallest feasible management unit because managing within-field variation (Dobermann et al 1995, 1997a) is difficult without georeferencing. Rice farmers always try to manage within-field variation in crop growth by applying fertilizers or pesticides according to their visual observations, but this is a subjective process that can hardly be made into a reproducible management strategy.

In this paper, we describe an approach for SSNM in irrigated rice systems in Asia, which evolved over a period of about five years. Initial research on separate topics was conducted in the early and mid-1990s. It was found that the indigenous N supply was quite variable among fields and not related to soil organic matter content (Cassman et al 1996a,b) so that plant-based strategies for real-time N management were needed to increase yields and N-use efficiency (Peng et al 1996a,b). Research on P and K concluded that nutrient imbalances may limit yield and N-use efficiency, that existing soil test methods had limited applicability to lowland rice, and that negative K input-output balances were a general characteristic of intensive rice cropping (Dobermann et al 1996b,c,d). A detailed survey in Central Luzon, Philippines (Dobermann and Oberthür 1997, Oberthür et al 1996), revealed large spatial variability in soil fertility that was not related to soil types or landscape features. It became obvious that blanket fertilizer recommendations given for large areas have serious limitations and that a new approach was required to provide an integrated management of all nutrients.

The original concept for SSNM was developed in 1996 (Dobermann et al 1996a, Dobermann and White 1999) and has been tested on 205 irrigated rice farms in China, India, Indonesia, the Philippines, Thailand, and Vietnam since 1997. An important part of field testing was to continuously collect data that could be used to improve the approach. Our goal was to develop a generic but flexible, location-specific SSNM approach (see Chapter 17) based on models that were calibrated using data collected on-farm across many sites. Although there was one general strategy, we made location-specific adjustments for each site, reviewed them after each crop, and modified them as required. In this trial-and-error learning process, rules of thumb were mixed with quantitative models to arrive at an acceptable SSNM strategy for each domain within a period of about three years of on-farm testing. It is important to mention here that this was a novel type of strategic on-farm research for which we had little previous experience to draw on. We made mistakes because things that seemed to work on paper turned out to be too complicated to be used in the field, or we had to consider newly emerging issues. However, if rigorous standardization and quality control are ensured, the models and technologies developed through such an approach will likely have greater extrapolation potential than those produced in the traditional top-down research and technology transfer process.

In this chapter, we describe the major principles of this SSNM strategy and details of its implementation. Chapters 6 to 12 contain further details related to each site. Chapter 17 describes how all the experiences and data gathered during the empirical field evaluation period resulted in a more generic nutrient decision support system that will, we hope, form the basis for future research and adoption of SSNM in Asia.

5.2 A general strategy for site-specific nutrient management

The classical approach for developing recommendations for fertilizer use is based on empirical response functions derived from factorial fertilizer trials conducted across different locations. A key problem is that many existing algorithms do not adequately account for nutrient interactions as the driving force for plant uptake and internal nutrient efficiency at higher yields (Sinclair and Park 1993, Witt et al 1999). We hypothesized that, particularly in high-yielding, irrigated systems, the ability to predict soil nutrient supply and plant uptake in absolute terms rather than relative yield response is the key to fine-tuning nutrient management. The QUEFTS model (Janssen et al 1990, Smaling and Janssen 1993) presents such an alternative because it describes, in four steps, relations between, in order, (1) chemical soil test values, (2) potential NPK supply from soils and fertilizer, (3) actual NPK uptake, and (4) grain yield, acknowledging interactions among the three macronutrients in all steps. It accounts for the climatic yield potential and is particularly suitable for irrigated systems in which the yield response to nutrients is less confounded by water stress.

We further hypothesized that SSNM must account for the principal differences among three major groups of nutrients (Dobermann and White 1999):

Nitrogen. The goal of N management is to maximize plant N-use efficiency and minimize N losses into the environment because N drives yield and is involved in rapid biological turnover processes that affect water and air quality. Managing the large spatial and temporal variability in N requires a combination of both predictive and interventive management to achieve the best congruence between crop N demand and soil N supply at all growth stages. For irrigated systems, an integrated approach featuring a predictive model combined with temporally dense, real-time information will be required to properly manage N site-specifically. Key requirements include (1) generic models for estimating crop N demand that account for yield potential and provide optimal internal N-use efficiency; (2) algorithms for estimating the potential indigenous N supply and its variation with climate, soil type, and cropping history; and (3) algorithms for N management at key growth stages to improve the congruence between N supply and crop demand based on indicators of actual plant N status.

Phosphorus and potassium. The goal of P and K management is to maintain nutrient supply at levels that do not limit crop growth, that ensure optimal N-use efficiency, that increase plant resistance to pests and lodging (K), and that avoid leaching (K) or runoff hazard (P). Predicting the amount to apply is more important than the issue of maximizing the recovery efficiency of fertilizer P and K. The three major scenarios in P and K management are (1) fertilizer application to overcome nutrient limitation, (2) fertilizer application to replenish P and K removal with grain and straw to maintain soil nutrient supplies, and (3) reduced or no fertilizer application where the nutrient supply is substantially greater than the plant demand. Key components of P and K management are (1) generic models for estimating crop P and K demand that provide optimal internal-use efficiency; (2) algorithms for predicting the potential indigenous P and K supply based on soil tests or plant indicators; (3) rules for timing and placement of P and K applications depending on soil type, tillage, crop establishment method, and population density; and (4) generic models to develop adequate nutrient management strategies considering long-term changes in soil nutrient reserves.

Sulfur, zinc, others. Diagnosis is the key for managing secondary and micronutrients. Once identified as a problem, deficiencies can be alleviated by single or regular preventive applications (Dobermann and Fairhurst 2000). The question of whether and how to apply is more important than the question of accurately prescribing how much to apply.

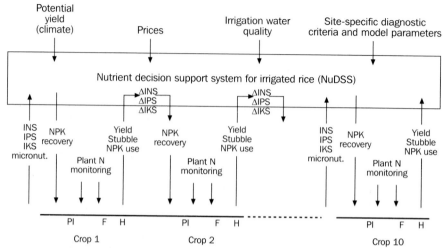

Fig. 5.1. Strategy for field-specific nutrient management in irrigated rice (Dobermann and White 1999). PI = panicle initiation, F= flowering, H = harvest.

Figure 5.1 shows the proposed general SSNM approach for irrigated rice, which focuses on managing field-specific spatial variation in indigenous NPK supply, temporal variability in plant N status occurring within one growing season, and medium-term changes in soil P and K supply resulting from the actual nutrient balance. The following steps are involved:

1. Field-specific estimation of the potential indigenous supplies of N (INS), P (IPS), and K (IKS, all in kg ha^{-1}) and diagnosis of other nutritional disorders in year one.

2. Field-specific recommendations for NPK use and alleviation of other nutritional problems.

3. Optimization of timing and amount of applied N. Decisions about timing and splitting of N applications are based on (1) 3–5 split applications following season-specific agronomic rules tailored to specific locations or (2) regular monitoring of plant N status up to the flowering stage, using a chlorophyll meter (Peng et al 1996b) or green leaf color charts (Balasubramaniam et al 1999).

4. Estimation of actual grain yield, stubble (straw) returned to the field, and amount of fertilizer used. Based on this, a P and K input-output balance is estimated and used to predict the change in IPS and IKS resulting from the previous crop cycle. The predicted IPS and IKS values are then used to develop fertilizer recommendations in the subsequent crop cycle.

5.3 Site-specific nutrient management in the RTDP on-farm trials

Estimation of field-specific fertilizer amounts using QUEFTS

The procedure shown in Figure 5.1 was applied in 205 farmers' fields for a succes-

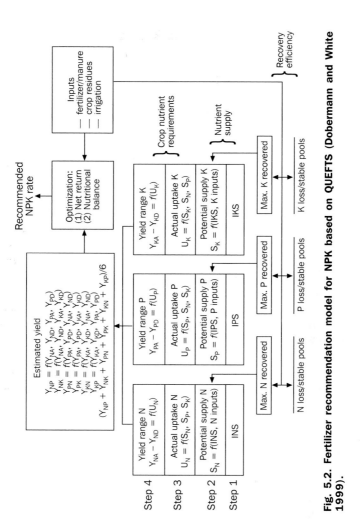

Fig. 5.2. Fertilizer recommendation model for NPK based on QUEFTS (Dobermann and White 1999).

sion of at least four crops (see Chapter 2 for details of the experimental work). We used a modification of the QUEFTS model (Janssen et al 1990, Witt et al 1999) for predicting the amount of fertilizer N, P, and K required for a specific yield target, including interactions among these three nutrients (Fig. 5.2). The recommendations for field-specific fertilizer applications were worked out using a simple spreadsheet model based on QUEFTS. A linear optimization procedure was used for finding the best combination of N, P, and K fertilizer rates to achieve the yield goal under the constraint of optimizing the internal efficiencies of N, P, and K in the plant. The model was improved in late 1998 when more on-farm data became available for the calibration of crucial model parameters (Janssen et al 1990, Witt et al 1999) . In the following, we describe the basic steps involved in calculating fertilizer recommendations and model improvements in 1997-99.

User-defined information needed to run the model for making a field-specific fertilizer recommendation for each field and crop cycle included
(1) potential yield and yield goal,
(2) definition of the relationship between grain yield and nutrient uptake,
(3) recovery efficiencies of fertilizer N, P, and K,
(4) field-specific estimates of the indigenous N, P, and K supply, and
(5) optimization constraints.

(1) Potential yield and yield goal. The potential yield (Y_{max}) can be defined as grain yield limited by climate and genotype only, with all other factors not limiting crop growth. Y_{max} fluctuates among sites, farms, and years (typically ±10%) because of climatic variation, differences in genotypes, and variation in planting dates. Y_{max} can be estimated using crop models such as ORYZA1 (Kropff et al 1993), WOFOST 7.1 (Boogaard et al 1998), SIMRIW (Horie et al 1995), or CERES-Rice (Singh et al 1991). A problem with this approach is that the model chosen must be validated and calibrated for the climate and the varieties grown at a particular site. It must also account for the major crop management effects on the development of the rice plant canopy, particularly differences between transplanted and direct-seeded plants. The mean simulated Y_{max} of current rice varieties grown across Asia was estimated to be 8.1 t ha^{-1} (ORYZA1, range 4 to 13 t ha^{-1}) or 8.5 t ha^{-1} (SIMRIW) (Matthews et al 1997). Typically, in the subhumid to humid subtropical and tropical regions of Asia, Y_{max} is around 9 to 10 t ha^{-1} in high-yielding (dry) seasons and 6 to 8 t ha^{-1} in low-yielding (wet) seasons, when solar radiation is lower because of greater cloud cover. Experimentally, Y_{max} can be measured only in maximum-yield trials with complete control of all growth factors other than solar radiation. In the absence of simulated or measured data, the only alternative is to estimate Y_{max} as the highest yield ever recorded at a particular site in an experiment with near-optimal growth conditions.

Table 5.1 shows Y_{max} values used in our SSNM trials and how they were derived. For specifying field-specific fertilizer rates using QUEFTS, season-specific yield goals were set in the range of 70% to 80% of Y_{max}. The rationale for this is that beyond that level internal efficiencies of nutrients in the plant decline (Witt et al 1999), making it difficult to achieve optimal efficiencies of N, P, and K. Moreover, practical experience indicates that yields of about 80% of Y_{max} seem to represent a ceiling for what can be achieved by the best farmers under field conditions (Cassman and Harwood 1995). In most cases, yields of 70% to 80% of Y_{max} are also associated with the highest profits, except when local rice prices significantly exceed world market prices. The latter is currently the case in the Philippines, where farmers would probably benefit more from aiming at yields closer to Y_{max} unless this would increase the risk of crop failure because of pests, diseases, or lodging. Typically, yield goals within a season and domain varied little among farms (CV 2% to 4%, range <1 t ha^{-1}). However, in fields with a low indigenous supply of one or more nutrients, the yield goal was lowered because even very high mineral fertilizer rates cannot fully substitute for lower attainable yields because of low inherent soil fertility. In such cases, our goal was to slowly build up soil fertility and raise the yield goal over time unless crucial soil properties indicated limited opportunities to improve soil fertility

Table 5.1. Estimates of the climatic yield potential (Y_{max}, t ha^{-1}) in high- (HYS) and low-yielding (LYS) cropping seasons at all sites.

Site	HYS	LYS	Comments
Maligaya, PhilRice	10.5	7.5	Based on crop modeling and field experiments
Suphan Buri, PTRRC	9.0	7.5	Little information available for the varieties grown there
Omon, CLRRI	9.0	7.0	Expert guess and field experiments, shorter growth duration
Sukamandi, RIR	8.0–9.0	6.5–7.5	HYS is the wet season, variation among villages because of different planting dates
Aduthurai, TNRRI & Thanjavur, SWMRI	9.8	8.3	Based on crop modeling and field experiments
Hanoi, NISF	8.5–10.0	6.5–7.5	Field experiments, variation among villages because of different planting dates
Jinhua, ZU	10.0	9.0	Based on crop modeling and field experiments

through fertilization in the long term. For example, a lower yield goal was chosen for the sandy soils of the New Cauvery Delta than for the more fertile soils of the Old Cauvery Delta in Tamil Nadu, India (Chapter 6). Under such conditions, it is safer to choose a yield target that is 10–20% higher than the yields currently achieved by farmers if a thorough analysis of yields and fertilizer practice indicates opportunities to increase productivity through improved nutrient management. This example demonstrates that local knowledge of farming conditions has to be integrated when choosing reasonable yield goals, particularly in cases where the climate-adjusted yield potential of a certain variety could never be reached because of other constraints (e.g., low fertility of sandy soils, acid sulfate soils, low rooting depth, etc.). Thus, suggested yield goals for a particular variety may vary within an area of similar climatic conditions depending on the spatial variability of soil fertility or other biophysical yield constraints.

(2) Definition of the relationship between grain yield and nutrient uptake. Crop nutrient requirements for a specific yield goal were quantified using the empirical modeling approach in QUEFTS (Janssen et al 1990, Janssen 1998). The relationship between grain yield and nutrient accumulation is described as a function of the climatic yield potential and the supply of the three macronutrients N, P, and K. In a situation where crop growth is not limited by water supply or pest infestations, biomass production is mainly driven by nutrient supply. Estimating crop nutrient demand must avoid deficiency and unproductive accumulation of N, P, or K in the plant. Y_{max} and nutrient interactions determine the internal-use efficiency of a nutrient (IE = kg grain yield per kg nutrient in aboveground plant dry matter at maturity). In a situation of balanced nutrition, the QUEFTS model assumes a linear relationship between grain yield and plant nutrient uptake (YN, YP, YK, Fig. 5.3) or constant internal

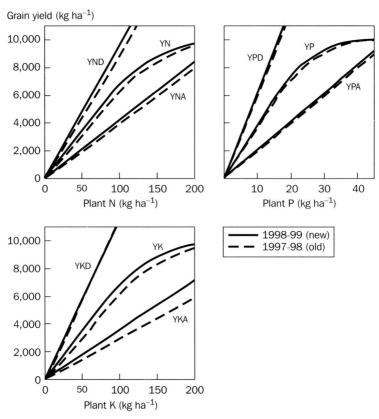

Fig. 5.3. The balanced N, P, and K uptake requirements (YN, YP, and YK) for targeted grain yields depending on the yield potential (Y_{max}) of 10 t ha⁻¹ as calculated by the QUEFTS model used in 1997-98 (old) and 1998-99 (new, Witt et al 1999). The regression lines to the left of each figure represent the boundary of maximum dilution of plant N, P, and K (YND, YPD, and YKD), while the lines to the right indicate the boundary of maximum accumulation (YNA, YPA, and YKA). The constants describing these borderlines are given in Table 5.2.

efficiencies until yield targets reach about 70–80% of Y_{max}. When yields approach the potential yield, the internal nutrient efficiencies decline as the relationship between grain yield and nutrient uptake enters a nonlinear phase (Fig. 5.3). To model this in a generic sense required the empirical determination of two boundary lines describing the minimum and maximum IEs of N, P, and K in the plant across a wide range of yields and nutrient statuses. In the field testing of SSNM, we initially used preliminary boundary lines derived from long-term experiments and on-farm experiments at some sites (Cassman et al 1997). The parameters corresponding to these borderlines are summarized in Table 5.2. Later on, a much larger database was used to derive improved generic boundary lines that are valid for any site in Asia at which modern rice varieties with a harvest index of about 0.45–0.55 are grown (Witt et al

Table 5.2. Envelope functions relating grain yield of rice (Y) at maximum accumulation (a) and dilution (d) of N, P, and K to their accumulation in the aboveground dry matter at maturity (U) for the QUEFTS model used in 1997-98 (old) and 1998-99 (new, Witt et al 1999).

Nutrient	QUEFTS models used in	
	1997-98 (old)	1998-99 (new)
Nitrogen	$Y_a = 40\ (U_N - 2)$	$Y_a = 42\ (U_N)$
	$Y_d = 90\ (U_N - 2)$	$Y_d = 96\ (U_N)$
Phosphorus	$Y_a = 200\ (U_P - 0.1)$	$Y_a = 206\ (U_P)$
	$Y_d = 600\ (U_P - 0.1)$	$Y_d = 622\ (U_P)$
Potassium	$Y_a = 30\ (U_K - 3)$	$Y_a = 36\ (U_K)$
	$Y_d = 120\ (U_K - 3)$	$Y_d = 115\ (U_K)$

1999). These new relationships have been used for SSNM since late 1998 (Table 5.2). The balanced plant nutrient requirements to produce 1,000 kg of grain yield were newly estimated with 14.7 kg N, 2.6 kg P, and 14.5 kg K valid for the linear phase of the relationship between yield and nutrient uptake (Witt et al 1999). The new borderlines describing the minimum and maximum IEs were estimated at 42 and 96 kg grain kg^{-1} N, 206 and 622 kg grain kg^{-1} P, and 36 and 115 kg grain kg^{-1} K, respectively (Table 5.2). Note that these models are built into the NuDSS (see Chapter 17) so that users do not need to enter such values to make a site-specific recommendation.

(3) Recovery efficiencies of fertilizer N, P, and K. First-crop recovery efficiencies of fertilizer N, P, and K had to be specified for all farms within a domain. Values used at all sites ranged from 0.4 to 0.6 kg kg^{-1} (N), 0.2 to 0.3 kg kg^{-1} (P), and 0.4 to 0.5 kg kg^{-1} (K). For N, we mostly assumed recovery efficiencies of 0.5 kg kg^{-1}, assuming that this level is achievable with a good plant-based N management strategy (see below). Estimates for recovery efficiencies of P and K were derived from other data available such as field experiments or "expert knowledge" and were adjusted slightly up or down to account for variation in soil types and cropping practices. For example, the common practice in direct-seeded rice is to broadcast P and K together with the first N dose on the surface at about 2 to 3 wk after sowing. Considering the high plant densities observed in broadcast-sown rice and rapid crop development during this period, much of the P and K fertilizer applied is probably not adsorbed by soil particles so that recovery efficiencies are larger than for incorporated fertilizer and transplanted rice. Naturally, the estimates of recovery efficiencies represent a considerable source of uncertainty. More research is needed to quantify differences in the recovery efficiency of P and K depending on methods of crop establishment, water management, and fertilizer application.

(4) Field-specific estimates of the indigenous N, P, and K supply. A critical issue is obtaining reliable estimates of the indigenous supply of the three macronutrients. The potential indigenous supply is defined as the cumulative amount of a nutrient originating from all indigenous sources that circulates through the soil solution surrounding the entire root system during one complete crop cycle (Janssen et al 1990). This is not measurable, but can be estimated by plant nutrient accumulation in a nutrient omission plot. For example, IPS can be measured as total plant P accumulation in a 0-P plot, which receives N, K, and other nutrients (see Chapter 2). An advantage of using plant indicators of nutrient supply is that they integrate nutrient supply from all indigenous sources under field conditions over time. They result in a measurement unit that is directly usable in fertilizer calculations (kg nutrient ha^{-1} crop^{-1}). Components include soil nutrient supply across the whole rooting depth, irrigation water, atmospheric deposition, biological N_2 fixation, or crop residues, but also factors affecting root uptake. A disadvantage is that plant-based indicators are affected by genotypic and environmental variation in harvest index, rooting patterns, and factors affecting nutrient uptake. An assumption is made that the effective rooting depth and exploitation of indigenous nutrient pools are similar for fertilized and nonfertilized crops.

Direct measurements of NPK uptake in nutrient omission plots are often not available for estimating INS, IPS, and IKS on a routine basis. An alternative is to develop simple empirical models that predict the potential supply from information such as soil tests and climate (Smaling and Janssen 1993) or crop biomass. For example, for a given nutrient combination (e.g., +PK, +NK, or +NP), the relationship between total plant uptake and grain yield is less scattered. Therefore, INS, IPS, and IKS can be estimated from grain yield measurements in small N, P, and K omission plots embedded in farmers' fields if other nutrients are fully supplied and the harvest index is approximately 0.5. For example, IPS (kg P ha^{-1}) can be estimated from grain yield in a P omission plot (Y_{0P}, t ha^{-1}) by comparing it also with the yield obtained with a full nutrient supply (GY_{NPK}) using (Dobermann and Fairhurst 2000) the following:

If $Y_{0P} \geq Y_{NPK}$ P supply is not limiting yield in 0P plot, IPS = $Y_{0P} \times 2.6$

If $Y_{0P} < Y_{NPK}$ P supply is limiting yield in 0P plot, IPS = $Y_{0P} \times 2.3$

A second alternative is to estimate INS, IPS, and IKS from soil tests if the latter provide a good index of the total soil nutrient release to plant roots during the whole growth period. If so, one can establish generic or more location-specific models for predicting INS, IPS, and IKS from a soil test by conducting controlled experiments with N, P, and K omission plots across a wide range of soil and environmental conditions. Data collected from the omission plots at all sites (see Chapter 2) will be used to explore such possibilities. However, for irrigated lowland rice systems, many existing soil tests are of limited use for this because they fail to account for the changes in soil nutrient supply caused by flooding (Cassman et al 1996a). Improvements may be possible by using dynamic soil tests such as *in situ* ion exchange resin techniques (Dobermann et al 1997b, 1998), but the current soil testing infrastructure in Asia is too weak to support such methods.

To work out field-specific fertilizer recommendations during the testing of the SSNM approach at all sites, we used estimates of INS, IPS, and IKS obtained from measurements of nutrient uptake in omission plots conducted in previous years (see Chapters 6–12 for details regarding each site).

(5) Optimization constraints. A linear optimization procedure was used to find the best combination of N, P, and K fertilizer rates to achieve the yield goal under the constraint of optimizing the internal efficiencies of N, P, and K in the plant. The model was constrained to arrive at a solution close to the situation of the most balanced nutrition, that is, where the ratio between predicted uptake and potential supply of each macronutrient was ≥0.95 (Janssen 1998). Other constraints included upper and lower possible limits of fertilizer rates. In the case of N, an upper limit (e.g., typically 180 to 200 kg N ha^{-1} in a dry season) was set to avoid excessive N rates that could cause pest problems or lodging. Upper limits of 40 to 45 kg P ha^{-1} and 150 to 200 kg K ha^{-1} were set so that fertilizer rates applied remained within economically reasonable ranges. In cases where the model did not arrive at an acceptable solution within these limits, the yield goal was reduced until an optimal solution was found. This approach follows the concept that crop yields on soils with moderate to high P or K status cannot be fully matched on comparable soils of low P or K status even when large amounts of fertilizer are applied (Johnston et al 1986). Lower limits of 10 kg P ha^{-1} and 30 kg K ha^{-1} were set as the minimum amount to be applied to replenish the net removal from the field.

Sensitivity analysis

The sensitivity of the QUEFTS model to changes in envelope functions used in 1997-98 and 1998-99 (Fig. 5.3, Table 5.2) was investigated for two scenarios: (1) grain yield as affected by yield potential and potential nutrient supply (Table 5.3) and (2) fertilizer recommendations as affected by yield potential, yield target, recovery efficiency, and indigenous nutrient supply (Table 5.4). In both scenarios, a yield potential of 10 t ha^{-1} was chosen and the indigenous nutrient supplies were set to 60 kg N ha^{-1}, 15 kg P ha^{-1}, and 90 kg K ha^{-1}. These nutrient supplies estimated from plant nutrient uptake in omission plots would be equivalent to a grain yield of about 4 t ha^{-1} in 0-N plots, and 6 t ha^{-1} in both 0-P and 0-K plots. Taking the nutrient interactions into account, QUEFTS predicted a yield of 4.5 t ha^{-1} for an unfertilized plot. Table 5.3 gives the fertilizer rates and assumed recovery efficiencies used in the analysis.

The changes in envelope functions had no effect on plant nutrient uptake as affected by yield potential and potential nutrient supply (nutrient supply from soil and fertilizer) as shown in Table 5.3, and this would also not change at other supply levels. The parabolic function used for the relationship between nutrient uptake and soil nutrient supply in QUEFTS is mainly sensitive to nutrient supply and yield potential as long as the constants given in Table 5.2 are not dramatically changed. The relatively small changes in envelope functions, however, resulted in greater internal efficiencies predicted with the new model, that is, nutrients taken up by the plant would be more efficiently transformed into grain yield depending on the yield level

Table 5.3. Predicted grain yield, plant nutrient uptake, internal nutrient efficiencies (IEN, IEP, IEK; kg grain yield per kg plant nutrient), and reciprocal internal efficiencies (RIEN, RIEP, RIEK; kg plant nutrient per 1,000 kg grain yield) as affected by the QUEFTS model parameters given in Table 5.2. Input parameters were Y_{max} = 10 t ha^{-1}; indigenous nutrient supplies of INS = 60 kg ha^{-1}; IPS = 15 kg ha^{-1}; IKS = 90 kg ha^{-1}; fertilizer rates of 113 kg N ha^{-1}, 25 kg P ha^{-1}, and 57 kg K ha^{-1}; assumed recovery efficiencies (RE) of applied fertilizer were REN = 0.45 kg kg^{-1} for N, REP = 0.25 kg kg^{-1} for P, and REK = 0.50 kg kg^{-1} for K.

	Unit	QUEFTS models used in		
		1997-98 (old)	1998-99 (new)	Difference
Predicted grain yield	t ha^{-1}	7.08	7.50	+0.42
Plant N	kg ha^{-1}	115	115	0
Plant P	kg ha^{-1}	20	20	0
Plant K	kg ha^{-1}	114	114	0
Internal efficiencies (IE)				
Of N (IEN)	kg kg^{-1}	62	65	+3
Of P (IEP)	kg kg^{-1}	349	368	+19
Of K (IEK)	kg kg^{-1}	62	66	+4
Reciprocal IE				
Reciprocal IEN	kg kg^{-1}	16.2	15.3	−0.9
Reciprocal IEP	kg kg^{-1}	2.9	2.7	−0.2
Reciprocal IEK	kg kg^{-1}	16.1	15.1	−1.0

(Fig. 5.3). As a consequence, the newly calibrated model predicted a 0.42 t ha^{-1} or 6% higher grain yield when the new envelope functions were used in the example given in Table 5.3.

The increase in predicted internal efficiencies had pronounced effects on fertilizer recommendations. We used a Microsoft Excel spreadsheet version of QUEFTS in combination with a solver module to simulate generic nutrient uptake curves representing optimal internal efficiencies at a certain yield potential (YN, YP, YK, Fig. 5.3). The nutrient uptake requirement to support a selected yield goal is based on these relationships. For the scenario shown in Table 5.4, nutrient requirements decreased by about 8% for N and P and by 11% for K for the chosen yield target of 7.5 t ha^{-1} when the new envelope functions were used. Taking the assigned fertilizer recovery efficiencies into account, QUEFTS predicted 15% lower fertilizer N and P requirements and 35% lower fertilizer K requirements in this scenario to support the targeted yield level. The reduction in applied fertilizer nutrients, particularly N and K in the second year of our experimental trials, was in part due to the introduction of the newly calibrated QUEFTS model, which substantially improved our fertilizer recommendations (see Chapters 6–12).

Table 5.4. Fertilizer and plant nutrient requirements in kg ha^{-1} for a grain yield target of 7.5 t ha^{-1} as predicted by QUEFTS using the two sets of model parameters given in Table 5.2. See Table 5.3 for input parameters (except fertilizer rates) and for an explanation of internal efficiencies (IEN, IEP, IEK) and reciprocal internal efficiencies (RIEN, RIEP, and RIEK).

	Unit	QUEFTS models used in		
		1997-98 (old)	1998-99 (new)	Difference
Fertilizer N	kg ha^{-1}	157	133	−24	(−15%)
Fertilizer P	kg ha^{-1}	30	25	−5	(−15%)
Fertilizer K	kg ha^{-1}	87	57	−30	(−35%)
Plant N	kg ha^{-1}	125	115	−10	(+8%)
Plant P	kg ha^{-1}	22	20	−2	(+8%)
Plant K	kg ha^{-1}	128	114	−14	(+11%)
Internal N eff. (IEN)	kg kg^{-1}	60	65	+5	(+8%)
Internal P eff. (IEP)	kg kg^{-1}	350	368	+18	(+5%)
Internal K eff. (IEK)	kg kg^{-1}	59	66	+7	(+8%)
Reciprocal IEN	kg t^{-1}	16.7	15.3	−1.4	(−8%)
Reciprocal IEP	kg t^{-1}	2.9	2.7	−0.2	(−7%)
Reciprocal IEK	kg t^{-1}	17.1	15.1	−2.0	(−12%)

The above examples highlight the sensitivity of the QUEFTS model to changes in envelope functions and the importance of a sufficiently large data set of high quality for calibration. The new envelope functions for rice were based on a database with more than 2,000 entries on the relationship between grain yield and nutrient uptake, and all data were derived using a standard methodology in plant sampling and processing (Witt et al 1999). The internal efficiencies predicted with the newly calibrated model showed good agreement with data obtained from farmers' fields. However, fertilizer recommendations are not only affected by model parameterization, but by all input parameters including yield potential, soil nutrient supply, and recovery efficiencies. In the following, we investigate the sensitivity of the model to the variation in these input parameters using a risk analysis and simulation add-in for Microsoft Excel (@Risk 4.0, Palisade Corp.).

Using the example of Table 5.3, it was assumed that input parameters were normally distributed so that the functions describing the variation were determined by means and standard deviations as given in Table 5.5. The average yield potential was set to 10 t ha^{-1} assuming a year-to-year variation of about ±10%. The variation in indigenous nutrient supplies and recovery efficiencies were based on typical values observed at the experimental sites (Chapters 6–12). Correlation coefficients establishing a relationship between certain input parameters in the simulation are given in Table 5.5. We assumed that recovery efficiencies are positively correlated with yield potential because greater plant uptake of applied fertilizer nutrients is expected in high-yielding seasons. The relationships among soil nutrient supplies of N, P, and K were based on data collected at all sites from 1997 to 1999 (A. Dobermann, un-

Table 5.5. Correlation coefficients used for the simulation given in Table 5.6.

	Y_{max}	INS	IPS	IKS
Y_{max}	1			
RFN	0.75			
RFP	0.50			
RFK	0.50			
INS	0	1		
IPS	0	0.80	1	
IKS	0	0.80	0.95	1

published). It appears that indigenous supplies of N, P, and K are related to each other and do not vary randomly in farmers' fields. For example, N is the most limiting nutrient in most rice fields in Asia, and differences in soil fertility status among sites seem to affect all three macronutrients to a certain extent. The correlation coefficients presented in Table 5.5 may not be true and are expected to vary depending on cropping history. The model was run for 1,000 iterations randomly assigning values for input parameters based on the distribution functions in each run. Simulation results appear in Table 5.6.

The simulation created a set of input parameters with a *variation* comparable to that observed at the experimental sites (see Chapters 6–12). For the given scenario, predicted grain yield varied from about 6.4 to 8.4 t ha^{-1} (10th and 90th percentiles), with a mean value of 7.4 t ha^{-1}. Average plant nutrient uptake was comparable to the requirements estimated using single values as input parameters (Table 5.4, new QUEFTS model). Likewise, average simulated internal efficiencies were similar to the optimal values achieved at balanced nutrition. The data set was subjected to a sensitivity analysis of input parameters affecting grain yield based on Spearman rank correlation coefficient calculations and multivariate stepwise regression (Table 5.7). The coefficients of the latter are calculated for each input variable measuring the sensitivity of the output to that particular input distribution. None of the input variables had a pronounced effect on yield on its own. The rank correlation coefficient is calculated between the selected output variable and the sample for each of the input distributions. Given a fixed rate of applied fertilizer nutrients, the indigenous nutrient supply parameters were the most critical input parameters for achieving high yields because of the strong relationship among the three macronutrients (Table 5.5). In other words, yields were higher when the soil nutrient supplies increased regardless of climate and recovery efficiencies. Yield was also affected by climate (yield potential), whereas the variation in fertilizer recovery efficiencies was of less importance for yield expression. These simulation results correspond well to a stepwise multiple regression analysis of the actual data set showing that 76% of the variation in grain yield in SSNM treatments was explained by indigenous N and P supplies, the number of N applications per crop, and a parameter expressing the degree of crop care (Dobermann et al 2002). In the on-farm experiments, fertilizer rates were ad-

Table 5.6. Simulated grain yield and plant nutrient uptake as predicted by the newly parameterized QUEFTS model used in 1998-99 (Witt et al 1999) based on the scenario given in Table 5.4. Fertilizer rates were fixed with 133 kg N ha^{-1}, 25 kg P ha^{-1}, and 57 kg K ha^{-1}, while input parameters given below were randomly assigned based on normal distribution functions determined by the mean and a CV of 20% (CV of 10% for yield potential). Minimum and maximum values were assigned to INS (25–95 kg ha^{-1}), IPS (5–25 kg ha^{-1}), IKS (40–140 kg ha^{-1}), REN (0.35–0.55 kg kg^{-1}), REP (0.1–0.4 kg kg^{-1}), and REK (0.35–0.65 kg kg^{-1}). Simulation results are based on 1,000 model runs.

Parameters	Unit	Mean	SD[a]	10th percentile	Lower quartile	Upper quartile	90th percentile
Input parameters							
Yield potential	kg ha^{-1}	10,000	1,000	8,715	9,323	10,674	11,280
Indigenous N supply (INS)	kg ha^{-1}	60	12	45	52	68	75
Indigenous P supply (IPS)	kg ha^{-1}	15	3	11	13	17	19
Indigenous K supply (IKS)	kg ha^{-1}	90	18	67	78	102	113
Recovery efficiency N (REN)	kg kg^{-1}	0.45	0.05	0.38	0.41	0.49	0.52
Recovery efficiency P (REP)	kg kg^{-1}	0.25	0.05	0.19	0.22	0.28	0.31
Recovery efficiency K (REK)	kg kg^{-1}	0.50	0.07	0.4	0.44	0.56	0.60
Output parameters							
Grain yield	kg ha^{-1}	7,386	760	6,431	6,886	7,916	8,366
Agronomic efficiency	kg kg^{-1}	21	4	16	19	23	25
Plant N	kg ha^{-1}	114	13	97	105	123	131
Plant P	kg ha^{-1}	20	3	16	18	22	24
Plant K	kg ha^{-1}	113	17	92	101	125	135
Internal N efficiency (IEN)	kg kg^{-1}	65	3	61	63	67	69
Internal P efficiency (IEP)	kg kg^{-1}	368	28	353	348	386	404
Internal K efficiency (IEK)	kg kg^{-1}	66	5	59	62	70	73
Reciprocal IEN	kg kg^{-1}	15.4	0.8	14.4	14.9	16.0	16.5
Reciprocal IEP	kg kg^{-1}	2.7	0.2	2.5	2.6	2.9	3.0
Reciprocal IEK	kg kg^{-1}	15.3	1.2	13.7	14.4	16.1	16.8

[a]SD = standard deviation.

Table 5.7. Sensitivity analysis of input parameters affecting grain yield for the simulation given in Table 5.6 based on Spearman rank correlation coefficient calculation and multivariate stepwise regression.

Rank	Parameter	Unit	Correlation coefficient	
			Spearman rank	Stepwise regression
1	Indigenous K supply	kg ha^{-1}	0.806	0.285
2	Indigenous P supply	kg ha^{-1}	0.792	0.316
3	Indigenous N supply	kg ha^{-1}	0.750	0.282
4	Yield potential	kg ha^{-1}	0.519	0.158
5	Recovery efficiency N	kg kg^{-1}	0.418	0.266
6	Recovery efficiency P	kg kg^{-1}	0.229	0.195
7	Recovery efficiency K	kg kg^{-1}	0.209	0.137

justed to match the deficit between uptake requirement and indigenous supply, so that including fertilizer rates in the analysis did not explain additional yield variation.

The on-farm experiments were not designed to estimate fertilizer responses, so that it is not possible to estimate the variation in "optimal" fertilizer rates for the given data set. However, the QUEFTS model can be used to estimate the variation in *recommended* fertilizer rates for a certain yield target based on the measured variation in input parameters. Table 5.8 shows the predicted fertilizer rates for the input parameters yield potential, indigenous nutrient supplies, and recovery efficiencies used in the simulation of Table 5.6. The required fertilizer rates ranged from 114 to 157 kg ha^{-1} for fertilizer N, from 18 to 35 kg ha^{-1} for fertilizer P, and from 36 to 81 kg ha^{-1} for fertilizer K (lower and upper quartiles). These values were based on a fixed yield target and may not fully represent the variation in optimal fertilizer rates since the optimal yield target is likely to vary from season to season according to the variation in yield potential. However, a substantial temporal and spatial variation in optimal fertilizer N rates was also observed in an analysis using actual data from a series of N-response experiments in the Philippines during 1965-88 (Dawe and Moya 1999). For two nearby sites at the IRRI experimental farm, year-to-year variation in optimal fertilizer N rates differed considerably by 30 kg N ha^{-1} in both dry and wet seasons. These results suggest that only a certain degree of precision can be expected from preseason fertilizer recommendations, which may well be in the range of ±15–25 kg for N, ±5–10 kg for P, and ±15–20 kg for K, corresponding to about ±0.5 t grain yield ha^{-1}.

The sensitivity analysis of the QUEFTS model indicated that the model is sufficiently robust to arrive at sensible ranges of fertilizer recommendations despite uncertainties in the recovery efficiencies used for the calculation. A greater precision is probably not achievable as optimal fertilizer N, P, and K rates vary for a particular season or site because of the variation in climate and indigenous nutrient supplies, and other parameters that cannot be controlled. However, a greater precision may

Table 5.8. Simulated fertilizer N, P, and K rates (in kg ha^{-1}) to achieve a yield target of 7.5 t ha^{-1} as predicted by QUEFTS for the simulated input data given in Table 5.6. Simulation results are based on 1,000 model runs.

Parameters	Mean	SD[a]	10th percentile	Lower quartile	Upper quartile	90th percentile
Fertilizer N	134	31	95	114	157	179
Fertilizer P	26	12	11	18	35	45
Fertilizer K	59	34	14	36	81	103

[a]SD = standard deviation.

also not be required. Opportunities exist to arrive at near-optimal fertilizer N rates through dynamic N management compensating for the uncertainties of preseason fertilizer N recommendations. Fertilizer P and K recommendations produced by QUEFTS can be further improved using a simple nutrient balance model that would take other nutrient inputs into account and produce long-term rather than short-term strategies for P and K fertilization (Chapter 17).

Dynamic nitrogen management

Nitrogen management strategies differed among sites to account for differences in (1) climatic seasons, (2) varieties and growth duration, (3) crop establishment methods, (4) water management, and (5) possible pest problems (Dobermann and Fairhurst 2000). Our general strategy was to

1. Obtain a rough estimate of the average amount of N needed to achieve the yield target for average weather conditions,
2. Depending on the indigenous N supply (INS), crop establishment method, and variety (hybrid or inbred), decide whether and how much N needs to be applied at very early growth stages, and
3. Dynamically adjust the amount of topdressed N applications based on the actual plant N status, which accounts for N demand driven by actual crop growth (weather, crop density, water, pests, supply of other nutrients, other constraints to growth).

Because actual growth conditions may significantly deviate from average conditions assumed in an empirical fertilizer recommendation model, N was adjusted in-season to further increase N-use efficiency. Strategies for splitting and timing of N applications differed among sites. Preplant N was applied in a few cases, most noticeably at all sites in China where hybrid rice is grown or in the early rice crop in North Vietnam. At other sites, preplant N was applied only on soils with very low INS (typically, if INS was less than 40 kg N ha^{-1}). A chlorophyll meter (SPAD 502, Minolta, Ramsey, NJ) was used for making decisions on topdressed N applications, but the decision criteria were empirically changed over time based on the experience accumulated in the previous cropping seasons and other field research conducted at each site. Several years of consecutive experimentation were required to develop an "optimal" N management scheme for each site through a process that combined trial and error with factual knowledge. For example, in the first year of SSNM, increased pest incidence was often noted because of too large or too late N applications. Examples included the attraction of "mobile" pests to our SSNM plots from surrounding farmers' fields (rats, stem borer, leaffolder) or increased disease incidence (bacterial leaf blight, sheath blight). Moreover, relying on a single SPAD threshold value to identify both the date and amount of topdressed N applications may not be practical because this requires weekly field measurements, may miss critical growth stages, and does not allow for a more gradual adjustment of rates. Therefore, over time, a more practical N management strategy was implemented at each site, targeting somewhat lower yields but also trying to avoid increased pest damage or lodging.

At Jinhua, for example, SSNM in the first two crops (1998) followed existing best N management practices. Nitrogen was applied in fixed split applications at preset growth stages (40% incorporated before transplanting, 20% topdressed at 7–14 DAT, and 40% topdressed at the PI stage). Weekly chlorophyll meter readings were collected during this period to gain understanding about location-specific SPAD ranges. Beginning in 1999, plant N status was monitored only at preset critical growth stages at which N must be applied, but the amount of N was varied based on the actual SPAD value (Chapter 12). This strategy accounted for field-specific variation in INS during early growth and variation in late-season N demand depending on the actual growth conditions. Late N at 55 DAT was applied only in cases with a good crop stand to support the extra yield potential by adding more N for grain filling (Perez et al 1996). For comparison, farmers in the JI domain typically applied all N in two splits of about 40% preplant and 60% within 7–10 DAT and only a few applied a third dose at later growth stages. At Omon, the final site-specific N management strategy was tailored to the needs of direct-seeded rice and included no preplant application (Chapter 10). Instead, a uniform initial N dose was broadcast on the soil surface within the first 2 weeks after sowing, shortly before the fields became permanently flooded. From then on, decisions about two to three more topdressed applications were based on SPAD readings using thresholds lower than those used for transplanted rice (Balasubramaniam et al 2000, Dobermann and Fairhurst 2000).

Adjustment of INS, IPS, and IKS after each crop

Precise management of P and K must account for long-term changes in IPS and IKS related to actual yield and fertilizer use (Fig. 5.1). Moreover, when the SSNM plots were introduced, the quality of the available data on INS, IPS, and IKS in each field varied. In many cases (PhilRice, TNRRI, CLRRI, RIR, PTRRC), only INS was directly measured before so that IPS and IKS had to be estimated with less accuracy. In other cases, initial measurements of INS, IPS, and IKS were less accurate because of methodological or weather problems. Therefore, values of INS, IPS, and IKS for each SSNM crop grown were regularly adjusted by (1) incorporating new data collected and (2) estimating the change caused by the nutrient input-output balance of the previous crop. Adjusted INS, IPS, and IKS values were then used as model inputs for making the field-specific fertilizer recommendation for the subsequent rice crop.

The change in IPS and IKS was estimated empirically, based on the actual nutrient balance after harvest. Chapter 2 describes details of the nutrient balance calculations for all sites. We used the following algorithms:

$$IPS_a = IPS_i + \Delta IPS$$
$$\text{if P balance} \quad >0 \quad \Delta IPS = \text{P balance} \times 0.1$$
$$\qquad\qquad\quad <0 \quad \Delta IPS = \text{P balance} \times 0.01$$

$$IKS_a = IKS_i + \Delta IKS$$
$$\text{if K balance} \quad >0 \quad \Delta IKS = \text{K balance} \times 0.3$$
$$\text{if K balance} \quad <0 \quad \Delta IKS = \text{K balance} \times 0.05$$

where IPS_i and IKS_i are the initial indigenous supplies (kg ha^{-1}) and IPS_a and IKS_a represent the adjusted values for the next crop. These relationships assume that, over short periods, in a situation of nutrient depletion (balance <0), nutrients in the depleted pools contributing to IPS and IKS are largely replenished by those from other soil pools so that the net loss of IPS or IKS is small (1–5% per crop). In a situation of balance >0, we assumed a next-season recovery fraction of 10% P and 30% K of the excessive residual nutrient remaining in the field. Note, however, that often SSNM recommendations had to be worked out before data from the previous crop became available, that is, the small adjustments based on the P and K balance usually lagged behind by one crop.

For INS, we assumed no significant change over the short to medium term that would be related to the N input-output balance. In an irrigated rice system that has been under double-cropping for at least 20 years (all sites in this study), we assumed that the measured INS represents a more or less steady-state level of soil N supply that is largely maintained by the N input from crop residues and the biological N$_2$ fixation in the soil-floodwater system. Long-term experiments have demonstrated that, after a transition period to a new, more intensive system, yields in 0-N plots tend to stabilize. Only major disturbances such as a change in soil tillage and straw management or unusually long and dry fallow periods (Dobermann et al 2000) would significantly affect INS. Situations like this occurred at several sites in 1998 (e.g., PhilRice, RIR, PTRRC) because of El Niño phenomenon. Current knowledge does not allow an accurate prediction of such short-term fluctuations in INS within an SSNM approach. However, because N management followed a real-time, plant-based approach, possible deviations of INS from the assumed value were accounted for by adjusting the topdressed N applications.

As an example, Table 5.9. shows the changes in model input values of INS, IPS, and IKS over time at NISF. In this case, measurements of INS made in the 1997 LR crop underestimated the true INS, which was taken into account in 1999 by using data from two or three crops with more consistent values. Phosphorus management slowly increased the predicted IPS in the SSNM because of a positive P balance. Management of K only maintained IKS, mainly because at this site all straw is removed, making it difficult to achieve a significantly positive K balance with afford-

Table 5.9. Average measured and model input values of INS, IPS, and IKS (kg ha^{-1}) in on-farm SSNM trials in the Red River Delta (N = 24 farms).

	INS	IPS	IKS
Measured			
1997 ER	60.1	14.1	68.3
1997 LR	43.4	15.9	64.0
1998 ER	60.1	12.0	86.2
1998 LR	58.7	–	–
Model input			
1998 ER[a]	60.1	14.1	68.3
1998 LR[b]	43.4	14.7	65.2
1999 ER[c]	60.1	15.6	68.8
1999 LR[d]	55.6	16.7	66.5

[a]INS = INS$_{97ER}$; IPS = IPS$_{97ER}$; IKS = IKS$_{97ER}$. [b]INS = INS$_{97LR}$; IPS = mean(IPS$_{97ER}$,IPS$_{97LR}$); IKS = mean(IKS$_{97ER}$,IKS$_{97LR}$). [c]INS = mean(INS$_{97ER}$,INS$_{98ER}$); IPS = mean(IPS$_{97ER}$,IPS$_{98ER}$) + ΔIPS$_{98ER}$; IKS = mean(IKS$_{97ER}$,IKS$_{98ER}$) + ΔIKS$_{98ER}$. [d]INS = mean(INS$_{97ER}$,INS$_{98ER}$, INS$_{98LR}$); IPS = mean(IPS$_{97ER}$,IPS$_{97LR}$,IPS$_{98ER}$) + ΔIPS$_{98ER}$ + ΔIPS$_{98LR}$; IKS = mean(IKS$_{97ER}$,IKS$_{97LR}$,IKS$_{98ER}$) + ΔIKS$_{98ER}$ + ΔIKS$_{98LR}$.

able K input. Obviously, a major future research need is to validate these model-predicted changes in IPS and IKS by placing omission plots into the SSNM plot. In retrospective, short-term adjustments of indigenous nutrient supplies were probably not necessary for calculating fertilizer rates given the variation of indigenous nutrient supply estimates. It may take several years until changes in indigenous nutrient supplies become significant and measurable.

Other crop management practices

In the majority of cases, all crop management in the SSNM other than fertilizers was the same as in the surrounding field operated by the same farmer and usually done by him/her. However, where problems were suspected or observed, measures to either control them in advance (prophylactic) or correct them were implemented under the guidance of the researchers. The main principle was to minimize possible negative effects of factors other than nutrients so that the performance of SSNM could be assessed properly. Below, we list the major examples, but it should be emphasized that many of these measures did not significantly confound the comparison between SSNM and FFP treatments because they affected both treatments similarly.

Seed quality and the availability of certified seeds were a problem at sites such as CLRRI and PhilRice. Therefore, farmers were given a choice of varieties and provided with certified seeds, which were sown in both SSNM and FFP areas. In

China and North India, attempts were made to achieve optimal planting density because the contract laborers tended to plant much wider than required. In cases with snail damage (PhilRice, Thailand, South Vietnam), farmers were encouraged to replant damaged patches. Occasionally, researchers reminded the farmers to irrigate their fields to avoid water stress. On calcareous soils with high pH (South and North India), blanket application of Zn was done, which is standard practice among most farmers there. High weed incidence occurred occasionally, mainly in the low-yielding season of direct-seeded areas (Central Luzon, Thailand, South Vietnam), and affected both SSNM and FFP treatments. Although hand-weeding was sometimes attempted, it was difficult in direct-seeded rice and would have also caused damage to the rice plants. However, weeds often occurred in patches and extreme spots were avoided for sampling.

Prophylactic pesticide sprays for controlling stem borer and sheath blight were originally proposed as standard measures in all SSNM plots. However, this was not followed consistently and pest control measures were mostly similar in SSNM and FFP plots. In some cases, the researchers provided guidance to farmers about identifying pests and what needed to be sprayed. Rats were a problem at four sites (RIR, PTRRC, CLRRI, and PhilRice), but, with the exception of RIR, no specific control measures were implemented. At Sukamandi, rat damage during the first crop grown was severe. In all subsequent crops, a plastic barrier (fence) was installed around the whole FFP field, which included SSNM, FFP, and omission plots. Despite these attempts, insufficient pest control remained a major problem throughout the 2-year period of SSNM testing reported here.

5.4 Future improvements of the SSNM approach

The success of the SSNM strategy proposed here largely depends on (1) how accurate indigenous supplies and recovery efficiencies can be estimated and (2) how accurate N management can be fine-tuned to real plant needs. Both determine approaches for simplifying the SSNM concept and adopting it to local needs (see Chapters 17 and 18). Key technical components to simplify or improve include guidelines on the use of nutrient omission plots to estimate indigenous nutrient supplies, site-specific schemes for N management, and P and K management scenarios. Research is currently ongoing to develop empirical models for predicting INS, IPS, and IKS. Grain yield measurement in omission plots instead of measuring plant nutrient accumulation or soil nutrient pools is a promising alternative for estimating INS, IPS, and IKS, at least within a few broad categories.

Simple guidelines are needed for setting a realistic preseason yield goal to guide the estimation of nutrient requirements, particularly for P and K, but also at sites where adoption of plant-based N management is less attractive. Empirical studies suggest that a yield goal should be within the yield of the past three to five years plus 10% to 20%, but less than 80% of the inferred climatic yield potential because difficult-to-control yield-limiting factors always exist in subtropical and tropical rice environments. Such rules of thumb may be sufficiently robust as a starting point in the iterative participatory development of improved recommendations.

The model used to develop the field-specific fertilizer prescriptions assigned the same weights to N, P, and K for their effect on yield. It is likely, however, that K can be diluted relatively more in the plant than N before a significant reduction in growth occurs. Model improvements should also include a more generic approach for estimating fertilizer recovery efficiencies, particularly for P and K, but simple nutrient balance models may offer practical alternatives to arrive at economic longer-term fertilization strategies (Chapter 17). Possibly, yield gains similar to those in the SSNM approach tested can be achieved with well-chosen blanket doses of P and K, but in combination with improved N management tactics.

The nutrient management strategies were an empirical attempt to use a pre-plant fertilizer model in combination with a decision aid for location-specific, in-season fine-tuning of N at critical growth stages. More research is required to (1) develop a generic, validated scientific basis for this, (2) replace the chlorophyll meter with simpler tools such as a leaf color chart or fine-tuned standard split application schemes, and (3) rigorously compare this approach with other forms of N management in participatory approaches.

References

Adhikari C, Bronson KF, Panaullah GM, Regmi AP, Saha PK, Dobermann A, Olk DC, Hobbs P, Pasuquin E. 1999. On-farm soil N supply and N nutrition in the rice-wheat system of Nepal and Bangladesh. Field Crops Res. 64:273-286.

Angus JF, St.-Groth CFD, Tasic RC. 1990. Between-farm variability in yield responses to inputs of fertilizers and herbicide applied to rainfed lowland rice in the Philippines. Agric. Ecosyst. Environ. 30:219-234.

Balasubramaniam V, Morales AC, Cruz RT, Abdulrachman S. 1999. On-farm adaptation of knowledge-intensive nitrogen management technologies for rice systems. Nutr. Cycl. Agroecosyst. 53:59-69.

Balasubramaniam V, Morales AC, Thiyagarajan TM, Nagarajan R, Babu M, Abdulrachman S, Hai LH. 2000. Adaptation of the chlorophyll meter (SPAD) technology for real-time N management in rice: a review. Int. Rice Res. Notes 25:4-8.

Boogaard HL, Van Diepen CA, Roetter RP, Cabrera JMCA, van Laar HH. 1998. WOFOST 7.1: user's guide for the WOFOST crop growth simulation model and WOFOST control center. Wageningen (Netherlands): DLO Winand Staring Centre. 144 p.

Cassman KG, Dobermann A, Sta. Cruz PC, Gines HC, Samson MI, Descalsota JP, Alcantara JM, Dizon MA, Olk DC. 1996a. Soil organic matter and the indigenous nitrogen supply of intensive irrigated rice systems in the tropics. Plant Soil 182:267-278.

Cassman KG, Gines HC, Dizon M, Samson MI, Alcantara JM. 1996b. Nitrogen-use efficiency in tropical lowland rice systems: contributions from indigenous and applied nitrogen. Field Crops Res. 47:1-12.

Cassman KG, Harwood RR. 1995. The nature of agricultural systems: food security and environmental balance. Food Policy 20:439-454.

Cassman KG, Peng S, Dobermann A. 1997. Nutritional physiology of the rice plant and productivity decline of irrigated lowland rice systems in the tropics. Soil Sci. Plant Nutr. 43:1111-1116.

Dawe D, Moya PF. 1999. Variability of optimal nitrogen applications for rice. In: Program report for 1998. Makati City (Philippines): International Rice Research Institute p 15-17.

Dobermann A, Adviento MAA, Pampolino MF, Nagarajan R, Stalin P, Skogley EO. 1998. Opportunities for in situ soil testing in irrigated rice. In: Proceedings of the 16th World Congress of Soil Science. Montpellier (France): ISSS, CIRAD. p Symposium 13A, 106.

Dobermann A, Cassman KG, Peng S, Tan PS, Phung CV, Sta. Cruz PC, Bajita JB, Adviento MAA, Olk DC. 1996a. Precision nutrient management in intensive irrigated rice systems. In: Proceedings of the International Symposium on Maximizing Sustainable Rice Yields Through Improved Soil and Environmental Management, 11-17 November 1996, Khon Kaen, Thailand. Bangkok (Thailand): Department of Agriculture, Soil, and Fertilizer Society of Thailand, Department of Land Development, ISSS. p 133-154.

Dobermann A, Cassman KG, Sta. Cruz PC, Adviento MAA, Pampolino MF. 1996b. Fertilizer inputs, nutrient balance, and soil nutrient-supplying power in intensive, irrigated rice systems. II. Effective soil K-supplying capacity. Nutr. Cycl. Agroecosyst. 46:11-21.

Dobermann A, Cassman KG, Sta. Cruz PC, Adviento MAA, Pampolino MF. 1996c. Fertilizer inputs, nutrient balance, and soil nutrient-supplying power in intensive, irrigated rice systems. III. Phosphorus. Nutr. Cycl. Agroecosyst. 46:111-125.

Dobermann A, Dawe D, Roetter RP, Cassman KG. 2000. Reversal of rice yield decline in a long-term continuous cropping experiment. Agron. J. 92:633-643.

Dobermann A, Fairhurst TH. 2000. Rice: nutrient disorders and nutrient management. Singapore, Makati City (Philippines): Potash and Phosphate Institute, International Rice Research Institute. 191 p.

Dobermann A, Goovaerts P, Neue HU. 1997a. Scale dependent correlations among soil properties in two tropical lowland ricefields. Soil Sci. Soc. Am. J. 61:1483-1496.

Dobermann A, Oberthür T. 1997. Fuzzy mapping of soil fertility: a case study on irrigated riceland in the Philippines. Geoderma 77:317-339.

Dobermann A, Pampolino MF, Adviento MAA. 1997b. Resin capsules for on-site assessment of soil nutrient supply in lowland rice fields. Soil Sci. Soc. Am. J. 61:1202-1213.

Dobermann A, Pampolino MF, Neue HU. 1995. Spatial and temporal variation of transplanted rice at the field scale. Agron. J. 87:712-720.

Dobermann A, Sta. Cruz PC, Cassman KG. 1996d. Fertilizer inputs, nutrient balance, and soil nutrient-supplying power in intensive, irrigated rice systems. I. Potassium uptake and K balance. Nutr. Cycl. Agroecosyst. 46:1-10.

Dobermann A, White PF. 1999. Strategies for nutrient management in irrigated and rainfed lowland rice systems. Nutr. Cycl. Agroecosyst. 53:1-18.

Dobermann A, Witt C, Dawe D, Abdulrachman S, Gines GC, Nagarajan R, Satawathananont S, Son TT, Tan CS, Wang GH, Chien NV, Thoa VTK, Phung CV, Stalin P, Muthukrishnan P, Ravi V, Babu M, Chatuporn S, Sookthongsa J, Sun Q, Fu R, Simbahan G, Adviento MAA. 2002. Site-specific nutrient management for intensive rice cropping systems in Asia. Field Crops Res. 74:37-66.

Donovan C, Wopereis MCS, Guindo D, Nebie B. 1999. Soil fertility management in irrigated rice systems in the Sahel and Savanna regions of West Africa. Part II. Profitability and risk analysis. Field Crops Res. 61:147-162.

Horie T, Nakagawa H, Centeno HGS, Kropff MJ. 1995. The rice crop model simulation model SIMRIW and its testing. In: Matthews RB, Kropff MJ, Bachelet D, van Laar HH, editors. Modeling the impact of climate change on rice production in Asia. Wallingford (UK): CABI, IRRI. p 51-66.

Janssen BH. 1998. Efficient use of nutrients: an art of balancing. Field Crops Res. 56:197-201.

Janssen BH, Guiking FCT, Van der Eijk D, Smaling EMA, Wolf J, van Reuler H. 1990. A system for Quantitative Evaluation of the Fertility of Tropical Soils (QUEFTS). Geoderma 46:299-318.

Johnston AE, Lane PW, Mattingly GEG, Poulton PR, Hewitt MV. 1986. Effects of soil and fertiliser P on yields of potatoes, sugar beet, barley and winter wheat on a sandy clay loam at Saxmundham. J. Agric. Sci. 106:155-167.

Kropff MJ, van Laar HH, ten Berge HFM. 1993. ORYZA1: a basic model for irrigated lowland rice production. Los Baños (Philippines), Wageningen (Netherlands): IRRI, CABO. 89 p.

Matthews RB, Kropff MJ, Horie T, Bachelet D. 1997. Simulating the impact of climate change on rice production in Asia and evaluating options for adoption. Agric. Syst. 54:399-425.

Oberthür T, Dobermann A, Neue HU. 1996. How good is a reconnaissance soil map for agronomic purposes? Soil Use Manag. 12:33-43.

Olk DC, Cassman KG, Simbahan GC, Sta. Cruz PC, Abdulrachman S, Nagarajan R, Tan PS, Satawathananont S. 1999. Interpreting fertilizer-use efficiency in relation to soil nutrient-supplying capacity, factor productivity, and agronomic efficiency. Nutr. Cycl. Agroecosyst. 53:35-41.

Peng S, Garcia FV, Gines HC, Laza RC, Samson MI, Sanico AL, Visperas RM, Cassman KG. 1996a. Nitrogen use efficiency of irrigated tropical rice established by broadcast wet-seeding and transplanting. Fert. Res. 45:123-134.

Peng S, Garcia FV, Laza RC, Sanico AL, Visperas RM, Cassman KG. 1996b. Increased N-use efficiency using a chlorophyll meter on high-yielding irrigated rice. Field Crops Res. 47:243-252.

Perez CM, Juliano BO, Liboon SP, Alcantara JM, Cassman KG. 1996. Effects of late nitrogen fertilizer application on head rice yield, protein content, and grain quality of rice. Cereal Chem. 73:556-560.

Pierce FJ, Nowak P. 1999. Aspects of precision agriculture. Adv. Agron. 67:1-85.

Pingali PL, Hossain M, Pandey S, Price L. 1998. Economics of nutrient management in Asian rice systems: towards increasing knowledge intensity. Field Crops Res. 56:157-176.

Sinclair TR, Park WI. 1993. Inadequacy of the Liebig limiting-factor paradigm for explaining varying crop yields. Agron. J. 85:742-746.

Singh U, Ritchie JT, Godwin DC. 1991. A user's guide to CERES-RICE: V2.10. Muscle Shoals, Ala. (USA): International Fertilizer Development Center. 132 p.

Smaling EMA, Janssen BH. 1993. Calibration of QUEFTS, a model predicting nutrient uptake and yields from chemical soil fertility indices. Geoderma 59:21-44.

Witt C, Dobermann A, Abdulrachman S, Gines HC, Wang GH, Nagarajan R, Satawathananont S, Son TT, Tan PS, Tiem LV, Simbahan GC, Olk DC. 1999. Internal nutrient efficiencies of irrigated lowland rice in tropical and subtropical Asia. Field Crops Res. 63:113-138.

Wopereis MCS, Donovan C, Nebie B, Guindo D, N'Diaye MK. 1999. Soil fertility management in irrigated rice systems in the Sahel and Savanna regions of West Africa. Part I. Agronomic analysis. Field Crops Res. 61:125-145.

Notes

Authors' addresses: A. Dobermann, Department of Agronomy and Horticulture, University of Nebraska, P.O. Box 830915, Lincoln, NE 68583-0915, USA, e-mail: adobermann2@unl.edu; C. Witt, International Rice Research Institute, Los Baños, Philippines.

Citation: Dobermann A, Witt C, Dawe D, editors. 2004. Increasing productivity of intensive rice systems through site-specific nutrient management. Enfield, N.H. (USA) and Los Baños (Philippines): Science Publishers, Inc., and International Rice Research Institute (IRRI). 410 p.

6 | Site-specific nutrient management in irrigated rice systems of Tamil Nadu, India

R. Nagarajan, S. Ramanathan, P. Muthukrishnan, P. Stalin, V. Ravi, M. Babu, S. Selvam, M. Sivanantham, A. Dobermann, and C. Witt

6.1 Characteristics of rice production in the Cauvery Delta, Tamil Nadu

Trends in rice production

Tamil Nadu is one of the most important states for rice production in India because of its favorable soil and climatic conditions. In 2000, the annual rice harvest area accounted for about 2.1 million ha, with an annual production of 10.8 million t of paddy rice (Department of Agriculture and Cooperation 2003). The state ranked fifth in rice production in the country (8.5% of the total) and its yields are among the highest in India (Punjab 5.2 t ha^{-1}, Tamil Nadu 5.1 t ha^{-1}, Andhra Pradesh 4.3 tha^{-1}).

Rice in Tamil Nadu is mainly grown in the Cauvery Delta Zone (CDZ), which lies in the eastern part of the state, including Thanjavur District (Fig. 6.1). The CDZ has a total land area of 1.45 million ha, which is equivalent to 11% of the state area. The CDZ is one of the seven agroclimatic zones of Tamil Nadu. This zone is a large flat alluvial terrain, gently sloping toward the east, but mostly with an elevation of just 20 to 25 m. A humid tropical monsoon climate brings an average annual rainfall of about 1,000 mm. Most of the rain falls during the monsoon season from September to December, which is also associated with lower solar radiation and temperature (Fig. 6.2).

The east-flowing water of the Cauvery River is diverted at the Grand Anicut gate into the Old Cauvery Delta (fed by the Cauvery and Vennar rivers) and the New Cauvery Delta fed by the Grand Anicut Canal, thereby forming large irrigation systems with different characteristics. The Old Delta has been under irrigated rice cultivation for centuries, whereas the New Delta represents about 50 years of irrigated rice cultivation. Soils are generally of alluvial origin, low in soil organic C and total N, and have neutral to slightly alkaline pH, but they differ significantly between the two deltas. The important cropping systems are rice-rice-fallow, rice-rice-pulses, rice-rice-sesame, and rice-rice-cotton. There are three major rice-growing seasons—kuruvai (KR), thaladi (TH), and samba. The kuruvai season (June to September) is a premonsoon dry season with short-duration rice (105–110 d) with high yield potential. Thaladi and samba are rainy and wet seasons with lower yield potential. Medium-duration rice (125–135 d) is grown during the thaladi season from October to

101

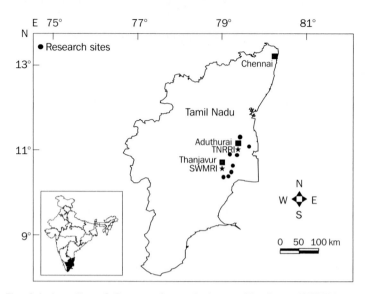

Fig. 6.1. Location of the experimental sites at Thanjavur (SWMRI) and Aduthurai (TNRRI), Tamil Nadu, India.

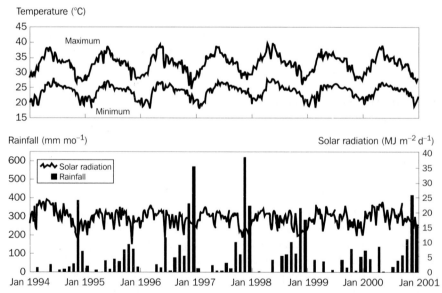

Fig. 6.2. Climatic conditions at Aduthurai, Tamil Nadu, India, from 1994 to 2000. Solar radiation and temperature data are 7-day moving averages; rainfall is monthly total.

Table 6.1. Area, yield, production of rough rice, and fertilizer use in the state of Tamil Nadu from 1961 to 2000 (5-year averages).

Item	1961-70	1971-80	1981-90	1991-2000
Rice area (10^6 ha)[a]	2.61	2.60	2.12	2.23
Irrigated rice area (%)	86.1	92.1	92.7	90.8
Rice yield (t ha^{-1})[a]	2.3	2.9	3.8	4.6
Rice production (10^6 t)[a]	6.00	7.59	7.93	10.29
Fertilizer use (1,000 t)	183	411	728	857
Nitrogen (1,000 t)	114	253	404	471
Phosphorus (1,000 t)	39	75	150	172
Potash (1,000 t)	30	83	174	214

[a]Irrigated and rainfed land. Data sources: IRRI (1994): World rice statistics 1993-1994. International Rice Research Institute (IRRI), Philippines; Government of India (2001): Economic survey 2000-2001; Department of Economics and Statistics (2000): Season and crop report 1959 to 2000; Department of Agriculture and Cooperation (2002). http://agricoop.nic.in, Ministry of Agriculture, Government of India, Chennai, India.

February as a second crop after the kuruvai season. In rotations with only one rice crop per year, long-duration rice (155–165 d) is grown in the samba season from August to January. Crops in between two rice crops (e.g., pulses, cotton, sesame) are grown under no-tillage conditions, using the residual moisture and nutrients in the Old Delta. In the New Delta, such summer crops often receive some irrigation.

The advent of semidwarf high-yielding varieties and improved production technologies led to a rapid increase in rice productivity during the initial period of the Green Revolution, and production in Tamil Nadu increased by 126,400 t year^{-1} from 1961 to 2000 (Table 6.1). During this period, farmers doubled their average yields with marked increases in the 1980s, which were, however, almost fully compensated in that decade by a 480,000-ha decrease in the total annual rice area (Table 6.1). There has been substantial variation in production in Tamil Nadu in the last 30 years. For instance, production in the 1980s ranged from 5.26 million t in 1982 to 9.35 million t in 1989. The variation in production in the 1970s and '80s was mainly caused by a large year-to-year variation in the area planted to rice, which could only partly be explained by water availability or the percentage of land under irrigation. About 90% of the rice area is irrigated, without major changes in the last 30 years, but year-to-year variation is large. In the 1990s, the percentage of irrigated rice land ranged from 81% to 98%, mainly depending on rainfall pattern (Fig. 6.2) and the timely release of water for the kuruvai crop through the Grand Anicut Canal in the major rice-growing area of the Cauvery Delta. Water availability is certainly of major concern for rice farmers in Tamil Nadu. The water release from the Cauvery River is now strictly regulated, and farmers mostly depend on groundwater pumped from bore wells for irrigation. Farmers with less capital can't afford bore wells and face great risk when taking up rice cultivation because rainfall during the monsoon may be insufficient in certain years to support the crop.

Yield increases in the last 40 years were accompanied by large increases in fertilizer use, particularly during the 1970s and '80s (Table 6.1). Unlike in many other rice areas of Asia, the use of both phosphate and potash fertilizer for more balanced plant nutrition is already more widespread in Tamil Nadu. Fertilizer recommendations are mainly based on blanket recommendations given for the whole state.

The production trends in Tamil Nadu shown in Table 6.1 look promising for meeting the increasing food demand for a population of 62 million people in 2001 growing at 1.19% per year. Population growth in Tamil Nadu was 11.19% from 1991 to 2000, which is the second lowest in India among the states with a population of more than 20 million (Athreya 2001). However, farmers may be forced to switch to other, not necessarily more profitable, crops in less favorable years with limited water availability. They will therefore have to improve the efficiency of all rice production inputs (water, nutrients, pesticides, labor, energy) to sustain high yield levels as well as soil quality (Nagarajan et al 1996). At issue is whether a more site-specific approach to nutrient management in rice can contribute to accelerating profitability increases in farming in both the Old and New Cauvery Deltas. Recent research conducted in Tamil Nadu suggested that variability in indigenous N supply (INS) was large among farms and that fertilizer use was not in close congruence with the variation in INS (Nagarajan et al 1997, Olk et al 1999, Stalin et al 1996). It was also found that soil N mineralization and crop N uptake dynamics varied widely among different soils, but were difficult to predict with commonly used static soil tests (Stalin et al 1996, Thiyagarajan et al 1997). Soils could, however, be classified into those that would require preplant N application and soils that wouldn't (Thiyagarajan et al 1997). Therefore, it was suggested to use a soil test or crop-based measure of INS to estimate the need for preplant N, but adjust topdressed N applications based on monitoring of crop N status or by using a simplified crop model (Stalin et al 1996, ten Berge et al 1997).

Much research has been conducted since then in Tamil Nadu to develop new N management strategies for irrigated rice. Studies included work on improved fixed-N splitting patterns (Stalin et al 1999) as well as research on developing real-time N management strategies based on monitoring leaf N status with a chlorophyll meter (SPAD) or leaf color charts (LCC). Research on SPAD and LCC mainly focused on fine-tuning critical thresholds and recommended application rates to local modern varieties, growth stages, climatic seasons, and crop management factors such as plant density (Babu et al 2000a,b, Balasubramanian et al 2000, Janaki et al 2000, Ramanathan et al 2000). On-farm evaluations generally showed that there was little need for preplant N and that N rates could be reduced significantly, but yield increases were less consistent. Studies were also conducted to compare soil test-based approaches with a Manage-N modeling approach and the chlorophyll meter concept (Stalin et al 2000), indicating that model-based N management was a promising strategy also.

This paper presents the initial results of on-farm testing of site-specific nutrient management (SSNM) conducted from 1997 to 2000 in the two major rice-growing regions of Tamil Nadu. Continuing the research on N management strategies, a major goal of this work was to develop a generic SSNM strategy that combines N management with that of P and K at a field- and season-specific level.

Current biophysical and socioeconomic farm characteristics

The two experimental domains of the RTDP project in Tamil Nadu were located around Aduthurai (11°1'N, 79°29'E) in the Old Cauvery Delta and around Thanjavur (10°47'N, 79°10'E) in the New Cauvery Delta (Fig. 6.1). Both domains represent more than 100,000 ha under intensive irrigated rice systems. To conduct a detailed analysis of current farm-level productivity and its variation among farms, on-farm monitoring experiments began in 1995 at Aduthurai and in 1997 at Thanjavur. The experimental approaches were the same as described in Chapter 2. Twenty-five farms within 15 to 20 km of Aduthurai and 18 farms within 20 km of Thanjavur were selected to represent dominant soil series, cropping systems, and socioeconomic farm characteristics in both domains. Data reported here refer to an initial period of farm monitoring (1995 to 1996 at Aduthurai, 1997 to 1998 at Thanjavur), which was followed by testing the SSNM strategy for four consecutive crops grown from 1997 to 1999 (Aduthurai) or 1998 to 2000 (Thanjavur).

Soils in the Old Delta are heavier in texture (clay loam to clay) and classified into Kalathur, Adanur, Padugai, and Alangudi series (Udorthentic Chromusterts, Vertic Ustropepts, and Typic Ustifluvents), whereas soils in the New Delta mostly belong to the Pattukkottai and Madukkur series (Typic Haplustalfs, 56% of total area) with sandy loam to clay loam texture and good drainage. These soil differences are also reflected in the general soil properties of rice farms at Aduthurai and Thanjavur (Table 6.2). Clay content averaged 39% at Aduthurai sites vis-à-vis less than 9% at Thanjavur. Compared with those at Thanjavur, soils at Aduthurai had higher cation exchange capacity (CEC) and contained more extractable potassium (K). In both domains, soil organic C content was typically 10 g kg^{-1} or less and pH was around 7. However, pH values of >7.5 were measured for about 25–30% of all sites in each domain. In both domains, median levels of extractable Olsen-P were relatively high (26 to 30 mg P kg^{-1}) and values below the commonly suggested critical level of 10 mg kg^{-1} did not occur. Micronutrient availability varies widely with soil type in the Cauvery Delta zone. Large surveys conducted in the 1980s suggested that about 80% of the soils in the Old Delta and 47% in the New Delta were deficient in Zn. Similarly, soil test levels of DTPA-extractable Fe were found to be below proposed critical levels for 60% (Old Delta) or 44% (New Delta) of all samples (Savithri et al 1999). In general, Zn availability is low on calcareous soils of the Cauvery Delta that have been under irrigated rice cultivation for centuries (Nagarajan and Manickam 1985), but Zn fertilization has become a common practice (Nagarajan and Manickam 1986, Savithri et al 1999) so that deficiencies in rice are less common. With the exception of a few farms, HCl-extractable Zn levels at the experimental sites (Table 6.2) were generally above the commonly used critical level of 1 mg kg^{-1} (Dobermann and Fairhurst 2000).

Table 6.2. General soil properties on rice farms at Aduthurai and Thanjavur, Tamil Nadu.

Soil properties	Min.	25%	Median	75%	Max.
Aduthurai, Old Cauvery Delta (25 farms)[a]					
Clay content (%)	34.0	38.0	39.0	40.0	41.0
Silt content (%)	23.0	24.0	26.0	26.0	27.0
Sand content (%)	31.0	33.0	35.0	36.0	41.5
Soil organic C (g kg ^{1})	7.1	9.0	10.1	11.0	12.1
Total soil N (g kg ^{1})	0.80	0.89	0.93	1.00	1.14
Soil pH (1:1 H_2O)	6.6	7.1	7.3	7.5	7.7
Cation exchange capacity (cmol$_c$ kg ^{1})	22.3	28.7	31.4	33.7	35.7
Exchangeable K (cmol$_c$ kg ^{1})	0.49	0.52	0.57	0.61	0.76
Exchangeable Na (cmol$_c$ kg ^{1})	1.10	1.19	1.33	1.63	2.19
Exchangeable Ca (cmol$_c$ kg ^{1})	15.43	20.10	21.39	23.33	27.17
Exchangeable Mg (cmol$_c$ kg ^{1})	6.73	8.37	9.40	10.57	11.43
Extractable P (Olsen-P, mg kg ^{1})	16.67	20.00	25.84	32.67	34.67
Extractable Zn (0.05N HCl, mg kg ^{1})	0.87	1.64	1.85	2.98	3.96
Thanjavur, New Cauvery Delta (18 farms)[b]					
Clay content (%)	4.55	7.88	8.65	9.20	10.10
Silt content (%)	2.45	2.80	2.90	3.10	3.65
Sand content (%)	86.60	87.83	88.45	88.85	92.30
Soil organic C (g kg ^{1})	5.8	6.7	7.0	7.8	8.5
Total soil N (g kg ^{1})	0.43	0.52	0.55	0.57	0.58
Soil pH (1:1 H_2O)	5.6	6.8	7.1	7.6	8.3
Cation exchange capacity (cmol$_c$ kg ^{1})	5.0	7.5	9.0	10.0	15.8
Exchangeable K (cmol$_c$ kg ^{1})	0.16	0.23	0.27	0.31	0.36
Exchangeable Na (cmol$_c$ kg ^{1})	0.22	0.27	0.29	0.33	0.38
Exchangeable Ca (cmol$_c$ kg ^{1})	4.90	5.87	6.80	8.05	11.75
Exchangeable Mg (cmol$_c$ kg ^{1})	2.40	3.05	3.40	3.62	3.95
Extractable P (Olsen-P, mg kg ^{1})	19.5	27.5	29.8	35.0	41.0

[a]Measured on initial soil samples collected before the 1996 thaladi crop. [b]Measured on initial soil samples collected before the 1997 kuruvai crop.

Figure 6.3 shows the variability of the indigenous N (INS), P (IPS), and K (IKS) supplies among farmers' fields as measured from 1997 to 1999. Differences among the four rice crops are mainly attributed to seasonal fluctuations in climate. In general, indigenous nutrient supplies in the thaladi rice season (wet season) were significantly lower than those for kuruvai rice (dry season), particularly at Aduthurai (Fig. 6.3A). However, within each season, ranges of indigenous N, P, and K supplies were large among the farms in each domain sampled. In addition to this, differences occurred between the two domains. Whereas the average IPS during the kuruvai seasons was similar in both domains (19 kg P ha^{-1} per crop), sites at Aduthurai had a higher INS (average of 58 vs 44 kg N ha^{-1}) as well as a much higher IKS (95 vs 66 kg K ha^{-1}) than those at Thanjavur. Assuming optimal nutrient requirements of 14.7 kg N, 2.6 kg P, and 14.5 kg K per 1,000 kg grain yield (Witt et al 1999), current average levels of indigenous nutrient supplies would be sufficient to achieve kuruvai rice yields of about 3 (Thanjavur) to 4 t ha^{-1} (Aduthurai) without N application, 7.3 t

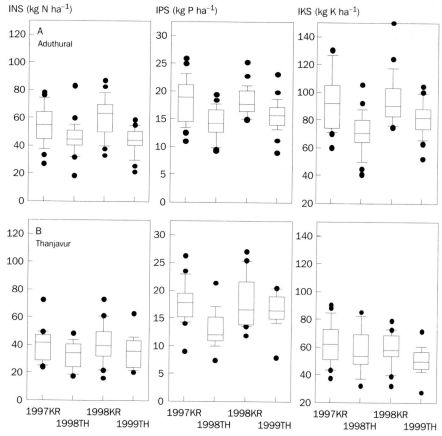

Fig. 6.3. Variability of the indigenous N (INS), P (IPS), and K (IKS) supply among farmers' fields at Aduthurai (25 farms) and Thanjavur (18 farms), Tamil Nadu, India (1994-2000). Median with 10th, 25th, 75th, and 90th percentiles as vertical boxes with error bars; outliers as bullets. KR = kuruvai, TH = thaladi.

ha^{-1} without P at both Thanjavur and Aduthurai, and 4.5 (Thanjavur) to 6.6 t ha^{-1} (Aduthurai) without K. The large differences between the Old and New Cauvery Deltas suggest that different SSNM strategies may be required for managing rice in each domain.

Demographic and economic characteristics of rice production on rice farms of Tamil Nadu were summarized in Table 6.3. The median farm size at Aduthurai was 2.8 ha vis-à-vis 10 ha at Thanjavur. However, the typical rice area in which the treatment plots were embedded in subsequent years was similar in both domains (2.2 vs 2.6 ha per farm). The average age of the farmers was 43 and they had typically attended school for 10 years. Farmers at Aduthurai usually applied more N and P fertilizer than those at Thanjavur, whereas K rates were similar. Pesticide use was generally low. Only farmers at Aduthurai applied insecticides at an average rate of 0.13

Table 6.3. Demographic and economic characteristics of rice production on rice farms at Aduthurai and Thanjavur, Tamil Nadu, India. Values shown are based on socioeconomic farm surveys conducted for whole farms.

Soil properties	Min.	25%	Median	75%	Max.
Aduthurai, 1997 kuruvai crop					
Total cultivated area (ha)	0.4	1.6	2.8	5.0	31.7
Age of household head (y)	27	35	43	56	70
Education of household head (y)	5	9	10	15	20
Household size (persons)	2	4	5	8	11
Rice area (ha)[a]	0.3	1.3	2.2	3.1	8.0
Yield (t ha^{-1})	3.8	5.1	5.9	7.1	7.7
N fertilizer (kg ha^{-1})	55	86	103	119	165
P fertilizer (kg ha^{-1})	12	14	23	26	50
K fertilizer (kg ha^{-1})	0	23	31	50	60
Insecticide (kg ai ha^{-1})	0	0	0.13	0.42	5.78
Herbicide (kg ai ha^{-1})	0	0	0	0	1.67
Other pesticides (kg ai ha^{-1})[b]	0	0	0	0	0
Total labor (8-h d ha^{-1})	130	147	163	179	225
Net return from rice (US$ ha^{-1} crop^{-1})	167	281	388	520	703
Thanjavur, 1998 kuruvai rice crop					
Total cultivated area (ha)	1.8	5.5	10.0	12.8	41.2
Age of household head (y)	35	41	43	60	73
Education of household head (y)	5	7	10	10	17
Rice area (ha)[a]	1.2	2.0	2.6	5.6	8.0
Household size (persons)	4	5	6	8	13
Yield (t ha^{-1})	3.3	4.3	5.2	5.5	5.6
N fertilizer (kg ha^{-1})	33	44	64	75	83
P fertilizer (kg ha^{-1})	8	11	18	22	38
K fertilizer (kg ha^{-1})	0	27	34	42	72
Insecticide (kg ai ha^{-1})	0	0	0	0.42	0.54
Herbicide (kg ai ha^{-1})	0	0	0	0	1.25
Other pesticides (kg ai ha^{-1})[b]	0	0	0	0	0.91
Total labor (8-h d ha^{-1})	66	74	81	86	111
Net return from rice (US$ ha^{-1} crop^{-1})	178	294	379	427	482

[a]Rice area in which the treatment plots were embedded in subsequent years. The total rice area may be even larger. [b]Includes fungicide, molluscicide, rodenticide, and crabicide.

and 0.17 kg ai ha^{-1} during the kuruvai and thaladi seasons, respectively. Farmers at Thanjavur usually did not use any insectides or herbicides. Weed control was mainly done by hand weeding. Labor input at Aduthurai (median of 163 8-h d ha^{-1} in kuruvai 1998 crop) was generally about twice as high as that at Thanjavur (median of 81 8-h d ha^{-1} in kuruvai 1998 crop). Field operations such as land preparation, pulling out and transporting seedlings, applying pesticides and fertilizers, harvesting, and threshing were mostly done by men, whereas women did the transplanting, weeding, cleaning, and drying of rice. Wages of men were typically higher than those of women, even for similar operations. Cost and return analysis revealed that median net returns were similar in both domains (US$379 to $388 ha^{-1} per crop) because yields at Aduthurai tended to be higher than at Thanjavur. Wide ranges in gross returns over fertilizer cost were observed among farms (Table 6.3) that reflect the variation in soil quality and quality of crop care.

Measurements of baseline agronomic characteristics of two consecutive rice crops sampled in 1995-96 at Aduthurai and 1997-98 at Thanjavur suggested slightly positive average P input-output balances in both domains (about +2 kg P ha^{-1} per crop), whereas the average K input-output balance was negative, typically –30 to –50 kg K ha^{-1} per crop. Average recovery efficiency of fertilizer-N (REN) was 0.54 kg kg^{-1} at Aduthurai and 0.45 kg kg^{-1} at Thanjavur. Average agronomic efficiency of fertilizer-N (REN) was 20.5 kg kg^{-1} at Aduthurai and 13.8 kg kg^{-1} at Thanjavur. Compared to other rice-growing regions in Asia, this represents a relatively high level of N-use efficiency, which is mainly caused by the moderate N use, frequent splitting of N applications, and generally good quality of crop management in both Tamil Nadu domains. Partial factor productivity of fertilizer-N (PFPN) averaged 55 to 60 kg kg^{-1}.

6.2 Effect of site-specific nutrient management on productivity and nutrient-use efficiency

Management of the SSNM plots

Beginning with the 1997 kuruvai crop at Aduthurai and the 1998 kuruvai crop at Thanjavur, an SSNM plot was established on all farms as a comparison to the farmers' fertilizer practice (FFP, see Chapter 2). The initial SSNM approach was tested over a period of four consecutive rice crops as described for other sites (see Chapter 5).

The size of the SSNM plots ranged from 500 to 1,000 m^2. All farmers planted conventional modern varieties, most frequently ADT38 (35% of all cases), ADT43 (20%), ADT42 (9%), ADT36 (8%), IR50 (6%), IR72 (5%), and TKM9 (5%). During the kuruvai season, the farmers used short-duration varieties (ADT36, ADT42, ADT43, IR50, TKM9, and IR72). In the kuruvai (KR) crop, farmers favor growing ADT43 because it is a fine-grain variety preferred by consumers, so that its market value is relatively high. The recommended planting density was 66 hills m^{-2} (15 × 10-cm spacing) with 2–3 plants per hill. During the thaladi seasons, farmers grew medium-duration varieties (e.g., ADT38) for which the recommended planting density was 50 hills m^{-2} (20 × 10 cm) with 2–3 plants per hill. However, the planting density in farmers' fields was typically below the recommended level in both the kuruvai and thaladi seasons.

Farmers controlled water, weeds, and pests in both FFP and SSNM plots following the commonly adopted methods. A shallow water depth of at most 2 cm was maintained until about 7 days after transplanting. Thereafter, soils were kept submerged at a water depth of 2.5 to 5 cm throughout the cropping period. The time of irrigation differed among sites depending on percolation rates and season. Once there was no standing water in the field, water was supplied after 1 day in the kuruvai season or 3 days in the thaladi season on the sandy to clay loam soils of the New

Cauvery Delta. On the clay loam to clay soils of the Old Cauvery Delta, there was no delay in irrigation in the kuruvai season but farmers kept their fields without standing water for 1–2 days before they reirrigated. Irrigation was withheld 15 days ahead of harvest.

Weeding was mostly done by hand and the quality of crop management was usually good in most fields. Pesticide use in both FFP and SSNM plots was generally low and a detailed pest impact assessment was performed for two seasons in 1998 (Sta. Cruz et al 2001).

Fertilizer applications for SSNM were prescribed on a field- and crop-specific basis following the approach described in Chapter 5. In the 1997 kuruvai crop at Aduthurai, average values of the INS measured in 1995 and 1996 were used as model input. Similarly, the average IPS and IKS were estimated from the plant P and K accumulation in the FFP sampled during this period, assuming recovery fractions of 0.2 kg kg^{-1} for fertilizer P and 0.4 kg kg^{-1} for fertilizer K. Beginning with the 1998 KR crop, estimates of INS, IPS, and IKS were continuously improved by incorporating the values measured in omission plots in the year before and by adjusting IPS and IKS according to the actual P and K input-output balance. At Thanjavur, SSNM recommendations were based on INS, IPS, and IKS measured in nutrient omission plots in 1997 and 1998.

Target yields for working out field-specific fertilizer rates were 7.2 to 7.5 t ha^{-1} in the thaladi season (WS) at both sites, and 8 t ha^{-1} (Aduthurai) and 7.0 to 7.5 t ha^{-1} (Thanjavur) for kuruvai rice (DS). Yield targets varied little (typically within 0.5 t ha^{-1}) among farms. The climatic yield potential was set to 8.3 t ha^{-1} for thaladi rice and at 9.8 t ha^{-1} for the kuruvai season based on crop simulations conducted for this region (Matthews et al 1995). First-crop recovery fractions of 0.50 to 0.55, 0.20, and 0.45 to 0.55 kg kg^{-1} were assumed for fertilizer N, P, and K, respectively. Previous on-farm research demonstrated a relatively high level of N-use efficiency caused by frequent splitting of N application and intensive crop care, thus explaining the high target values for N recovery efficiency assumed for SSNM. Urea and ammonium sulfate were used as N sources, single superphosphate for P, and muriate of potash for K. All P and 50% of the K fertilizer were incorporated in the soil before sowing or planting. Another 50% of the K rate was topdressed at panicle initiation (PI). No farmyard manure was applied.

The N management strategy gradually changed in response to the observations made in each cropping season. In the first season, in the 1997 KR crop at Aduthurai, preplant N at 20 kg N ha^{-1} was applied at sites with an INS less than 60 kg N ha^{-1}, but none at others. At 14 days after transplanting (DAT), all SSNM plots received 30 kg N ha^{-1}. After 14 DAT, chlorophyll meter (SPAD) readings were taken weekly and 30 to 45 kg N ha^{-1} applied whenever the SPAD value was below 35. This strategy resulted in three to six N applications per site at an average total N rate of about 130 kg N ha^{-1}. Late-season pest incidence was relatively high and the strategy was changed in subsequent years to improve the congruence between N supply and crop N demand. Beginning with the 1998 thaladi crop, no preplant N was applied at both Aduthurai and Thanjavur, but N management aimed at applying an amount close to that predicted by the QUEFTS model. The following N management scheme was used in the 1998 kuruvai season:

FN 0	Preplant		No preplant N
FN 1	14 DAT		30 kg N ha^{-1}
FN 2	21–35 DAT	If SPAD >35	No N
	(tillering)	If SPAD <35	FN 2 = 2/7 × (FN − 30)
FN 3	42–49 DAT	If SPAD >35	No N
	(panicle initiation)	If SPAD <35	FN 3 = 3/7 × (FN − 30)
		If SPAD <35 and FN 2 = 0	FN 3 = 5/7 × (FN − 30)
FN 4	56–65 DAT	If SPAD >35	No N
	(flowering)	If SPAD <35	FN 4 = 2/7 × (FN − 30)
		If SPAD <35 and FN 3 = 0	FN 4 = 3/7 × (FN − 30)

where FN was the model-predicted total N rate (kg N ha^{-1}). Starting with the 1999 KR season, this procedure was replaced by a more standardized N schedule, which led to the proposed schedule for tropical transplanted rice (Dobermann and Fairhurst 2000). This included no preplant N application, 25% applied at 14–20 DAT, and two more topdressings at 30–35 DAT and at 40–45 DAT (PI stage), with rates depending on the ranges of SPAD values. For example, the following schedule was implemented in the 2000 KR season:

FN 0	Basal		0%	
FN 1	14–20 DAT		25%	
FN 2	30–35 DAT	If SPAD >36		20% reduction in standard rate
		If SPAD 34–36	30%	Standard rate
		If SPAD <34		20% increase in standard rate
FN 3	40–45 DAT	If SPAD >36		20% reduction in standard rate
	(panicle initiation)	If SPAD 34–36	45%	Standard rate
		If SPAD <34		20% increase in standard rate
FN 4	60–65 DAT	If SPAD >36		No N
	(flowering)	If SPAD 34–36		Additional 15% of total N
		If SPAD <34		Additional 23% of total N

The average number of N applications per crop varied from 3.3 to 4.2 at Aduthurai and from 3.1 to 3.6 at Thanjavur.

Effect of SSNM on grain yield and nutrient uptake

Significant increases in grain yield and nutrient uptake because of SSNM were observed in both experimental domains. The average grain yield increase of SSNM over FFP across all four crop cycles was 0.49 t ha^{-1} at Aduthurai (8%, $P = 0.003$) and 0.63 t ha^{-1} at Thanjavur (13%, $P = 0.006$, Table 6.4). At Aduthurai, the yield difference increased significantly from 0.22 t ha^{-1} in the first year to 0.78 t ha^{-1} in the second year (crop-year effect $P = 0.000$). Yield increases in the first two crops grown

Table 6.4. Effect of site-specific nutrient management on grain yield and nitrogen-use efficiency of rice at Aduthurai (1997-99) and Thanjavur (1998-2000), Tamil Nadu, India.

Parameter	Levels[a]	Treatment[b] SSNM	Treatment[b] FFP	Δ^c	$P > \|T\|^c$	Effects[d]	$P > \|F\|^d$
Aduthurai, Old Cauvery Delta							
Grain yield (GY)	All	6.45	5.96	0.49	0.003	Village	0.025
(t ha ^{1})	Year 1	6.19	5.97	0.22	0.424	Year[e]	0.000
	Year 2	6.72	5.94	0.78	0.000	Season[e]	0.813
	Kuruvai	6.97	6.46	0.51	0.035	Year × season[e]	0.879
	Thaladi	5.95	5.47	0.48	0.012		
Agronomic efficiency	All	15.8	13.7	2.1	0.096	Village	0.127
of N (AEN)	Year 1	15.1	15.2	−0.1	0.984	Year	0.006
(kg grain kg ^{1} N)	Year 2	16.5	12.3	4.2	0.005	Season	0.263
	Kuruvai	17.4	14.4	3.0	0.107	Year × season	0.398
	Thaladi	14.3	13.1	1.2	0.478		
Recovery efficiency	All	0.43	0.39	0.04	0.084	Village	0.500
of N (REN)	Year 1	0.43	0.43	0.01	0.869	Year	0.049
(kg plant N kg ^{1} N)	Year 2	0.43	0.35	0.08	0.012	Season	0.547
	Kuruvai	0.48	0.43	0.05	0.135	Year × season	0.110
	Thaladi	0.39	0.35	0.03	0.310		
Thanjavur, New Cauvery Delta							
Grain yield (GY)	All	5.63	5.00	0.63	0.006	Village	0.287
(t ha ^{1})	Year 1	5.37	4.74	0.63	0.029	Year	0.889
	Year 2	5.96	5.33	0.63	0.078	Season	0.445
	Kuruvai	6.15	5.62	0.53	0.082	Year × season	0.484
	Thaladi	5.03	4.29	0.74	0.003		
Agronomic efficiency	All	14.6	13.2	1.4	0.392	Village	0.501
of N (AEN)	Year 1	14.0	11.5	2.5	0.228	Year	0.271
(kg grain kg ^{1} N)	Year 2	15.3	15.3	0.0	0.997	Season	0.357
	Kuruvai	14.5	12.6	1.9	0.425	Year × season	0.179
	Thaladi	14.6	13.9	0.7	0.735		
Recovery efficiency	All	0.46	0.45	0.02	0.611	Village	0.036
of N (REN)	Year 1	0.45	0.40	0.05	0.231	Year	0.241
(kg plant N kg ^{1} N)	Year 2	0.48	0.51	−0.03	0.581	Season	0.008
	Kuruvai	0.49	0.45	0.04	0.376	Year × season	0.803
	Thaladi	0.43	0.44	−0.01	0.795		

[a]All = all four crops grown from 1997 KR to 1999 TH (Aduthurai) or 1998 KR to 2000 TH (Thanjavur); Year 1 = 1997 KR and 1998 TH at Aduthurai and 1998 KR and 1999 TH at Thanjavur; Year 2 = 1998 KR and 1999 TH at Aduthurai and 1999 KR and 2000 TH at Thanjavur; kuruvai = two dry-season crops; thaladi = two wet-season crops. [b]FFP = farmers' fertilizer practice; SSNM = site-specific nutrient management. $^c\Delta$ = SSNM − FFP. $P > \|T\|$ = probability of a significant mean difference between SSNM and FFP. [d]Source of variation of analysis of variance of the difference between SSNM and FFP by farm; $P > \|F\|$ = probability of a significant F-value. [e]Year refers to two consecutive cropping seasons.

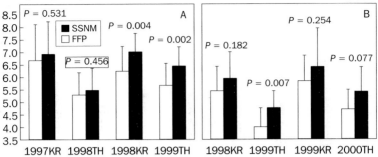

Fig. 6.4. Grain yield in the farmers' fertilizer practice (FFP) and site-specific nutrient management (SSNM) plots at Aduthurai (A) and Thanjavur (B), Tamil Nadu, India, 1997-2000 (bars: mean; error bars: standard deviation).

were not significant, but improvements in N management were probably the main reason for the improved performance of SSNM thereafter (Fig. 6.4). At Thanjavur, no such crop-year effect was observed probably because the SSNM treatment started one year later than at Aduthurai and included the refined N management scheme from the beginning. At both sites, similar yield increases were observed in kuruvai and thaladi rice crops (nonsignificant crop-season effect). Very large yield increases because of SSNM of more than 1 t ha^{-1} were observed for 24% of all cases at both Aduthurai and Thanjavur. Negative yield differences between SSNM and FFP were not frequent (20% at Aduthurai, 9% at Thanjavur), suggesting little risk associated with the SSNM approach. Grain yield was lowest in the 1998 thaladi crop at Aduthurai and in the 1999 thaladi crop at Thanjavur, mainly because of unfavorable rainy weather (Fig. 6.4).

The average increase in plant nitrogen uptake of SSNM over FFP across all four crops was significant, with 11.6 kg ha^{-1} at Aduthurai (13%, $P = 0.001$, Fig. 6.5) and 17.2 kg ha^{-1} at Thanjavur (22%, $P = 0.000$). For grain yield, relative differences in N uptake at Aduthurai increased from just 7% in year 1 to almost 19% in year 2 (Fig. 6.5, crop-year effect $P = 0.001$), whereas they were of similar magnitude in both years at Aduthurai and generally did not differ among dry- and wet-season crops. Similar trends were observed for plant P and K uptake. The average increase in plant P uptake across all four crops was highly significant and 2.4 kg ha^{-1} at Aduthurai (12%) vs 4.9 kg ha^{-1} at Thanjavur (26%). Average plant K uptake increased by 10.4 kg ha^{-1} at Aduthurai (11%) and by 12.4 kg ha^{-1} at Thanjavur (19%). In both domains, crop-year and crop-season effects on K uptake were not significant (Fig. 6.5).

Effect of SSNM on nitrogen-use efficiency

Increases in N-use efficiency through SSNM were small in both domains (Table 6.4). Across all four crops grown, average AEN was 15.8 kg kg^{-1} in the SSNM treatment at Aduthurai vis-à-vis 13.7 kg kg^{-1} with FFP management ($P = 0.096$). However, differences were not significant only in the first year, when N management in the

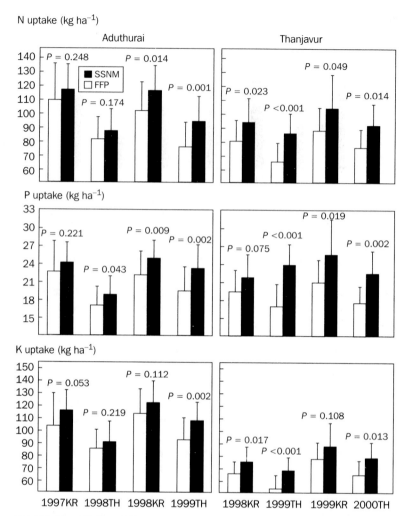

N uptake (kg ha^{-1})

P uptake (kg ha^{-1})

K uptake (kg ha^{-1})

Fig. 6.5. Plant N, P, and K accumulation in the farmers' fertilizer practice (FFP) and site-specific nutrient management (SSNM) plots at Aduthurai and Thanjavur, Tamil Nadu, India (bars: mean; error bars: standard deviation).

SSNM was not fine-tuned yet. In the second year, AEN increased from 12.3 kg kg^{-1} in the FFP to 16.5 kg kg^{-1} with SSNM (34%, $P = 0.005$), suggesting a gradual improvement of the N management scheme over time (crop-year effect $P = 0.006$). Similarly, a significant increase in REN (by 22%) was observed in the second year. At Thanjavur, however, AEN and REN were similar in both FFP and SSNM throughout the 2-year period. The average values of REN (0.45–0.46 kg kg^{-1}) were relatively high in both treatments.

At both sites, crop-season effects were not statistically significant, suggesting similar levels of N-use efficiency in kuruvai (DS) and thaladi (WS) crops. In both domains, higher fertilizer-N rates in the SSNM treatment caused a decrease in PFPN vis-à-vis FFP. Average PFPN with FFP was 58 kg kg^{-1} at Aduthurai and 57 kg kg^{-1} at Thanjavur, whereas it was 53 kg kg^{-1} and 45 kg kg^{-1} in SSNM, respectively. However, these levels, which can be considered close to those typically achieved in high-yielding, well-managed irrigated rice (Dobermann and Fairhurst 2000), and a lower PFPN are not necessarily a negative consequence of technology change (Dawe and Dobermann 1999).

The relatively small differences in N-use efficiency between SSNM and FFP were mainly caused by similar frequencies and timing of N applications in both treatments. Field observations indicated that many farmers changed their N applications in response to practices observed in the SSNM. Evidence for this is provided by an increase in the average number of N applications in the FFP from 3.1 (Thanjavur) or 3.3 (Aduthurai) before intervention (FFP baseline data) to 3.9 and 3.3 during the period of SSNM testing, respectively. This resembles, on average, 3.9 N applications in the SSNM treatments at Aduthurai or 3.8 N applications at Thanjavur.

In general, N-use efficiencies achieved in both domains were among the highest observed for all irrigated rice domains included in the RTDP project. Although the comparison of FFP and SSNM was somewhat confounded by the apparent changes in farmers' N applications, the results suggest further potential for improving the site-specific N management strategy. The approach tested throughout most of the experimental period combined fixed split applications with SPAD-based decisions at critical growth stages. It can probably be improved by introducing a more rigorous scheme for real-time decisions, which also allows for reducing the N rate in seasons with unfavorable weather. At Aduthurai, for example, the algorithm used resulted in the application of the full amount of fertilizer N for the predetermined yield target even in cropping seasons with unpredictably low yield potential such as the 1998 thaladi season (Fig. 6.4). Of particular importance is to avoid attracting insect pests such as leaffolder because of a late application of high N doses, as observed in the 1997 kuruvai season at Aduthurai.

Effect of SSNM on fertilizer use and profit

Site-specific nutrient management usually increased the amounts of N and K applied, whereas fertilizer-P rates were similar to those in the FFP (Table 6.5). On average, about 15 kg N ha^{-1} (13%) more were applied in SSNM at Aduthurai than in the FFP. At Thanjavur, the average difference was 34 kg N ha^{-1} per crop (35%). Differences in N rates were generally largest in the thaladi season, indicating that the N management scheme of the SSNM approach was not fully developed yet to account for the typically lower yields of thaladi rice (Fig. 6.4). Average fertilizer-K rates in SSNM were 70 kg K ha^{-1} at Aduthurai or 80 kg K ha^{-1} on the coarser-textured soils at Thanjavur, almost twice the average amounts applied by the farmers. Further research has to be conducted to clarify optimal K rates for longer periods.

Throughout all four crops grown, average total fertilizer cost in the SSNM treatment was about \$15 to \$30 ha^{-1} higher than in the FFP. However, because of

Table 6.5. Effect of site-specific nutrient management on fertilizer use at Aduthurai (1997-99) and Thanjavur (1998-2000), Tamil Nadu, India.

Parameter	Levels[a]	Treatment[b] SSNM	Treatment[b] FFP	Δ^c	$P > \lvert T \rvert^c$	Effects[d]	$P > \lvert F \rvert^d$
Aduthurai, Old Cauvery Delta							
N fertilizer (FN)	All	127.0	112.1	14.9	0.000	Village	0.002
(kg ha^{c1})	Year 1	125.4	112.0	13.4	0.027	Year[e]	0.585
	Year 2	128.7	112.3	16.4	0.002	Season[e]	0.097
	Kuruvai	130.5	121.2	9.3	0.105	Year × season[e]	0.746
	Thaladi	123.6	103.4	20.2	0.000		
P fertilizer (FP)	All	25.8	23.7	2.1	0.195	Village	0.004
(kg ha ^{1})	Year 1	29.2	23.5	5.7	0.027	Year	0.009
	Year 2	22.2	23.8	−1.6	0.411	Season	0.370
	Kuruvai	27.4	26.6	0.7	0.756	Year × season	0.213
	Thaladi	24.3	20.9	3.5	0.115		
K fertilizer (FK)	All	69.7	37.5	32.1	0.000	Village	0.005
(kg ha^{-1})	Year 1	70.6	42.7	27.9	0.000	Year	0.219
	Year 2	68.7	32.2	36.6	0.000	Season	0.042
	Kuruvai	75.7	37.3	38.4	0.000	Year × season	0.592
	Thaladi	63.9	37.8	26.1	0.000		
Thanjavur, New Cauvery Delta							
N fertilizer (FN)	All	129.1	95.5	33.6	0.000	Village	0.036
(kg ha ^{1})	Year 1	126.3	97.8	28.5	0.000	Year	0.241
	Year 2	132.6	92.6	39.9	0.000	Season	0.008
	Kuruvai	123.5	99.4	24.0	0.001	Year × season	0.803
	Thaladi	135.6	90.7	44.9	0.000		
P fertilizer (FP)	All	18.7	16.8	1.9	0.392	Village	0.283
(kg ha ^{1})	Year 1	21.2	19.4	1.8	0.602	Year	0.737
	Year 2	15.6	13.8	1.8	0.382	Season	0.939
	Kuruvai	18.7	16.4	2.3	0.313	Year × season	0.481
	Thaladi	18.8	17.4	1.4	0.730		
K fertilizer (FK)	All	79.8	36.0	43.8	0.000	Village	0.199
(kg ha ^{1})	Year 1	73.6	38.9	34.7	0.000	Year	0.028
	Year 2	87.6	32.5	55.1	0.000	Season	0.019
	Kuruvai	76.7	41.6	35.1	0.000	Year × season	0.280
	Thaladi	83.4	29.2	54.2	0.000		

[a]All = all four crops grown from 1997 KR to 1999 TH (Aduthurai) or 1998 KR to 2000 TH (Thanjavur); year 1 = 1997 KR and 1998 TH at Aduthurai and 1998 KR and 1999 TH at Thanjavur; year 2 = 1998 KR and 1999 TH at Aduthurai and 1999 KR and 2000 TH at Thanjavur; kuruvai = two dry-season crops; thaladi = two wet-season crops. [b]FFP = farmers' fertilizer practice; SSNM = site-specific nutrient management. $^c\Delta$ = SSNM – FFP. $P > \lvert T \rvert$ = probability of a significant mean difference between SSNM and FFP. [d]Source of variation of analysis of variance of the difference between SSNM and FFP by farm; $P > \lvert F \rvert$ = probability of a significant F-value. [e]Year refers to two consecutive cropping seasons.

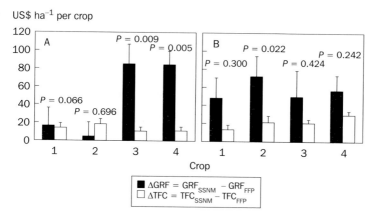

US$ ha^{-1} per crop

ΔGRF = GRF$_{SSNM}$ − GRF$_{FFP}$
ΔTFC = TFC$_{SSNM}$ − TFC$_{FFP}$

Fig. 6.6. Difference in total fertilizer cost (ΔTFC) and profit (ΔGRF) between the farmers' fertilizer practice (FFP) and site-specific nutrient management (SSNM) on rice farms at Aduthurai (A) and Thanjavur (B), Tamil Nadu (four successive crops, means ± standard error). The P values shown indicate the probability of a significant mean difference in gross return above fertilizer cost (GRF) between the FFP and SSNM in each season.

significant yield increases, SSNM was highly profitable in most crops grown (Fig. 6.6). At Aduthurai, the average profit increase over the FFP (ΔGRF) was $47 ha^{-1} per crop (7%), but it increased from $11 ha^{-1} in the first year (n.s.) to $85 ha^{-1} in the second year (13%, $P = 0.000$, significant crop-year effect). At Thanjavur, the average profit increase over the FFP was $56 ha^{-1} per crop (10%), with no significant differences between years or cropping seasons.

Factors affecting the performance of SSNM

Climate as well as crop management affected the performance of SSNM and explained why the yield target was not always achieved. However, on average, relatively few stresses occurred and the SSNM strategy was improved over time. In the first year of SSNM, average yields were only 73% of the target yield at Thanjavur and 79% at Aduthurai. This proportion increased to 84% at Thanjavur and 88% at Aduthurai. At Aduthurai, average yield was high and close to the optimal line describing the relationship between grain yield and plant N accumulation. Yield losses observed there were mainly caused by insects. Average yield increases of 13% were achieved at Thanjavur, but the yield goal achievement was somewhat lower and the internal N-use efficiency (amount of grain produced per unit N taken up) was suboptimal because of the occasional occurrence of water shortages or insect damage in some fields (Dobermann et al 2002).

Observations from the Aduthurai sites provide more detailed insights into the factors affecting on-farm research on developing a new technology such as SSNM. The poorer performance in the first year at Aduthurai mainly resulted from (1) use of a not fully calibrated QUEFTS model that overestimated nutrient needs (Dobermann and Witt, Chapter 5, this volume), (2) an inefficient N management algorithm, (3)

Table 6.6. Influence of crop management on grain yield, N-use efficiency, and profit increase by site-specific nutrient management (SSNM) at Aduthurai, Tamil Nadu. Farms were grouped into farms with no severe crop management problems (SSNM+) and farms in which one or more severe constraints (water, pests, crop establishment) occurred (SSNM–).[a]

Crop	Grain yield (t ha^{-1})		AEN (kg grain kg^{-1} N)		REN (kg N kg^{-1} N)		Δ Profit (US$ ha^{-1})	
	SSNM+	SSNM–	SSNM+	SSNM–	SSNM+	SSNM–	SSNM+	SSNM–
1997 kuruvai (DS)								
Mean	7.75 a	5.79 b	24.7 a	11.3 b	0.61 a	0.37 b	22.6 a	7.8 a
N	14	10	14	10	14	10	14	10
1998 thaladi (WS)								
Mean	5.66 a	4.89 a	11.3 a	11.4 a	0.37 a	0.36 a	21.9 a	–39.6 a
N	19	6	19	6	19	6	19	6
1998 kuruvai (DS)								
Mean	7.13 a	6.68 a	17.8 a	9.4 b	0.47 a	0.39 a	108.2 a	20.0 b
N	17	6	17	6	17	6	17	6
1999 thaladi (WS)								
Mean	6.77 a	5.99 b	19.1 a	14.9 a	0.45 a	0.34 b	86.3 a	82.0 a
N	14	10	14	10	14	10	14	10

[a]Within each row (season), means of SSNM+ and SSNM– followed by the same letter are not significantly different using LSD (0.05%).

pest incidence induced by late N applications in the 1997 kuruvai crop, and (4) very rainy weather affecting the 1998 thaladi crop. The comparison of SSNM with a "true" FFP was also confounded because farmers often copied SSNM practices in their own FFP plots adjacent to the SSNM. Many farmers used similar amounts of fertilizer-N in the FFP as in the SSNM and a similar timing of N applications, and, compared to previous years, farmers' P application was higher, much more variable, and usually similar to the SSNM. As a result, grain yield, AEN, and REN in the FFP were higher than in the 1995 and 1996 kuruvai seasons and differences from the SSNM plots were small during the 1997-98 period. In addition, the quality of crop management varied.

In the first crop, 1997 KR rice, we distinguished 14 farms with good management (SSNM+) and 10 farms on which water scarcity at critical growth stages (panicle initiation), weed problems, stem borer damage, or leaffolder damage (late infestation on flag leaves) caused severe yield losses (SSNM–, Table 6.6). Late leaffolder incidence in SSNM– plots was mostly caused by too high plant N status attracting leaffolders from the surrounding farmland. Average leaffolder damage (% of flag-leaf area damaged) was 43% (ranging from 4% to 94%) in SSNM compared with just 16% in the FFP (2–36%) at the same sites. Average stem borer damage was 14% whiteheads (range 4–41%) in SSNM– and 11% (4–25%) in the FFP. The higher amount of N applied in the SSNM (153 vs 118 kg N ha^{-1} in the fields without severe problems) and the prolonged N application based on SPAD readings contributed to attracting these pests to the SSNM plots. Yields in SSNM+ plots averaged 7.8 t ha^{-1}

vis-à-vis 5.8 t ha^{-1} in SSNM– plots with poorer management and differences in AEN and REN were of similar magnitude (Table 6.6). Moreover, variability in yield, N-use efficiency, and profit increase among farms was higher among SSNM– farms than for well-managed farms.

In subsequent cropping seasons, differences between SSNM+ and SSNM– farms at Aduthurai were smaller than in the 1997 kuruvai season and mostly not statistically significant (Table 6.6), mainly because of improvements in the SSNM approach and field management. However, heavy rains and a nonoptimal N management strategy prevented a good response to the SSNM in the 1998 thaladi crop (Fig. 6.4). This was mainly caused by an insufficient number of panicles per m^2 (reduced tillering) and number of spikelets per panicle. Average grain yield was 5.5 t ha^{-1} in SSNM vs 5.3 t ha^{-1} in the FFP, but yields ranged widely from 3.6 to 7.1 t ha^{-1}.

In contrast, climate was favorable in the 1998 kuruvai crop and a more conservative N management strategy was used in SSNM, which focused on preventing excessive plant N buildup during later growth stages to reduce susceptibility to pest attack (leaffolder) during the reproductive and grain-filling stages, but still maintaining N above critical levels. Basal N was not applied in any of the SSNM plots. Incidence of leaffolder and stem borer in SSNM was small and similar to that of the FFP. The more balanced nutrition because of higher rates of K used in SSNM led to a significant reduction in brown planthopper incidence in the SSNM compared with the FFP. As a result of all these factors, a high average grain yield of 7 t ha^{-1} was obtained with SSNM and yield increases were 0.8 t ha^{-1} (13%) vis-à-vis the FFP. Yield increases were mainly attributed to a 10% increase in the average number of panicles per m^2 (526 in SSNM vs 480 in FFP), whereas grain filling and 1,000-grain weight did not differ. Moreover, yield variability decreased to a CV of only 10% in the SSNM compared with 16% in FFP or 20% in FFP and SSNM in the 1997 kuruvai season, suggesting that field-specific management effectively evened out differences in soil fertility among farms.

6.3 Future opportunities for improvement and adoption of SSNM

Site-specific nutrient management must become an integral component of a wider, integrated crop management approach. Major differences among soil type, water management, and pest control appear to be of greatest importance for improving crop response to nutrients in the Cauvery Delta. Detailed pest monitoring was conducted at all Thanjavur sites in 1998 (kuruvai and thaladi seasons). The most frequently observed pest-related problems were grain discoloration, brown spot, sheath blight, leaffolder, and stem borer (Sta. Cruz et al 2001), but more understanding of nutrient × pest interactions is required to fine-tune SSNM. Earlier research showed that N and K applications significantly affect the incidence of pests such as green leafhopper (Raju et al 1996), but more work is needed to clarify K effects on yields and other pests and separate these effects from those caused by improved N timing and N-use efficiency.

The SSNM strategy also requires further simplification to make it fit for wider-scale delivery to farmers. While the omission plot strategy proved to be a cheap and reliable alternative to soil testing, a few broad categories of indigenous nutrient supplies may be sufficient to develop sensible recommendations for larger areas. Estimates of indigenous nutrient supplies would be based on grain yield rather than nutrient uptake measurements in a limited number of nutrient omission plots per domain (Dobermann et al 2003a,b). Existing, recently developed soil maps for the Cauvery Delta can then be used to delineate and verify such recommendation domains with similar soil indigenous nutrient supplies. While improved blanket recommendations for reasonably large areas will probably be sufficiently robust for P and K, the algorithm used for applying fertilizer N needs simplification, including the replacement of the SPAD meter with the cheaper leaf color chart. A stronger emphasis will have to be given to farmer participatory approaches to identify the N management strategy most acceptable to farmers prior to wider-scale dissemination. Extension workers will need to be trained so that they can share their newly acquired knowledge with farmers. The new strategy could then be promoted through farmers' group meetings, monthly zonal meetings of extension personnel, and mass media such as newspapers and radio. Regional and state support will be required to implement SSNM on a larger scale, including funding for promotional material and the manufacture and distribution of LCCs.

In view of current water and labor shortages at the time of transplanting, some farmers will likely change from transplanting to direct seeding in the near future. Crop establishment technologies that are currently under investigation include broadcasting of dry or pregerminated rice seeds and a newly developed drum seeder. SSNM technology will need to be adjusted if farmers change their crop management practices.

6.4 Conclusions

Transplanted irrigated rice has been grown for centuries in the Cauvery Delta of Tamil Nadu, but production increases have slowed down in recent years. Rice is grown in medium-size fields (0.5 to >1 ha) with high labor input, relatively high and balanced fertilizer use, and little use of pesticides.

Site-specific nutrient management resulted in large increases in grain yield and profit in two distinctly different regions of the Cauvery Delta zone, typically on the order of 8% to 13%. These increases were achieved in comparison with relatively high average yields in the FFP treatment (6.0 t ha^{-1} at Aduthurai and 5.0 t ha^{-1} at Thanjavur), which mainly reflect the generally high level of crop care. Relative increases in plant nutrient uptake were larger than yield increases, suggesting that yields and profits can be further raised by eliminating other constraints such as pests or water stress during early growth. Moreover, increases in yield and nutrient uptake were larger on the coarser-textured soils of the New Cauvery Delta than on the heavier soils around Aduthurai. Yield and profit increases because of SSNM are likely to become even larger with further development of the SSNM approach, including improved N management approaches, reduction in N rates in low-yielding seasons, optimal use of fertilizer K rates, and better integration of nutrient and pest management.

References

Athreya V. 2001. Census 2001. The Tamil Nadu picture. In: Frontline. 18 (09): 28 Apr.–11 May 2001 (online). Available at www.flonnet.com (last update 2001; accessed 3 Feb. 2003).

Babu M, Nagarajan R, Mohandass S, Susheela C, Muthukrishnan P, Subramanian M, Balasubramanian V. 2000a. On-farm evaluation of chlorophyll meter-based N management in irrigated transplanted rice in the Cauvery Delta, Tamil Nadu, India. Int. Rice Res. Notes 25(2):28-30.

Babu M, Nagarajan R, Ramanathan SP, Balasubramanian V. 2000b. Optimizing chlorophyll meter threshold values for different seasons and varieties in irrigated lowland rice systems of the Cauvery Delta zone, Tamil Nadu, India. Int. Rice Res. Notes 25(2):27.

Balasubramanian V, Morales AC, Thiyagarajan TM, Nagarajan R, Babu M, Abdulrachman S, Hai LH. 2000. Adaptation of the chlorophyll meter (SPAD) technology for real-time N management in rice: a review. Int. Rice Res. Notes 25(1):4-8.

Dawe D, Dobermann A. 1999. Defining productivity and yield. IRRI Discussion Paper Series No. 33. Makati City (Philippines): International Rice Research Institute. 13 p.

Department of Agriculture and Cooperation. 2003. Statistics at a glance (online). Available at http://agricooop.nic.in/welcome.html (last update 2002; accessed 30 Jan. 2003). New Delhi, India: Ministry of Agriculture, Government of India.

Dobermann A, Fairhurst TH. 2000. Rice: nutrient disorders and nutrient management. Singapore, Makati City: Potash and Phosphate Institute, International Rice Research Institute. 191 p.

Dobermann A, Witt C, Dawe D, Gines GC, Nagarajan R, Satawathananont S, Son TT, Tan PS, Wang GH, Chien NV, Thoa VTK, Phung CV, Stalin P, Muthukrishnan P, Ravi V, Babu M, Chatuporn S, Kongchum M, Sun Q, Fu R, Simbahan GC, Adviento MAA. 2002. Site-specific nutrient management for intensive rice cropping systems in Asia. Field Crops Res. 74:37-66.

Dobermann A, Witt C, Abdulrachman S, Gines HC, Nagarajan R, Son TT, Tan PS, Wang GH, Chien NV, Thoa VTK, Phung CV, Stalin P, Muthukrishnan P, Ravi V, Babu M, Simbahan GC, Adviento MA. 2003a. Soil fertility and indigenous nutrient supply in irrigated rice domains of Asia. Agron. J. 95:913-923.

Dobermann A, Witt C, Abdulrachman S, Gines HC, Nagarajan R, Son TT, Tan PS, Wang GH, Chien NV, Thoa VTK, Phung CV, Stalin P, Muthukrishnan P, Ravi V, Babu M, Simbahan GC, Adviento MA, Bartolome V. 2003b. Estimating indigenous nutrient supplies for site-specific nutrient management in irrigated rice. Agron. J. 95:924-935.

Janaki P, Thiyagarajan TM, Balasubramanian V. 2000. Effect of planting density on chlorophyll meter-based N management in transplanted rice. Int. Rice Res. Notes 25(2):24-27.

Matthews RB, Kropff MJ, Bachelet D, van Laar HH. 1995. Modeling the impact of climate change on rice production in Asia. Wallingford (UK): CAB International, International Rice Research Institute.

Nagarajan R, Manickam TS. 1985. Zinc fixing capacity of rice soils. Madras Agric. J. 72:572-576.

Nagarajan R, Manickam TS. 1986. Optimising zinc application in rice soils of Tamil Nadu. Madras Agric. J. 73:1-6.

Nagarajan R, Muthukrishnan P, Mohanty SK, Nambiar KKM, Badrachalam A, Panda D, Samantaray RN, Bhatacharya H, Olk DC. 1996. Continuous rice cropping: its implication on soil health and environmental sustainability. Paper presented at the India-IRRI Dialogue, 27-29 September 1996. Delhi: ICAR, IRRI.

Nagarajan R, Muthukrishnan P, Stalin P, Olk DC, Cassman KG, Ranganathan TB. 1997. Indigenous soil N supply and use efficiency in irrigated rice: experience from Tamil Nadu, India. In: Soil quality management and agroecosystem health for East and Southeast Asia. Cheju Island, Korea. p 200-215.

Olk DC, Cassman KG, Simbahan GC, Sta. Cruz PC, Abdulrachman S, Nagarajan R, Tan PS, Satawathananont S. 1999. Interpreting fertilizer-use efficiency in relation to soil nutrient-supplying capacity, factor productivity, and agronomic efficiency. Nutr. Cycl. Agroecosyst. 53:35-41.

Raju N, Nagarajan R, Rajendran R, Kareem AA. 1996. Role of nitrogen (N) & potassium (K) in the incidence of green leafhopper (GLH) *Nephotettix virescens* (Distant) in rice. J. Potassium Res. 12:313-318.

Ramanathan SP, Nagarajan R, Balasubramanian V. 2000. Assessment of chlorophyll meter-based N application at critical growth stages of irrigated transplanted rice. Int. Rice Res. Notes 25:34-35.

Savithri P, Perumal R, Nagarajan R. 1999. Soil and crop management technologies for enhancing rice production under micronutrient constraints. Nutr. Cycl. Agroecosyst. 53:83-92.

Sta. Cruz PC, Simbahan GC, Hill JE, Dobermann A, Zeigler RS, Pham Van Du, dela Peña FA, Samiayyan K, Suparyono, Nguyen Van Tuat, Zheng Zhong. 2001. Pest profiles at varying nutrient input levels. In: Peng S, Hardy B, editors. Rice research for food security and poverty alleviation. Proceedings of the International Rice Research Conference, 31 March-3 April 2000, Los Baños, Philippines. Los Baños (Philippines): International Rice Research Institute. p 431-440.

Stalin P, Dobermann A, Cassman KG, Thiyagarajan TM, ten Berge HFM. 1996. Nitrogen supplying capacity of lowland rice soils in southern India. Commun. Soil Sci. Plant Anal. 27:2851-2874.

Stalin P, Thiyagarajan TM, Ragarajan R. 1999. Nitrogen application strategy and use efficiency in rice. Oryza 36:322-326.

Stalin P, Thiyagarajan TM, Ramanathan S, Subramanian M. 2000. Comparing management techniques to optimize fertilizer N application in rice in the Cauvery Delta of Tamil Nadu, India. Int. Rice Res. Notes 25(2):25-26.

ten Berge HFM, Thiyagarajan TM, Quinghua Shi, Wopereis MCS, Drenth H, Jansen MJW. 1997. Numerical optimization of nitrogen application to rice. Part I. Description of MANAGE-N. Field Crops Res. 51:29-42.

Thiyagarajan TM, Stalin P, Dobermann A, Cassman KG, ten Berge HFM. 1997. Soil N supply and plant N uptake by irrigated rice in Tamil Nadu. Field Crops Res. 51:55-64.

Witt C, Dobermann A, Abdulrachman S, Gines HC, Wang GH, Nagarajan R, Satawathananont S, Son TT, Tan PS, Tiem LV, Simbahan GC, Olk DC. 1999. Internal nutrient efficiencies of irrigated lowland rice in tropical and subtropical Asia. Field Crops Res. 63:113-138.

Notes

Acknowledgments: This research has been the joint task of many colleagues. We wish to thank R. Sakunthala, V. Gnanabharathi, G. Sasikumar, S. Sridevi, S. Antony Samy, and S. Sujatha at SWMRI and D. Kabilar, M. Selvakumar, S. Natarajan, R. Jayaseelan, S. Arumugam, and S. Selvaganabathy at TNRRI for their long-term involvement in the field and laboratory research. Drs. J. Chandrasekran, S. Sridharan, K. Samiayyan, S. Monan, and P. Mathikumar have been involved in other aspects of this research, particularly pest assessment, and we wish to acknowledge their contributions. The project was coordinated by Dr. R. Nagarajan, TNRRI and SWMRI, from 1994 to 2000. Dr. S. Ramanathan, TNRRI, assumed leadership of the project in May 2000 and contributed much to the section on future opportunities for improvement and adoption of SSNM. The authors are much obliged to Dr. M. Subramaniam, former director of the Tamil Nadu Rice Research Institute, Dr. A. Kareem, former vice chancellor of Tamil Nadu Agricultural University, and Dr. R.K. Singh, liaison scientist, IRRI-India Office, for supporting this research. We are particularly grateful to all farm families involved at our site, and to Mr. Greg Simbahan, Ms. Arlene Adviento, Mr. Rico Pamplona, Ms. Olivyn Angeles, Ms. Julie Mae Criste Cabrera-Pasuquin, Mr. Edsel Moscoso, Dr. David Dawe, and Ms. Pie Moya (IRRI). The authors are grateful to the Swiss Agency for Development and Cooperation (SDC) for its support.

Authors' address: R. Nagarajan, M. Babu, S. Selvam, Soil and Water Management Research Institute (SWMRI), Thanjavar, Tamil Nadu; R. Nagarajan, Department of Soil Science and Agricultural Chemistry, Anbil Dharmalingam Agricultural College and Research Institute, Tiruchirappalli, Tamil Nadu; S. Ramanathan, P. Muthukrishnan, P. Stalin, V. Ravi, M. Sivanantham, Tamil Nadu Rice Research Institute (TNRRI), Aduthurai, Tamil Nadu, e-mail: ranagarajan@rediffmail.com; A. Dobermann, C. Witt, International Rice Research Institute, Los Baños, Philippines; A. Dobermann, University of Nebraska, Lincoln, Nebraska, USA.

Citation: Dobermann A, Witt C, Dawe D, editors. 2004. Increasing productivity of intensive rice systems through site-specific nutrient management. Enfield, N.H. (USA) and Los Baños (Philippines): Science Publishers, Inc., and International Rice Research Institute (IRRI). 410 p.

7 | Site-specific nutrient management in irrigated rice systems of Central Thailand

S. Satawathananont, S. Chatuporn, L. Niyomvit, M. Kongchum,
J. Sookthongsa, and A. Dobermann

7.1 Characteristics of rice production in Central Thailand

Trends in rice production

The Pathum Thani Rice Research Center (PTRRC) is the RTDP lead center in Thailand, but the experimental domain is located in Suphan Buri Province of Central Thailand, about 100 km northwest of Bangkok (Fig. 7.1). Suphan Buri has a tropical climate with warm and humid weather (average temperature 28 °C), except during the winter months of December until February, when temperature and humidity are somewhat lower (Fig. 7.2). Annual rainfall is about 1,400 mm and soils are mostly fertile alluvial soils (Alfisols, Inceptisols). Surface irrigation water is supplied from

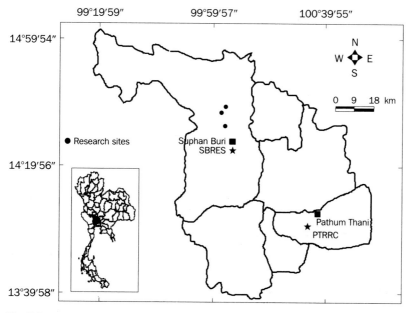

Fig. 7.1. Location of the experimental sites at Suphan Buri, Central Plain, Thailand.

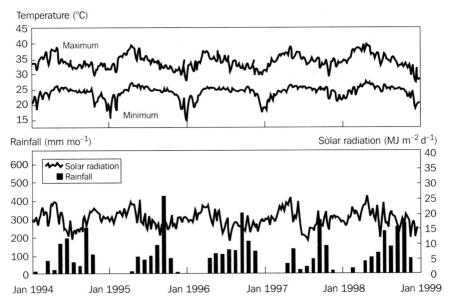

Temperature (°C)

Rainfall (mm mo⁻¹)

Solar radiation (MJ m⁻² d⁻¹)

Jan 1994   Jan 1995   Jan 1996   Jan 1997   Jan 1998   Jan 1999

Fig. 7.2. Climatic conditions at Suphan Buri, Thailand, from 1994 to 1998. Solar radiation and temperature data are 7-day moving averages; rainfall is monthly total. Solar radiation was obtained from sunshine hours using the standard coefficients for humid tropical conditions of 0.29 for a and 0.45 for b (Frère and Popov 1979) for Suphan Buri (14°28′N, 100°07′E).

a reservoir and the Chao Phraya River. Rice is mostly grown in double- and triple-crop monoculture systems. The major seasons are the dry season (DS, November to February), early wet season (EWS, March to June), and wet season (WS, July to October). The climatic yield potential at this site increases in the order WS < EWS < DS.

Until about 30 years ago, farmers grew photosensitive traditional rice varieties only during the rainy season. They rarely applied organic manure or chemical fertilizers and rice yields were mostly below about 1.8 t ha⁻¹. After harvest, most land was left fallow. Later on, farmers began applying chemical fertilizer on traditional cultivars and on the first two semidwarf high-yielding varieties released in 1969. However, farmers in Suphan Buri started double rice cropping only after the popular variety RD7 was released in 1975, about a decade after the first availability of irrigation water in this area. In 1981, the promotion of wet seeding began through a government project and within a few years most farmers changed their crop establishment from transplanting to wet seeding. Through the use of high-yielding germplasm, an increase in fertilizer application, and other improvements in crop management, the mean rice yield in Suphan Buri increased from 3.4 t ha⁻¹ in the late 1980s to 4.3 t ha⁻¹ in the late 1990s (Table 7.1). During this period, WS yields increased from 2.9 to 3.9 t ha⁻¹ and DS yields from 4.2 to 4.9 t ha⁻¹. In 1990 and 1991, yields were low (about 2.5 t ha⁻¹) because of widespread damage caused by brown planthopper (BPH), attributed to the widespread consecutive use of variety SPR60. Yields increased

Table 7.1. Changes in rice production in Suphan Buri Province, Thailand.

Item	1985-89	1990-94	1995-99
Wet season			
Rice area (1,000 ha)	175.1	153.4	141.6
Rice yield (t ha $^{-1}$)	2.9	3.0	3.9
Rice production (1,000 t)	504.9	465.2	546.3
Dry season			
Rice area (1,000 ha)	104.6	82.7	128.8
Rice yield (t ha $^{-1}$)	4.2	3.9	4.9
Rice production (1,000 t)	440.3	317.9	629.4
Wet and dry season			
Rice area (1,000 ha)	279.7	236.0	270.4
Rice yield (t ha $^{-1}$)	3.4	3.3	4.3
Rice production (1,000 t)	945.2	783.1	1,175.7
Fertilizer consumption (total NPK, 1,000 t)	–	36,223	40,562

Source: Agricultural Economics Office, Department of Agriculture, Bangkok, 1985-99.

steadily by about 100 kg ha^{-1} y^{-1} (2.8% y^{-1}) since 1985 and yield increases were similar in the DS and WS.

Rice production in Suphan Buri increased by about 30,600 t or 3.2% y^{-1} since 1985 although production fluctuated much, mainly because of variation in the water supply affecting the harvested area (Table 7.1). The harvested area of DS rice was particularly low in the early 1990s, whereas the rice area in the WS declined within a decade from about 175,100 to 141,600 ha. Only 10% of the WS crops were grown without irrigation in the 1990s and yields reached about 60% of the irrigated rice yields. Because of abundant rainfall in 1995 and 1996 in conjunction with a favorable rice market and the full availability of combine harvesters, farmers in Suphan Buri have modified their cropping systems in recent years. Rice is now grown in an almost continuous mode with few distinctive seasons. Planting times vary widely among various fields in the same neighborhood. Cropping systems range from two to two and a half or three rice crops per year. It remains unclear how long farmers will continue this practice. The consequences for soil fertility and pest population dynamics are largely unknown.

Current biophysical and socioeconomic farm characteristics

The experimental domain is located around Suphan Buri City (14°28′N, 100°10′E) and includes 24 farms in Muang, Sriprachan, and Donchedi districts (Fig. 7.1), all located within 30 km of the SBRES research station at Suphan Buri. Detailed agronomic and socioeconomic farm monitoring using the procedures described in Chapter 2 started on all farms with the 1994-95 DS crop and this continues until today.

Farm sizes range from 1 to 8 ha, with a median of 4.8 ha (Table 7.2). Farmholders in this area currently have a low educational level (median is school grade 4), even though the standard government education system was extended to grade 6 many years ago and to grade 9 in 1993.

Labor use for growing rice at Suphan Buri is very low because many operations have been mechanized because of the increasing labor shortage. On average, total labor input is only about 14 person-days per hectare of rice, which is the lowest among all sites in the RTDP project. Labor-intensive operations such as transplanting and manual harvest have been completely replaced by wet seeding and combine harvest, also because farm sizes are relatively large and labor is drawn into other farm and off-farm activities not related to rice. In rice, about one-third of the total labor input is used for wet-land preparation using a rototiller. The first plowing is done a few days after land soaking with simultaneous incorporation of approximately half the biomass of the leftover rice stubble (the other half is burned).

Pregerminated seeds at 190–220 kg ha^{-1} are broadcast on drained but wet soil. About three days later, preemergence herbicide is sprayed. To enhance germination and prevent snail damage, fields remain without standing water for about 3 weeks and may dry out substantially during this period. About 3 weeks after sowing, fields are irrigated and kept flooded using rainwater or surface irrigation water until shortly before harvest. Broadcast fertilizer application starts only shortly after the first flooding. Farmers in the area are unlikely to follow official fertilizer recommendations. The only exception is the recommendation not to apply any K so that K fertilizer has never been used on rice (Table 7.2). Farmers use locally available fertilizer sources, which are introduced by fertilizer company agents and/or farmers' organizations. Currently, the predominant fertilizers are prilled urea (46-0-0), ammonium phosphate (16-20-0), and diammonium phosphate (18-46-0). Most farmers apply fertilizers in three splits in both the dry and wet seasons.

Pest problems are often a major constraint to increasing rice yields at Suphan Buri. Compared with that at other sites in Asia, pesticide use is high (Table 7.2). Farmers usually spray preemergence herbicides and molluscicides during germination and emergence of rice. Furadan is occasionally applied to control stem borers. Other chemicals are applied in attempts to control leaffolders, rice bugs, bacterial leaf blight, or sheath blight. The varieties used currently are resistant to blast and ragged stunt virus so that these diseases have been rarely observed lately. However, in areas with asynchronous planting, where farmers grow the same variety (e.g., SPR1) with practically all growth stages found in the same neighborhood, BPH damage can become severe.

Detailed agronomic background data were collected for three consecutive rice crops harvested from 1995 to 1996 (see Chapter 2). Most soils in the sampling domain are derived from riverine alluvial deposits. Soil texture is clay loam to clay in the Muang and Sriprachan subdomains and sandy clay loam at Donchedi. Based on the existing soil map produced by the Department of Land Development in 1998, the soils at our sites are classified as Ustic Endoaquerts, Aeric Endoaqualfs, or Plinthic Paleaquults (Soil Survey Staff 1994). Soil fertility varied widely among farms (Table

Table 7.2. Demographic and economic characteristics of rice production on 24 farms at Suphan Buri, Thailand. Values shown are based on socioeconomic farm surveys conducted for whole farms.

Production characteristics	Min.	25%	Median	75%	Max.
Total cultivated area (ha)	0.96	2.24	4.8	5.44	7.96
Age of household head (y)	29	42	45	52	70
Education of household head (y)	2	4	4	4	10
Household size (persons)	2	4	4	5	12
1998 dry-season crop					
Rice area (ha)[a]					
Yield (t ha^{-1})	3.16	3.53	4.19	5.02	5.94
N fertilizer (kg ha^{-1})	94	105	113	136	161
P fertilizer (kg ha^{-1})	5	8	24	29	51
K fertilizer (kg ha^{-1})	0	0	0	7	17
Insecticide (kg ai ha^{-1})	0.34	0.41	1.05	1.86	2.70
Herbicide (kg ai ha^{-1})	0.02	0.67	0.90	1.27	3.85
Other pesticides (kg ai ha^{-1})[b]	0	0.11	0.32	0.82	1.20
Total labor (8-h d ha^{-1})	6	8	14	16	26
Net return from rice ($ ha^{-1} crop^{-1})	75	207	258	421	478
1998 wet-season crop					
Rice area (ha)[a]	0.80	1.44	1.76	2.56	4.00
Yield (t ha^{-1})	2.88	4.69	5.41	5.90	7.13
N fertilizer (kg ha^{-1})	39	63	91	107	146
P fertilizer (kg ha^{-1})	5	14	20	26	30
K fertilizer (kg ha^{-1})	0	0	0	0	17
Insecticide (kg ai ha^{-1})	0.10	0.42	0.66	1.65	2.21
Herbicide (kg ai ha^{-1})	0.10	0.70	0.84	1.01	1.54
Other pesticides (kg ai ha^{-1})[b]	0	0.02	0.20	0.40	0.78
Total labor (8-h d ha^{-1})	6	9	13	15	22
Net return from rice ($ ha^{-1} crop^{-1})	111	437	537	625	831

[a]Rice area in which the treatment plots were embedded in subsequent years. The total rice area may be even larger. [b]Includes fungicide, molluscicide, rodenticide, and crabicide.

7.3). Soil organic C ranged from 9 to 23 g kg^{-1} and CEC from 7 to 30 cmol$_c$ kg^{-1}. Extractable Olsen-P was relatively high (>13 mg P kg^{-1}) on all farms, suggesting high P-fertilizer use during the past 20 years of intensive rice cultivation, perhaps because of the widespread use of compound fertilizers. Extractable soil K and Zn were above the commonly used critical level of 0.2 cmol K kg^{-1} and 1 mg Zn kg^{-1} on most farms (Dobermann and Fairhurst 2000).

Figure 7.3 illustrates the large temporal and spatial variation in indigenous N (INS), P (IPS), and K supply (IKS) measured from 1997 to 1998. Fourfold ranges in indigenous nutrient supply were measured in all four crops sampled. These measurements were affected by the very large variation in planting dates and differences in the length of the fallow period because farmers adopted an asynchronous planting schedule. For example, the INS (= plant N accumulation in a 0-N plot) was high and

Table 7.3. General soil properties on 24 rice farms, Suphan Buri, Thailand.

Soil properties	Min.	25%	Median	75%	Max.
Clay content (%)	9.8	28.9	33.7	36.9	46.1
Silt content (%)	1.5	6.1	15.4	25.6	37.6
Sand content (%)	29.0	38.4	49.6	62.4	86.4
Soil organic C (g kg ^{1})	9	14	16	18	23
Total soil N (g kg ^{1})	0.8	1.2	1.3	1.6	2.1
Soil pH (1:1 H_2O)	5.0	5.2	5.4	5.5	6.2
Cation exchange capacity (cmol$_c$ kg ^{1})	6.6	10.8	12.9	21.4	30.2
Exchangeable K (cmol$_c$ kg ^{1})	0.16	0.23	0.30	0.35	0.53
Exchangeable Na (cmol$_c$ kg ^{1})	0.48	0.57	0.83	0.99	1.37
Exchangeable Ca (cmol$_c$ kg ^{1})	2.45	3.36	4.23	4.94	7.07
Exchangeable Mg (cmol$_c$ kg ^{1})	0.72	1.00	1.22	1.75	2.61
Extractable P (Olsen-P, mg kg ^{1})	13.28	18.51	25.98	34.77	53.58
Extractable Zn (0.05N HCl, mg kg ^{1})	1.73	2.63	3.70	4.24	4.93

[a]Measured on soil samples collected before the 1996 dry season.

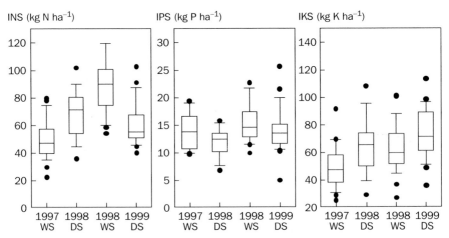

Fig. 7.3. Variability of the indigenous N (INS), P (IPS), and K (IKS) supply among 23 farmers' fields at Suphan Buri, Thailand (1997-99). Median with 10th, 25th, 75th, and 90th percentiles as vertical boxes with error bars; outliers as bullets.

variable in the 1998 WS crop because the data set included 15 farms with planting dates from 5 to 25 June 1998 and eight farms on which rice was planted only from 18 August to 18 September 1998 and harvested under more or less DS conditions. The greater INS in 1998 DS and WS crops was probably also due to longer dry fallow periods caused by climatic factors (El Niño) and delayed water release, whereas planting in the 1997 WS and 1999 DS followed more closely the normal cropping seasons. Soil drying is known to increase soil N mineralization in a succeeding rice crop because of enhanced breakdown of soil organic matter and soil microbial biomass (Dobermann et al 2000, Olk et al 1998, Witt et al 1998, 2000). At Suphan Buri, residual moisture and high temperature enhance crop residue decomposition during the fallow period after harvest of the DS crop until planting of EWS rice. In 1998, air temperature during March-April was particularly high, often exceeding 40 °C (Fig. 7.2). In addition, heterotrophic free-living N_2-fixing microorganisms contribute much to the INS and their activity may be enhanced by the increased soil carbon mineralization.

Despite the high Olsen-P levels, using a plant-based measure such as IPS (= plant P accumulation in a 0-P plot) suggested only moderate levels of soil P supply to rice in all four crops. Similarly, median IKS (= plant K accumulation in a 0-K plot) was low (1997 WS) to moderate only (1998 DS-1999 DS), suggesting that soil K extracted with $1N$ NH_4-acetate (Table 7.3) may overestimate the true soil K-supplying capacity on these soils. On several farms, IKS was very low (<40 kg K ha^{-1}, Fig. 7.3), indicating potential K deficiency. In summary, over the short term, the current average INS of about 50 kg N ha^{-1}, IPS of about 14 kg P ha^{-1}, and IKS of about 70 kg K ha^{-1} are only sufficient for achieving rice yields of about 3.5 t ha^{-1} without applying N, 5.5 t ha^{-1} without applying P, and 5 t ha^{-1} without applying K. These estimates assume optimal balanced nutrient requirements of 14.7 kg N, 2.6 kg P, and 14.5 kg K per 1,000 kg grain yield (Witt et al 1999).

The baseline agronomic characteristics obtained in 1995-96 showed that average rice yields in the farmers' fertilizer practice (FFP) were low in both the WS and DS crops, among the lowest within the RTDP project (Olk et al 1999). Yields varied widely among farms (CV about 20%). Typically, the high plant densities used led to only about 40 filled spikelets per panicle and plant N and K uptake were low (Table 7.4). Mean fertilizer N rates were 93 kg N ha^{-1} in the 1995 WS and 111 kg N ha^{-1} in the 1996 DS crops, but variation was large (CV = 24–30%) among farms and N-use efficiencies were low. The average agronomic N efficiency (AEN) was 5 kg kg^{-1} in the 1995 WS and 12 kg kg^{-1} in the 1996 DS. Average recovery efficiencies of N (REN) were 13% (WS) and 22% (DS), suggesting large gaseous N losses. Total plant N accumulation was relatively low in both seasons, although the average internal N efficiency of 64 kg grain kg^{-1} plant in the DS suggested favorable conditions for grain filling. Unbalanced fertilizer application may have contributed to low N uptake and N-use efficiencies because farmers did not apply K fertilizer. This is of particular concern for the lighter-textured soil of many farms in the Donchedi subdomain. Nutrient balance estimates indicated a negative K input-output balance of about –30 kg K ha^{-1} crop^{-1}, whereas the P balance was positive (Table 7.4).

7.2 Effect of SSNM on productivity and nutrient-use efficiency

Management of SSNM plots

A fixed-location SSNM plot was established within each of the 24 farmers' fields in the 1997 WS and its performance was compared with the FFP for four consecutive rice crops grown from the 1997 WS to 1999 DS (see Chapter 2). The size of the SSNM plots was 20×20 m during 1997-98 and 25×40 m in the 1999 DS. All crop

Table 7.4. Baseline agronomic characteristics of rice production on 24 farms at Suphan Buri, Central Thailand. Values shown are means and standard deviations (SD) of the farmers' fertilizer practice for two consecutive rice crops monitored before SSNM plots were established.

Agronomic characteristics	1995 WS crop		1996 DS crop	
	Mean	SD	Mean	SD
Grain yield (t ha^{-1})	3.46	0.73	4.48	0.79
Harvest index	0.46	0.05	0.46	0.04
No. of panicles per m^2	374	94	605	78
Total no. of spikelets per m^2	18,987	3,845	29,264	4,252
Total no. of spikelets per panicle	52	8	49	6
Filled spikelets (%)	80	5	83	5
Fertilizer N use (kg ha^{-1})	93.0	22.1	110.6	32.9
Fertilizer P use (kg ha^{-1})	21.9	11.5	23.8	8.7
Fertilizer K use (kg ha^{-1})	0.0	0.0	0.9	3.1
N uptake (kg ha^{-1})	63.3	13.6	70.2	15.0
P uptake (kg ha^{-1})	14.1	3.5	20.0	4.6
K uptake (kg ha^{-1})	57.6	13.4	68.4	15.9
Input-output N balance (kg ha^{-1} crop^{-1})	−2.3	10.1	−1.1	15.0
Input-output P balance (kg ha^{-1} crop^{-1})	10.7	11.4	7.4	10.4
Input-output K balance (kg ha^{-1} crop^{-1})	−26.9	6.0	−29.2	7.7
Partial productivity of N (kg kg^{-1})	39.2	11.9	41.4	13.0
Agronomic efficiency of N (kg kg^{-1})	4.9	4.7	11.6	11.4
Recovery efficiency of N (kg kg^{-1})	0.13	0.12	0.22	0.17
Physiological efficiency of N (kg kg^{-1})	31.9	18.3	45.6	27.4

management practices in the SSNM plots were the same as in the FFP area, except for fertilizer application and, to some degree, pest management.

Variety SPR1 has been used by most farmers since 1997. The major characteristics of SPR1 are 110–115 d growth duration, a moderately tall stature, a harvest index of about 0.5, and resistance to diseases such as blast, bacterial leaf blight, and ragged stunt. Other varieties occasionally grown included RD23, Rachini, Chainat, or SPR50. Pregerminated seeds at 190 to 220 kg ha^{-1} were broadcast on drained wet soil in the whole field. About 1 week after sowing, treatment plots were established and separated from the FFP by bunds. We tried to implement a prophylactic pest management strategy in SSNM, whereas most farmers would spray their FFP area based on visual plant symptoms of pest incidence. However, no complete pest control was achieved in either strategy and pests caused significant yield losses on most farms and in most seasons, also in the SSNM plots. Other problems encountered in 1997 to 1999 were related to climate and water management. Water shortages during the DS occurred occasionally and heavy rains caused flooding in some periods of the WS crops, particularly in the 1998 WS (La Niña phenomenon).

Fertilizer applications for SSNM were prescribed on a field- and crop-specific basis following the approach described in Chapter 5 and descriptions given elsewhere (Dobermann and White 1999, Witt et al 1999). In the 1997 WS, average values of the INS measured in three crops grown from 1995 to 1996 were used as model input. Similarly, the average IPS and IKS were estimated from the plant P and K accumulation in the FFP sampled during this period, assuming recovery fractions of 0.2 kg kg^{-1} for fertilizer P and 0.4 kg kg^{-1} for fertilizer K. Note that farmers applied no K so that FFP treatments were essentially NP treatments and plant K accumulation measured in the FFP was roughly equivalent to IKS. Estimates of IPS and IKS were adjusted after each crop grown by accounting for the actual P and K input-output balance. The adjusted values were then used as model input for the succeeding crop. Target yields were set to 6 t ha^{-1} in the 1997 WS, 6.5 t ha^{-1} in the 1998 DS, and 5.8 t ha^{-1} in the 1998 WS and 1999 DS. First-crop recovery fractions of 0.5, 0.2, and 0.40–0.45 kg kg^{-1} were assumed for fertilizer N, P, and K, respectively. The climatic yield potential was set to 7.5 t ha^{-1} for the WS and 9 t ha^{-1} for the DS, but no crop simulation data were available for this.

In the SSNM, fertilizer N was applied in two to three splits in the WS and three to four splits in the DS, focusing on critical growth stages such as early tillering (ET), midtillering (MT), panicle initiation (PI), and, occasionally, first flowering (FF). The actual amount and timing for each split were based on SPAD readings, but the N management was modified to take into account the observations of the previous seasons (Chapter 5).

In the 1997 WS, N was applied in SSNM as follows:

FN 1	10 DAS	If INS >50	0 kg N ha^{-1}
		If INS <50	20 kg N ha^{-1}
FN 2	20 DAS (ET)		30 kg N ha^{-1}
FN 3	30–40 DAS	If SPAD >33	0 kg N ha^{-1}
		If SPAD <33	40 kg N ha^{-1}
FN 4	40–60 DAS (PI)	If SPAD >33	0 kg N ha^{-1}
		If SPAD <33	50 kg N ha^{-1}

This was changed in the 1998 DS to

FN 1	20 DAS (ET)	If INS >50	30 kg N ha^{-1}
		If INS <50	40 kg N ha^{-1}
FN 2	35 DAS (MT)		40 kg N ha^{-1}
FN 3	50 DAS (PI)	If SPAD >33	40 kg N ha^{-1}
		If SPAD <33	50 kg N ha^{-1}
FN 4	60 DAS	If SPAD >33	0 kg N ha^{-1}
	If SPAD <33	20 kg N ha^{-1}	

and further modified in the 1998 WS and 1999 DS:

FN 1	20 DAS (ET)	If INS >50	30 kg N ha^{-1}
		If INS <50	40 kg N ha^{-1}
FN 2	35 DAS (MT)	If SPAD >33	30 kg N ha^{-1}
		If SPAD <33	40 kg N ha^{-1}

FN 3	50 DAS (PI)	If SPAD >35	30 kg N ha^{-1}
		If SPAD 33–35	40 kg N ha^{-1}
		If SPAD <33	50 kg N ha^{-1}

All P fertilizer was broadcast on the soil surface at 20 days after sowing (DAS) before irrigation started. Potassium was split into 50% at 20 DAS and 50% at 50 DAS (1997-98) or 100% at 20 DAS (since 1999 DS). Fertilizer sources used were urea (46% N), single superphosphate (8.3% P), and muriate of potash (50% K).

For comparison, farmers used mostly urea, ammonium phosphate (AP), and diammonium phosphate (DAP) as N and P sources in the FFP. They usually applied fertilizers in two to three splits, mostly at ET and PI or at ET, MT, and PI. Generally, each split was approximately a half-and-half mix of urea and AP or DAP. However, there was very little consistency among farmers. For example, in the 1997 WS, two farmers in Pho-Phraya village applied fertilizer at 20 and 50 DAS. However, in the same village, another farmer applied it at 28 and 72 DAS and another one at 21, 72, and 92 DAS. No farmer used organic fertilizer, except for incorporating about half of the remaining stubble biomass two times a year.

Effect of SSNM on grain yield and nutrient uptake

Mean grain yields in SSNM and FFP were 4.8 to 4.9 t ha^{-1} (Table 7.5), but ranged from 4.1 t ha^{-1} in the 1998 DS to about 5.2 t ha^{-1} in the 1998 WS and 1999 DS (Fig. 7.4). Averaged for all four crops grown, there was no significant increase in grain yield or N and P accumulation by using SSNM as compared with the FFP (Table 7.5). The only exception was a significant increase in plant K accumulation by 11 kg K ha^{-1} per crop (15%, $P = 0.015$). Crop-season effects were mostly not statistically significant, indicating no difference in the performance of SSNM between DS and WS crops. Village × crop interactions were not significant, but significant differences in plant P accumulation occurred among the three subdomains (villages).

There are some indications that the modifications made in the N management of the SSNM in year 2 (1998 WS and 1999 DS) have improved its performance vis-à-vis year 1 (1997 WS and 1998 DS). In 1998-99, grain yield and plant N, P, and K accumulation were about 1 t ha^{-1} larger than in 1997-98 (Fig. 7.4) or in 1995-96 (Table 7.4). Although yield differences between SSNM and FFP remained not significant, increases in N accumulation (11.6 kg ha^{-1}, 11%, $P = 0.056$) and K accumulation (12.3 kg ha^{-1}, 15%, $P = 0.028$) became significant (Table 7.4). In the 1998 WS, average grain yield in the SSNM was 5.4 t ha^{-1} versus 5.1 t ha^{-1} in the FFP (Fig. 7.4), yields of >7 t ha^{-1} were measured in the SSNM of four farms, and yield increases over the FFP exceeded 1 t ha^{-1} in eight cases. This all suggests potential for SSNM if other yield-determining factors can be controlled properly.

Average rice yield in the FFP was 5.1 t ha^{-1} in the 1998 WS and 5.3 t ha^{-1} in the 1999 DS (Fig. 7.4) vis-à-vis yields of 3.5 to 4.6 t ha^{-1} in all five rice crops sampled from the 1995 to 1998 DS. We do not know the exact reasons for the significant yield increase, but it coincides with (1) the introduction of SSNM, (2) changes in planting dates, and (3) climatic events that may have favored higher yields. Data from the

Table 7.5. Effect of site-specific nutrient management on agronomic characteristics at Suphan Buri, Thailand (1997-99).

Parameters	Levels[a]	Treatment[b] SSNM	Treatment[b] FFP	Δ^c	$P > \|T\|^c$	Effects[d]	$P > \|F\|^d$
Grain yield (GY)	All	4.90	4.81	0.10	0.595	Village	0.687
(t ha^{-1})	Year 1	4.42	4.31	0.11	0.569	Year[e]	0.703
	Year 2	5.29	5.20	0.09	0.737	Season[e]	0.361
	DS	4.77	4.86	−0.09	0.733	Year × season[e]	0.796
	WS	5.00	4.77	0.23	0.361	Village × crop	0.473
Plant N uptake (UN)	All	98.2	92.5	5.8	0.231	Village	0.185
(kg ha^{-1})	Year 1	74.5	76.0	−1.4	0.721	Year	0.051
	Year 2	117.4	105.8	11.6	0.056	Season	0.612
	DS	104.0	99.5	4.5	0.505	Year × season	0.464
	WS	94.3	87.6	6.6	0.313	Village × crop	0.391
Plant P uptake (UP)	All	17.8	16.7	1.1	0.215	Village	0.042
(kg ha^{-1})	Year 1	14.7	14.5	0.1	0.864	Year	0.000
	Year 2	20.3	18.5	1.9	0.142	Season	0.039
	DS	16.6	15.2	1.4	0.255	Year × season	0.924
	WS	18.6	17.8	0.9	0.459	Village × crop	0.731
Plant K uptake (UK)	All	80.9	70.5	10.5	0.015	Village	0.139
(kg ha^{-1})	Year 1	64.7	56.5	8.2	0.073	Year	0.467
	Year 2	94.1	81.8	12.3	0.028	Season	0.928
	DS	85.1	74.4	10.6	0.062	Year × season	0.478
	WS	78.1	67.8	10.3	0.090	Village × crop	0.571
Agronomic efficiency	All	9.4	8.7	0.7	0.655	Village	0.321
of N (AEN)	Year 1	8.9	6.8	2.1	0.309	Year	0.187
(kg grain kg^{-1} N)	Year 2	9.8	10.3	−0.5	0.834	Season	0.058
	DS	10.1	11.8	−1.7	0.503	Year × season	0.727
	WS	8.9	6.6	2.3	0.221	Village × crop	0.636
Recovery efficiency	All	0.29	0.24	0.05	0.121	Village	0.487
of N (REN)	Year 1	0.21	0.20	0.01	0.802	Year	0.184
(kg plant N kg^{-1} N)	Year 2	0.36	0.27	0.09	0.086	Season	0.758
	DS	0.36	0.30	0.07	0.260	Year × season	0.990
	WS	0.24	0.20	0.04	0.279	Village × crop	0.841
Partial productivity	All	45.3	50.1	−4.8	0.151	Village	0.379
of N (PFPN)	Year 1	46.2	40.7	5.5	0.163	Year	0.026
(kg grain kg^{-1} N)	Year 2	44.5	57.7	−13.2	0.009	Season	0.892
	DS	41.0	44.1	−3.1	0.403	Year × season	0.062
	WS	48.3	54.3	−6.0	0.226	Village × crop	0.507

[a]All = all four crops grown from 1997 WS to 1999 DS; Year 1 = 1997 WS and 1998 DS; Year 2 = 1998 WS and 1999 DS; WS = 1997 WS and 1998 WS; DS = 1998 DS and 1999 DS. [b]FFP = farmers' fertilizer practice; SSNM = site-specific nutrient management. [c]Δ = SSNM − FFP. $P > \|T\|$ = probability of a significant mean difference between SSNM and FFP. [d]Source of variation of analysis of variance of the difference between SSNM and FFP by farm; $P > \|F\|$ = probability of a significant F-value. [e]Year refers to two consecutive cropping seasons.

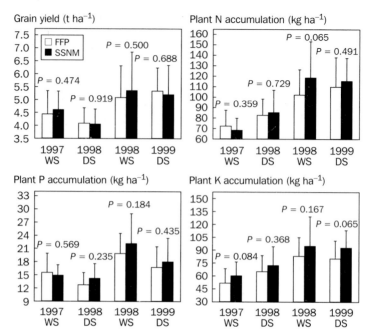

Fig. 7.4. Grain yield and plant N, P, and K accumulation in the farmers' fertilizer practice (FFP) and site-specific nutrient management (SSNM) plots at Suphan Buri, Thailand (1997-99; bars: mean; error bars: standard deviation).

1998 WS include eight farms with an early WS crop and 15 farms with a WS crop planted at the normal time. There was no clear evidence that farmers have copied the SSNM approach in their fields, but there was probably much interference by the researchers with regard to the overall crop management. After the first year of testing SSNM, measures were taken to ensure a better quality of land preparation and water and pest management in the SSNM plots, but perhaps also in the FFP area surrounding them. This is probably also one explanation for the small differences between the SSNM and FFP treatments.

Effect of SSNM on nitrogen-use efficiency

Overall, SSNM did not significantly increase any of the N-use efficiency parameters and there was little difference in this performance among years or climatic cropping seasons (Table 7.5). The only exception was a nearly significant crop $\times$ season effect for AEN ($P = 0.058$), suggesting that the difference between SSNM and the FFP measured in the WS (+2.3 kg kg^{-1}) was larger than the difference in DS crops (-1.7 kg kg^{-1}). Village or village $\times$ crop effects were all not significant.

However, regardless of SSNM and FFP treatments, the mean AEN and REN were generally low and highly variable among farms in the first three crops grown (Fig. 7.5), similar to what was measured in 1995-96 (Table 7.4). Physiological N

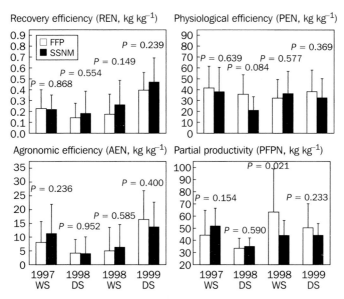

Recovery efficiency (REN, kg kg^{-1})

Physiological efficiency (PEN, kg kg^{-1})

Agronomic efficiency (AEN, kg kg^{-1})

Partial productivity (PFPN, kg kg^{-1})

Fig. 7.5. Fertilizer nitrogen-use efficiencies in the farmers' fertilizer practice (FFP) and site-specific nutrient management (SSNM) plots at Suphan Buri, Thailand (1997-99; bars: mean; error bars: standard deviation).

efficiency (PEN) in both SSNM and FFP treatments was mostly 30 to 40 kg grain kg^{-1} N. The average internal efficiency (IEN, kg grain per kg N taken up) in SSNM was only 46 to 49 kg kg^{-1} in the three crops grown in 1998 to 1999, suggesting the presence of other biotic or abiotic stresses (Witt et al 1999). An example is the SSNM treatment of the 1998 DS crop, in which PEN was only 20 kg kg^{-1} because the SSNM plots suffered from severe rat and leaffolder damage.

Values of AEN and REN increased significantly in the 1999 DS. In SSNM, mean AEN increased to 14 kg kg^{-1} (>20 kg kg^{-1} on six farms) and REN increased to 0.47 kg kg^{-1} (>0.45 kg kg^{-1} on 11 farms). A similar increase took place in the FFP (Fig. 7.5).

Effect of SSNM on fertilizer use and profit

On average, similar amounts of N but less P (−3 kg P ha^{-1}, −15%) and more K fertilizer (+43 kg K ha^{-1}) were applied in SSNM vis-à-vis FFP (Table 7.6). Total fertilizer cost was about $40 ha^{-1} crop^{-1} greater in SSNM than in the FFP ($P = 0.000$) and average profits in the FFP tended to be larger than in SSNM, although not statistically significant (Fig. 7.6). Only in the 1998 WS was significantly more N applied in SSNM (121 kg N ha^{-1}) than in the FFP (93 kg N ha^{-1}), but this was also the crop with the largest yield increase achieved by SSNM (Fig. 7.4).

Note the declining trend of P and K rates used in SSNM (Fig. 7.6). This is because the SSNM strategy aimed at building up IPS and IKS and because the actual yields were often below the yield targets set for estimating the fertilizer amounts. The

Table 7.6. Effect of site-specific nutrient management on fertilizer use, fertilizer cost, and gross returns above fertilizer costs from rice production at Suphan Buri, Thailand (1997-99).

Parameters	Levels[a]	Treatment[b] SSNM	Treatment[b] FFP	Δ[c]	$P > \|T\|$[c]	Effects[d]	$P > \|F\|$[d]
N fertilizer (FN)	All	111	109	2.0	0.642	Village	0.380
(kg ha[-1])	Year 1	101	116	-14.9	0.016	Year[e]	0.005
	Year 2	119	103	15.6	0.006	Season[e]	0.553
	DS	117	119	-2.5	0.648	Year × season[e]	0.006
	WS	107	102	5.0	0.380	Village × crop	0.217
P fertilizer (FP)	All	18.2	21.4	-3.2	0.021	Village	0.033
(kg ha[-1])	Year 1	20.4	24.0	-3.5	0.150	Year	0.854
	Year 2	16.4	19.4	-3.0	0.043	Season	0.850
	DS	17.2	20.9	-3.7	0.097	Year × season	0.876
	WS	18.9	21.8	-2.9	0.110	Village × crop	0.666
K fertilizer (FK)	All	44.8	1.4	43.3	0.000	Village	na
(kg ha[-1])	Year 1	59.2	2.2	57.0	0.000	Year	na
	Year 2	33.0	0.8	32.3	0.000	Season	na
	DS	35.9	1.5	34.4	0.000	Year × season	na
	WS	50.9	1.4	49.5	0.000	Village × crop	na
Fertilizer cost	All	109.8	69.7	40.1	0.000	Village	0.033
(US$ ha[-1])	Year 1	123.4	76.2	47.2	0.000	Year	0.018
	Year 2	98.8	64.4	34.4	0.000	Season	0.007
	DS	101.8	72.9	28.9	0.000	Year × season	0.196
	WS	115.3	67.4	47.9	0.000	Village × crop	0.003
Gross returns above	All	531	559	-27	0.265	Village	0.367
fertilizer costs	Year 1	455	488	-33	0.192	Year	0.857
(US$ ha[-1])	Year 2	594	616	-23	0.511	Season	0.810
	DS	521	562	-41	0.257	Year × season	0.592
	WS	538	557	-18	0.589	Village × crop	0.743

[a]All = all four crops grown from 1997 WS to 1999 DS; Year 1 = 1997 WS and 1998 DS; year 2 = 1998 WS and 1999 DS; WS = 1997 WS and 1998 WS; DS = 1998 DS and 1999 DS. [b]FFP = farmers' fertilizer practice; SSNM = site-specific nutrient management. [c]Δ = SSNM – FFP. $P > \|T\|$ = probability of a significant mean difference between SSNM and FFP. [d]Source of variation of analysis of variance of the difference between SSNM and FFP by farm; $P > \|F\|$ = probability of a significant F-value. [e]Year refers to two consecutive cropping seasons.

estimated increase in IPS and IKS over time was taken into account for subsequent rice crops so that fertilizer P and K rates appeared to level off over time, thus improving the economics of SSNM.

Factors affecting the performance of SSNM

A more detailed discussion is needed to explain the poor performance of SSNM at Suphan Buri. We usually observed excellent vegetative growth of rice in SSNM, with clear improvements over the FFP, but severe pest incidence often started around the PI stage and, together with other factors, caused large yield losses, particularly in the first year. Average actual yields in SSNM were 78% of the model-predicted target yield in the 1997 WS and 60% in the 1998 DS, but increased to 92% in the 1998

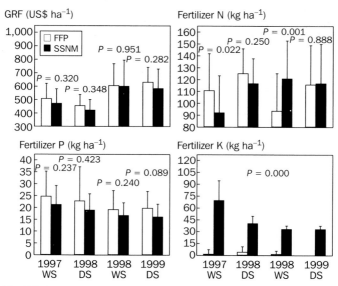

Fig. 7.6. Gross returns above fertilizer costs (GRF) and fertilizer use in the farmers' fertilizer practice (FFP) and site-specific nutrient management (SSNM) plots at Suphan Buri, Thailand (1997-99; bars: mean; error bars: standard deviation).

WS and to 90% in the 1998 DS. The latter suggests improvements in crop management over time.

Nevertheless, on many farms, the performance of SSNM was negatively affected by (1) climatic factors such as El Niño phenomenon, (2) water management, (3) suboptimal N management, (4) insufficient control of weeds, insects, and diseases, and (5) premature harvest. Although the direct climatic effect of El Niño is difficult to quantify, various observations suggest that it caused significant yield losses in the 1998 DS. Monthly maximum temperatures during February to April 1998 ranged from 39 to 41 °C, resulting in only 65% filled spikelets, vis-à-vis 36 to 39 °C in February to April of 1996, with more than 80% filled spikelets (Fig. 7.2). The 1998 DS yields were about 0.8 t ha^{-1} lower than in normal years, which explains why the yield target set for SSNM could not be achieved.

In DS crops, drought problems were observed at critical growth stages on some farms, whereas inundation occurred occasionally in the WS. The quality of land preparation was sometimes poor, resulting in poor crop establishment and increased weed growth. The poor performance of SSNM in the 1997 WS was mainly due to an incorrect N management strategy. Too much N was applied at early growth stages, which was not in congruence with the farmers' water management practice, which kept the field dry or saturated during the first 3 weeks after sowing. This led to large N losses but also excessive N accumulation during early growth stages, which was then followed by N deficiency during reproductive growth because no N was applied at PI if the SPAD value was above 33.

Table 7.7. Influence of crop management on grain yield, N-use efficiency, and profit increase by site-specific nutrient management (SSNM) at Suphan Buri, Thailand. Farms were grouped into farms with no severe crop management problems (SSNM+) and farms in which one or more severe constraints (water, pests, crop establishment) occurred (SSNM−).[a]

	Grain yield (t ha^{-1})		AEN (kg grain kg^{-1} N)		REN (kg plant N kg^{-1} N)		ΔProfit (US$ ha^{-1})	
	SSNM+	SSNM−	SSNM+	SSNM−	SSNM+	SSNM−	SSNM+	SSNM−
1997 wet season								
Mean	5.20 a	4.28 b	18.8 a	7.3 b	0.27 a	0.19 a	−13.7 a	−45.5 a
N	8	14	8	14	8	14	8	14
1998 dry season								
Mean	4.57 a	3.80 b	6.1 a	3.2 a	0.31 a	0.12 a	47.2 a	−71.5 b
N	4	8	4	8	4	8	4	8
1998 wet season								
Mean	6.53 a	4.09 b	11.7 a	0.7 b	0.41 a	0.11 b	75.5 a	−89.2 b
N	12	11	12	11	12	11	12	11
1999 dry season								
Mean	5.42 a	4.98 a	13.5 a	14.3 a	0.47 a	0.48 a	−48.0 a	−44.7 a
N	10	9	10	9	10	9	10	9

[a]Within each row (season), means of SSNM+ and SSNM− followed by the same letter are not significantly different using LSD (0.05%).

In all seasons, significant yield losses were caused by pests, mainly weeds, rats, and leaffolders, but also unknown "diseases" such as "orange leaf." Because of the higher plant N status, moving pests such as rats and leaffolders became particularly attracted to the SSNM plots so that pest incidence was often greater than in the surrounding FFP. A site-specific problem was premature harvest. Many farmers harvested their field on the date for which they had contracted a combine. This was probably also one reason for the small differences between SSNM and the FFP. Improved plant nutrition, particularly higher plant N status, caused plants in the SSNM to mature 3 to 5 d later than in the FFP, but harvest samples were often collected on the same day.

As described in Chapter 2, farms were grouped into farms with good crop management (SSNM+) and farms with poor crop management and severe problems in the SSNM plots (SSNM−). In all four crops sampled, a large proportion of farms were classified as SSNM− (Table 7.7), reflecting the numerous problems described. Mean grain yield differences between SSNM+ and SSNM− ranged from 0.44 (1999 DS) to 2.44 t ha^{-1} (1998 WS) and were statistically significant in three out of four seasons. They probably reflect the average yield losses caused by factors other than nutrients, mainly pests. Farms with better overall crop management had a significantly larger AEN in the 1997 and 1998 WS crops. In the 1998 WS, average REN was 0.41 kg kg^{-1} in SSNM+ vs 0.11 kg kg^{-1} in SSNM−. In addition, SSNM+ farms had large profit increases ($75 ha^{-1}) vis-à-vis a profit decrease of $89 ha^{-1} in SSNM−.

7.3 Future opportunities for adoption of SSNM

Our data suggest that, with good crop management, SSNM may be profitable at Suphan Buri and yields may exceed 7 t ha^{-1}, a level rarely achieved now. We noticed improvements in the performance of SSNM in the second year, but we will need at least one more year before drawing firm conclusions about the future potential of this technology in Thailand. Any dissemination of SSNM in this area must be part of an overall improvement of integrated crop management because complex interactions of land preparation, crop establishment, water management, weeds, nutrients, and pests occur.

Simplifications of the SSNM approach are required, such as better options for using soil P and K tests in the fertilizer recommendations as well as N management based on cheap leaf color charts rather than an expensive chlorophyll meter. We also need a better adjustment of the nutrient decision-support system to local conditions. This includes a better understanding of nutrient dynamics in interaction with water management in the wet-seeded rice system. Another critical issue is insufficient knowledge about the climatic and genetic yield potential of the predominantly long-grain varieties and its relationship with planting date. Planting date has become even more important in recent years because of the overlapping cropping seasons.

Once these problems are solved, we should start studies on how to disseminate SSNM in the rice-growing areas of Thailand. The existing extension system may be the main backbone for this, but village extension workers as well as farmers will need thorough training.

7.4 Conclusions

Suphan Buri Province is one of the most intensively irrigated rice domains in Central Thailand. Compared with other irrigated rice areas in Asia, labor input, farmers' educational background, and rice yields are low. Farmers use sufficient N and P fertilizer, but no potash. Nitrogen-use efficiencies have remained low. Pesticide use in this wet-seeded rice system is very high and includes sprays for controlling weeds, snails, insects, and diseases. Despite the high pesticide use, pests have remained a common problem. In any particular season, profits from rice production varied about 7-fold among the 24 farms studied.

Nutrient omission plots indicated that the INS varied widely among farms and cropping seasons. Climatic events such as El Niño or changes in cropping patterns caused large differences in the length of fallow periods and planting dates, which affected the INS. Contrary to simple routine soil tests, plant-based measurements of the IPS and IKS suggested only moderate levels of both, which were sufficient for achieving rice yields of 5 to 5.5 t ha^{-1} without applying P or K. Revision of the current fertilizer recommendations and actual practices appears to be required. Improved N management and applying K fertilizer appear to be key steps toward increasing rice yields at Suphan Buri.

During 1997 to 1999, we tested a new approach for field-specific, dynamic nutrient management. There was no significant increase in grain yield, plant N and P accumulation, N-use efficiency, and profit averaged for all four crops grown. However, poor crop management or climatic events caused severe yield reductions, particularly in the first two crops grown. Improvements in crop and N management were introduced in the second year and the actual yields reached about 90% of the target yield used for estimating the amounts of fertilizer to be applied. On many farms, factors such as poor land preparation, crop establishment, and water and pest management caused a suboptimal yield response to nutrients.

Further field testing of SSNM is required in Thailand before firm conclusions about its potential can be drawn. Without major improvements in the overall soil and crop management, particularly land preparation, weed control, and pest management, benefits from fine-tuning nutrient management may remain insignificant. Better adaptation of the SSNM approach to local conditions such as wet seeding, long-grain cultivars, and seasonally overlapping planting time is required. Yields of 6 t ha^{-1} in the WS and 7 t ha^{-1} in the DS appear to be realistic medium-term targets for such an approach.

References

Dobermann A, Dawe D, Roetter R, Cassman KG. 2000. Reversal of rice yield decline in a long-term continuous cropping experiment. Agron. J. 92:633-643.

Dobermann A, Fairhurst T. 2000. Rice: nutrient disorders and nutrient management. Singapore and Los Baños: Potash & Phosphate Institute (PPI), Potash & Phosphate Institute of Canada (PPIC), and International Rice Research Institute (IRRI). 191 p.

Dobermann A, White PF. 1999. Strategies for nutrient management in irrigated and rainfed lowland rice systems. Nutr. Cycl. Agroecosyst. 53:1-18.

Frère M, Popov GF. 1979. Agrometeorological crop monitoring and forecasting. FAO Plant Production and Protection Paper 17. Rome (Italy): FAO.

Olk DC, Cassman KG, Mahieu N, Randall EW. 1998. Conserved chemical properties of young soil humic acid fractions in tropical lowland soils under intensive irrigated rice cropping. Eur. J. Soil Sci. 49:337-349.

Olk DC, Cassman KG, Simbahan GC, Sta. Cruz PC, Abdulrachman S, Nagarajan R, Tan PS, Satawathananont S. 1999. Interpreting fertilizer-use efficiency in relation to soil nutrient-supplying capacity, factor productivity, and agronomic efficiency. Nutr. Cycl. Agroecosyst. 53:35-41.

Soil Survey Staff. 1994. Keys to soil taxonomy. 6th ed. Washington, D.C. (USA): USDA, SCS. 306 p.

Witt C, Cassman KG, Olk DC, Biker U, Liboon SP, Samson MI, Ottow JCG. 2000. Crop rotation and residue management effects on carbon sequestration, nitrogen cycling, and productivity of irrigated rice systems. Plant Soil 225:263-278.

Witt C, Cassman KG, Ottow JCG, Biker U. 1998. Soil microbial biomass and nitrogen supply in an irrigated lowland rice soil as affected by crop rotation and residue management. Biol. Fert. Soils 28:71-80.

Witt C, Dobermann A, Abdulrachman S, Gines HC, Wang GH, Nagarajan R, Satawathananont S, Son TT, Tan PS, Tiem LV, Simbahan GC, Olk DC. 1999. Internal nutrient efficiencies of irrigated lowland rice in tropical and subtropical Asia. Field Crops Res. 63:113-138.

Notes

Authors'addresses: S. Satawathananont, Pathum Thani Rice Research Center, Thanyaburi, Pathum Thani 12110, Thailand, e-mail: c/o irri-bangkok-t@cgiar.org; S. Chatuporn, J. Sookthongsa, Suphan Buri Rice Experiment Station, Rice Research Institute, Department of Agriculture, Thailand; L. Niyomvit, Technical and Planning Division, Department of Agriculture, Bangkok, Thailand; M. Kongchum, International Rice Research Institute, Bangkok, Thailand; A. Dobermann, International Rice Research Institute, Los Baños, Philippines, and University of Nebraska, Lincoln, Nebraska, USA.
Acknowledgments: This research has been the joint task of many colleagues. We wish to thank Ms. Samorn Okveja, Ms. Uraiwan Nasapot, and Mr. Sangob Punpeng for their long-term involvement in our socioeconomic farm monitoring. We thank Mrs. Wasana Inthalaeng, Mr. Samran Inthalaeng, Mrs. Orathai Teppayasirikamol, Ms. Thayarat Maneesaeng, and Mrs. Nareerat Ngernsawang of PTRRC for doing all laboratory work of the RTDP project. We are grateful to Mr. Supoj Noksakul for managing the on-farm trials at Suphan Buri. The authors are much obliged to Mr. Suthep Limthongkul, director of the Rice Research Institute and INMnet steering committee member. Mr. Vichai Hiranyupakorn, director of PTRRC; Mr. Nikool Rangsichol, director of SBRES; and Dr. Boriboon Somrith, liaison scientist, IRRI-Thailand Office, are acknowledged for supporting our work. We are particularly grateful to all 26 farm families involved at our site and to Christian Witt, Kenneth Cassman, Gregorio C. Simbahan, Ma. Arlene Adviento, Rico Pamplona, Olivyn Angeles, Julie Mae Criste Cabrera-Pasuquin, Edsel Moscoso, David Dawe, and Pie Moya (IRRI) for their help in various parts of this project.
Citation: Dobermann A, Witt C, Dawe D, editors. 2004. Increasing productivity of intensive rice systems through site-specific nutrient management. Enfield, N.H. (USA) and Los Baños (Philippines): Science Publishers, Inc., and International Rice Research Institute (IRRI). 410 p.

8 | Site-specific nutrient management in irrigated rice systems of Central Luzon, Philippines

H.C. Gines, G.O. Redondo, A.P. Estigoy, and A. Dobermann

8.1 Characteristics of rice production in Central Luzon, Philippines

Trends

The Philippine RTDP site is located in the province of Nueva Ecija, Central Luzon, Philippines (Fig. 8.1). Central Luzon occupies 18,230 km^2 and is composed of six provinces. The region is a flat alluvial floodplain bounded by two mountain ranges, the Sierra Madre mountains to the east and the Zambales mountains to the west. Central Luzon is the most intensively cropped lowland rice area in the Philippines and accounts for the largest share of the national rice production. Nueva Ecija Province has a total land area of 5,284.3 km^2. Its terrain is almost level, but elevation declines gradually from about 80 m above sea level in northeastern Nueva Ecija to

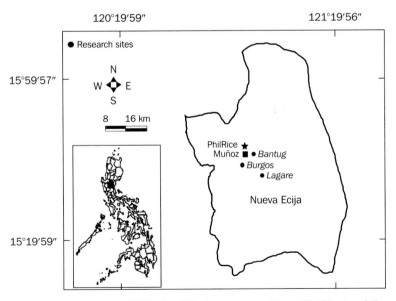

Fig. 8.1. Location of the Philippine Rice Research Institute (PhilRice) and the experimental sites near the city of Muñoz in Nueva Ecija Province, Philippines.

145

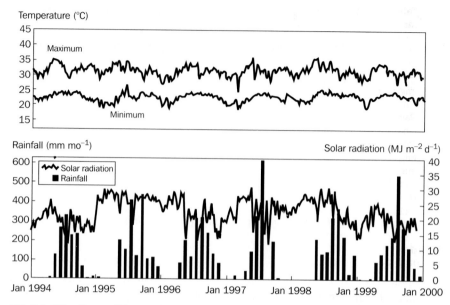

Fig. 8.2. Climatic conditions at Maligaya, Nueva Ecija Province, Philippines, from 1994 to 1999. Solar radiation and temperature data are 7-day moving averages; rainfall is monthly total. Solar radiation was obtained from sunshine hours using the standard coefficients for dry tropical conditions of 0.25 for a and 0.45 for b (Frère and Popov 1979) for Maligaya (15°42′N, 120°56′E).

about 30 m in the southeast. The soils in Nueva Ecija are primarily formed from alluvial deposits, originating from the adjacent mountains, frequently rejuvenated with volcanic sediments (Alicante et al 1948). Vertisols (Ustic Endoaquerts and Ustic Epiaquerts) and Inceptisols (Ustic Tropaquepts) with medium to heavy texture (clay loam to clay) and an aquic soil moisture regime predominate in almost the entire study area. Most of these soils are fertile and very suitable for growing irrigated rice (Gines 1982, Kawaguchi and Kyuma 1977). Coarser-textured soils (Alfisols and Inceptisols) occur along the Talavera and Pampanga rivers and are mainly used for growing vegetables and fruits.

Rice is the predominant crop grown in Nueva Ecija and accounts for about 93% of the agricultural land area. Most of this is irrigated lowland rice. A year-round growing season is possible in large parts of Nueva Ecija because of the moist tropical monsoon climate and the irrigation water provided by the Upper Pampanga River Irrigation System. Rainfall averages about 1,500 mm per year, most of this occurring from May to October (Fig. 8.2). The mean annual temperature is 27.2 °C and the temperature usually peaks in March to May. Farmers typically grow a dry-season (DS, December to April, climatic yield potential of 10–11 t ha^{-1}) and a wet-season crop (WS, July to October, yield potential <8 t ha^{-1}) in monoculture. Other crops planted occasionally on rice or on nonrice land are maize, onion, mango, eggplant, banana, tomato, peanut, mongo, garlic, coffee, coconut, cassava, camote, calamansi,

Table 8.1. Changes in area, production, and yield of rice in Nueva Ecija Province, Philippines, 1970-99.[a]

Item	1970-74	1975-79	1980-84	1985-89	1990-94	1995-99
Irrigated rice area (1,000 ha)	101	127	161	165	181	202
Rainfed rice area (1,000 ha)	86	37	51	49	59	40
Total rice area (1,000 ha)[b]	188	164	212	214	241	242
Irrigated rice area (%)	54	78	76	77	75	84
Irrigated rice production (1,000 t)	215	401	569	567	726	731
Rainfed rice production (1,000 t)	142	60	122	114	182	103
Total rice production (1,000 t)[b]	359	462	692	681	908	834
Irrigated rice yield (t ha[1])	2.13	3.14	3.51	3.41	4.01	3.62
Rainfed rice yield (t ha[1])	1.63	1.65	2.39	2.30	3.06	2.50
Average rice yield (t ha[1])[b]	1.90	2.83	3.24	3.17	3.78	3.44

[a]Source: PhilRice. Provincial rice statistics. [b]Including upland rice.

and cabbage. Small farms (1–3 ha) and field sizes (0.1–0.5 ha) predominate in the study region. Wet direct seeding of rice is predominant in the DS (since the mid-1980s), transplanting in the WS. Recently, however, many farmers have switched to direct seeding in the WS, mainly because of the increasing labor cost or unavailability of labor.

Table 8.1 presents the changes in irrigated rice production in the province of Nueva Ecija from 1970 to 1999. The change from single cropping of rice to intensive double-cropping systems took place in the early 1970s after the introduction of semi-dwarf, short-duration, high-yielding varieties and the completion of the Upper Pampanga River Irrigation System. The irrigated rice area increased at 2.4% y^{-1} until the late 1990s, whereas the area under rainfed rice remained more or less constant in the last 25 years. Irrigated rice area increased from 46% of the total area in 1970 to 82% in 1995, and there has been little further expansion in recent years. Average rice yields in Nueva Ecija increased from 1.9 t ha^{-1} in the early 1970s to 3.4–3.8 t ha^{-1} in the 1990s, but yield growth declined from about 7% y^{-1} in the 1970s to virtual stagnation in both irrigated and rainfed rice since the early 1980s. The production of irrigated rice increased from 215,000 t in the early 1970s to about 730,000 t in the early 1990s because of a combination of yield and area increases. For comparison, total rice production, including irrigated, rainfed, and upland rice, increased from about 360,000 t in the early 1970s to about 910,000 t in the early 1990s. In the last 20 years, however, total annual production growth was only 1.8% mainly because of a slowdown in both area and yield growth in irrigated rice. The production of rainfed rice fluctuated much since 1970 because of variation in both area and yield, and rainfed rice contributed 12–20% to the overall production in the last 25 years.

Similar trends were observed for the whole Central Luzon region. The harvested rice area increased from 246,300 ha in 1970 to 435,900 ha in 1997. Rice production increased from about 0.6×10^6 t in 1970 to about 1.7×10^6 t in 1997 because of increases in yield and in irrigated area. Regional rice yields increased from 2.4 t ha^{-1} in 1970 to 3.9 t ha^{-1} in 1997.

Since 1990, fertilizer consumption has increased at annual growth rates larger than the yield growth rate, but there has also been a tendency to apply more P and K as part of efforts directed at achieving a more balanced nutrition. From 1990 to 1997, fertilizer-N consumption in Nueva Ecija increased from 12,050 to 19,570 t, fertilizer P_2O_5 from 3,840 to 7,100 t, and fertilizer K_2O from 2,020 to 4,390 t (G. Gines, PhilRice, personal communication). However, soil nutrient depletion, particularly that of P, K, and Zn, has become widespread because of more than 25 years of intensive rice cropping with heavy emphasis on applying nitrogen. A survey of 384 rice fields in Nueva Ecija indicated that 54% to 64% of them had low available soil P, K, or Zn (Dobermann and Oberthür 1997).

Although there have been significant changes in cropping intensity, yield, and crop management practices during the past 35 years in Central Luzon, a recent reanalysis of productivity and yield trends suggests that total factor productivity has not declined during this period (Tiongco and Dawe 2000), contrary to what was previously believed (Cassman and Pingali 1995). However, better data are required to fully assess such productivity trends. Recent studies in Central Luzon also indicated large potential for increasing yields by improving the congruence between soil and fertilizer-N supply and crop N demand (Cassman et al 1996, Peng et al 1996a). Therefore, the main objectives of our study were to quantify trends and variability of productivity and soil fertility and to develop a new approach for site-specific nutrient management (SSNM) for the specific conditions of transplanted and direct-seeded rice cultivation in Central Luzon.

Current biophysical and socioeconomic farm characteristics

The experimental domain is located around the PhilRice headquarters near Muñoz (15°43′N, 120°54′E) and includes 27 farms in three villages, Burgos, Bantug, and Lagare (Fig. 8.1), all located within 20 km of PhilRice. Biophysical and socioeconomic data collection started on all farms with the 1994-95 DS crop and continues until today (see Chapter 2). Farm sizes range from 0.5 to 6 ha, with a median of 2 ha, and almost all land is used for growing rice (Table 8.2). Farmers in our sample had a median age of 55 with 7 years of school education. About 75% of them rent their land and pay about 20% of the harvest for this. In general, because of the rising labor cost, many farmers in this area started adopting direct seeding 15 to 20 years ago. Currently, about 70% of the DS rice is broadcast on wet soil (remaining 30% transplanted), whereas WS rice is about 60% transplanted and 40% wet-seeded. Total labor input averages about 50 8-hour person-days ha^{-1}, but varies widely among seasons and farms (Table 8.2).

Table 8.2. Demographic and economic characteristics of rice production on 26 farms of Nueva Ecija, Philippines. Values shown are based on socioeconomic farm surveys conducted for whole farms.

Production characteristics	Min.	25%	Median	75%	Max.
Total cultivated area (ha)	0.50	1.10	2.00	2.87	6.28
Age of household head (y)	26	41	55	61	76
Education of household head (y)	1	6	7	10	14
Household size (persons)	3	5	6	6	12
1998 dry-season crop					
Rice area (ha)[a]	0.50	1.10	2.00	2.80	6.08
Yield (t ha^{-1})	3.72	5.12	6.02	6.52	9.33
N fertilizer (kg ha^{-1})	62	117	141	152	192
P fertilizer (kg ha^{-1})	2	9	17	19	26
K fertilizer (kg ha^{-1})	0	12	23	31	56
Insecticide (kg ai ha^{-1})	0	0	0	0.003	0.73
Herbicide (kg ai ha^{-1})	0	0.20	0.29	0.40	0.71
Other pesticides (kg ai ha^{-1})[b]	0	0	0.03	0.09	12.25
Total labor (8-h d ha^{-1})	17	28	46	65	108
Net return from rice (US$ ha^{-1} crop^{-1})	420	747	920	1,129	1,546
1998 wet-season crop					
Rice area (ha)[a]	0.50	1.10	1.88	2.87	6.28
Yield (t ha^{-1})	2.13	2.88	3.98	4.99	6.99
N fertilizer (kg ha^{-1})	37	59	82	95	139
P fertilizer (kg ha^{-1})	2	7	12	18	31
K fertilizer (kg ha^{-1})	0	12	18	33	41
Insecticide (kg ai ha^{-1})[b]	0	0	0	0.01	1.15
Herbicide (kg ai ha^{-1})	0	0.18	0.27	0.45	1.10
Other pesticides (kg ai ha^{-1})[b]	0	0	0	0.06	0.54
Total labor (8-h d ha^{-1})	28	43	59	73	125
Net return from rice (US$ ha^{-1} crop^{-1})	128	312	535	677	982

[a]Rice area in which the treatment plots were embedded in subsequent years. The total rice area may even be larger.
[b]Includes fungicide, molluscicide, rodenticide, and crabicide.

To enhance germination and prevent snail damage, fields remain without standing water for 1 to 2 weeks after sowing or planting, with weed control mainly done by applying preemergence herbicides in combination with proper water management. Thereafter, fields are irrigated and kept flooded using rainwater or surface irrigation water until shortly before harvest. Because the area has a unimodal rainfall pattern (Fig. 8.2), the WS crop (June-October) is rainfed, but supplemented by irrigation from canals or tube wells when precipitation is erratic. Broadcast fertilizer application starts shortly before or after the first flooding and varies widely among farmers. Farmers rarely follow the official fertilizer recommendations (Fujisaka 1993). Prilled urea (46-0-0), ammonium phosphate (16-20-0), and other compound fertilizers such

Table 8.3. General soil properties on 21 rice farms of Nueva Ecija, Philippines.[a]

Soil properties	Min.	25%	Median	75%	Max.
Clay content (%)	20.0	27.0	33.0	38.0	59.0
Silt content (%)	37.0	50.0	54.0	56.0	62.0
Sand content (%)	4.0	9.0	11.0	20.0	28.0
Soil organic C (g kg[-1])	8.4	11.6	12.6	14.4	16.5
Total soil N (g kg[-1])	0.77	0.94	1.15	1.27	1.64
Soil pH (1:1 H_2O)	5.7	5.9	6.2	6.4	6.7
Cation exchange capacity (cmol$_c$ kg[-1])	14.7	21.2	24.6	28.4	39.5
Exchangeable K (cmol$_c$ kg[-1])	0.11	0.17	0.20	0.22	0.55
Exchangeable Na (cmol$_c$ kg[-1])	0.30	0.38	0.43	0.49	0.74
Exchangeable Ca (cmol$_c$ kg[-1])	9.44	14.60	17.50	22.40	24.90
Exchangeable Mg (cmol$_c$ kg[-1])	4.44	6.26	6.64	7.27	14.70
Extractable P (Olsen-P, mg kg[-1])	1.00	3.10	3.80	5.80	7.30
Extractable Zn (0.05N HCl, mg kg[-1])	0.40	0.64	0.72	1.20	2.20

[a]Measured on soil samples collected before the 1996 dry season.

as "14-14-14" are the dominant nutrient sources. Most farmers do not apply farm-yard manure or other organic materials. Nitrogen fertilizers are mostly applied in two to three splits in both dry and wet seasons. Pesticide use varies (Table 8.2), but most farmers spray preemergence herbicides. Farmers in some villages such as Bantug use an IPM approach and no insecticides are applied. Fungicide use is not common. At harvest, rice is cut by hand, leaving behind about 30 to 40 cm of stubble. Straw is piled up in the fields at threshing sites and burned in heaps.

Soil data in our sample confirmed the results from other studies in Nueva Ecija (Dobermann and Oberthür 1997). Soil organic C ranged from 8 to 17 g kg[-1] and CEC from 15 to 40 cmol$_c$ kg[-1]. Extractable Olsen-P was low on all farms (<8 mg P kg[-1]) and the exchangeable soil K level was <0.20 cmol$_c$ kg[-1] on 50% of all farms. Soil test levels indicate low Zn status (<1 mg Zn kg[-1]) on more than 50% of all farms, but visual symptoms of plant Zn deficiency were not observed. Moreover, soil fertility varied widely among farms (Table 8.3).

We measured grain yield and plant N accumulation in unfertilized plots on all 27 farms in all rice crops grown since the 1995 DS to test the hypothesis that the indigenous N supply (INS = plant N accumulation in a 0-N plot) is declining under intensive rice cropping. Available data for a time series of nine successive rice crops show no evidence for a decline in the median INS over time (Fig. 8.3) within this relatively short period, and distinct but relatively small seasonal differences. The average INS in the WS was 60 kg N ha[-1] vis-à-vis only 54 kg N ha[-1] in DS crops (standard deviation was about 15 kg N ha[-1] in all WS and DS crops). One possible explanation for this is that WS crops included more transplanted rice fields, whereas DS rice was mostly wet-seeded (Table 8.4). We hypothesize that extraction of native soil N is greater in transplanted rice than in wet-seeded rice because of a greater rooting depth. Another explanation may be that hot temperatures and greater soil drying during the DS to WS fallow period (Fig. 8.2) cause a flush of N mineralization

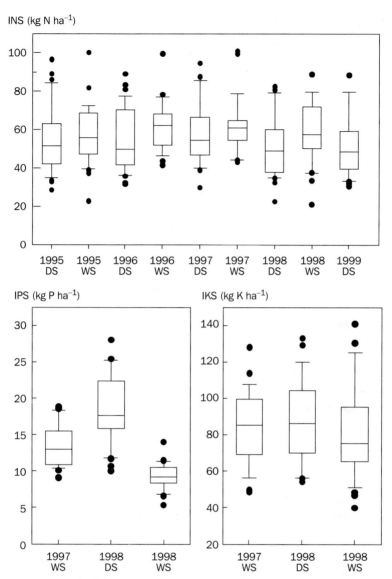

Fig. 8.3. Variability of the indigenous N (INS), P (IPS), and K (IKS) supply among 27 farmers' fields in Nueva Ecija, Philippines (1995-99). Median with 10th, 25th, 75th, and 90th percentiles as vertical boxes with error bars; outliers as bullets; WS = wet season; DS = dry season.

Table 8.4. Frequency of transplanting and wet seeding of rice in the 27 farmers' fields during 1997 to 1999.

Crop	Transplanted	Direct-seeded
1997 WS	17	10
1998 DS	6[a]	21
1998 WS	16	11
1999 DS	10[b]	17

[a]All at Burgos. [b]Eight of those at Burgos.

in the succeeding WS crop because of faster decomposition of soil organic matter (Dobermann et al 2000, Olk et al 1998). However, differences in INS between seasons are probably too small to be considered in practical decision making regarding fertilizer-N applications.

During 1997 to 1998, P and K omission plots were established to also obtain estimates of the indigenous P (IPS = plant P accumulation in a 0-P plot) and K supply (IKS = plant K accumulation in a 0-K plot). In general, indigenous N, P, and K supply varied by about threefold among the 27 farmers' fields (Fig. 8.3). Based on data from three cropping seasons, the IKS was about 70 to 100 kg K ha^{-1} on 50% of all farms. However, an IKS of less than 60 kg K ha^{-1} was measured on about one-third of all farms, suggesting that severe soil K depletion had occurred in the past. The current average INS of about 55 kg N ha^{-1}, IPS of about 17 kg P ha^{-1}, and IKS of about 80 kg K ha^{-1} are sufficient for achieving DS rice yields of about 3.7 t ha^{-1} without applying N, 6.5 t ha^{-1} without applying P, and 5.5 t ha^{-1} without applying K. These estimates assume optimal balanced nutrient requirements of 14.7 kg N, 2.6 kg P, and 14.5 kg K per 1,000 kg of grain yield (Witt et al 1999). Our data indicate that potassium has become the second most yield-limiting nutrient, mainly because farmers in Central Luzon applied little K fertilizer in the past and the surface water used for irrigation is low in nutrients such as K. This confirms observations of strong declines in available soil K and yield in the +NP treatment of the long-term rice experiment at PhilRice (Cassman et al 1997, De Datta et al 1988). Visual field observations indicated that K deficiency occurred in the 1998 and 1999 DS in several direct-seeded farmers' fields, particularly in those at Bantug with shallow plow layer, high plant density, and high yield potential.

The baseline agronomic data measured in 1995-96 showed distinct differences between WS and DS crops and large variability in performance among farmers (Olk et al 1999). Table 8.5 shows data for two crops sampled in 1995 and 1996. Generally, because of the large proportion of direct seeding at seed rates of up to 200 kg ha^{-1}, average plant densities were high (e.g., 677 panicles m^{-2} in the 1996 DS), but panicles had only about 50 spikelets. In both crops sampled, average grain-filling percentage was low (about 75%). As a result, average farm yield was only 3.5 t ha^{-1} in the 1995 WS and 5.7 t ha^{-1} in the 1996 DS. For comparison, yields in the NPK treatment of the long-term fertility experiment located at PhilRice were 4.8 t

Table 8.5. Baseline agronomic characteristics of rice production on 27 farms of Nueva Ecija, Philippines. Values shown are means and standard deviations (SD) of the farmers' fertilizer practice for two consecutive rice crops monitored before SSNM plots were established.

Agronomic characteristics[a]	1995 WS crop		1996 DS crop	
	Mean	SD	Mean	SD
Grain yield (t ha^{-1})	3.49	0.77	5.67	0.74
Harvest index	0.40	0.04	0.43	0.05
No. of panicles m^{-2}	582	115	677	201
Total no. of spikelets m^{-2}	27,427	6,790	30,699	7,928
Total no. of spikelets panicle^{-1}	49	13	48	15
Filled spikelets (%)	73	5	75	8
Fertilizer N use (kg ha^{-1})	84.3	33.0	123.2	35.1
Fertilizer P use (kg ha^{-1})	13.2	8.4	14.2	7.0
Fertilizer K use (kg ha^{-1})	14.7	10.8	22.2	14.7
N uptake (kg ha^{-1})	73.2	18.4	96.5	18.5
P uptake (kg ha^{-1})	9.6	2.4	13.3	2.7
K uptake (kg ha^{-1})	51.3	12.0	82.6	37.6
Input-output N balance (kg ha^{-1} crop^{-1})	9.1	11.9	9.2	15.5
Input-output P balance (kg ha^{-1} crop^{-1})	4.5	8.6	2.4	6.6
Input-output K balance (kg ha^{-1} crop^{-1})	−17.9	14.2	−28.7	24.3
Partial productivity of N (kg kg^{-1})	45.2	19.9	50.1	17.2
Agronomic efficiency of N (kg kg^{-1})	8.4	6.9	18.3	10.6
Recovery efficiency of N (kg kg^{-1})	0.22	0.14	0.35	0.13
Physiological efficiency of N (kg kg^{-1})	28.5	22.3	47.7	17.8

ha^{-1} in the 1995 WS and 6.8 t ha^{-1} in the 1996 DS. Whereas farmers applied relatively high amounts of N fertilizer, P rates averaged only about 14 kg P ha^{-1} crop^{-1} and K rates about 20 kg K ha^{-1} crop^{-1} (Tables 8.2 and 8.5). Plant P and K uptake were relatively low, suggesting that both nutrients limited yield on many farms. Moreover, a negative K input-output balance occurred on most farms (−18 kg K ha^{-1} WS crop^{-1} and −29 kg K ha^{-1} DS crop^{-1}), confirming similar results obtained from a larger sample of farmers in Nueva Ecija (Dobermann and Oberthür 1997). Nitrogen-use efficiencies were low in the 1995 WS and varied widely among farms in both seasons. The average agronomic N efficiency (AEN) was 8.4 kg kg^{-1} in the 1995 WS (CV = 82%) and 12 kg kg^{-1} in the 1996 DS (CV = 58%). On average, only 22% (WS) and 35% (DS) of the fertilizer N applied was recovered in the aboveground plant biomass, suggesting large gaseous N losses. The low average physiological N-use efficiency (PEN) of less than 30 kg grain kg^{-1} N indicates severe climatic and pest stresses occurring in the WS.

8.2 Effect of SSNM on productivity and nutrient-use efficiency

Management of the SSNM plots

An SSNM plot was established first in the 18 farmers' fields at Burgos and Bantug (Fig. 8.1) in the 1997 DS, but not at Lagare because no DS crop was grown because of a repair of the irrigation system. The 1997 DS crop at Burgos and Bantug served as a test of the SSNM procedure for all RTDP project sites and results are not reported here. Permanent SSNM plots were managed for all 27 farms and compared to the FFP for four consecutive rice crops grown from the 1997 WS to 1999 DS (see Chapter 2). The size of the SSNM plots was 400–500 m^2 during 1997-98, but increased to 1,000–1,300 m^2 in the 1999 DS. All crop management practices in SSNM plots were the same as in the FFP area, except fertilizer application and, to some degree, pest management as described below.

Varieties used in SSNM and FFP plots were chosen by the farmers. In the four crops sampled, 16 different semidwarf, high-yielding varieties were grown. The most frequently grown varieties were IR64 (40%), PSBRc28 (30%), PSBRc54 (9%), IR74 (5%), and IR60 (4%). Interestingly, PSBRc28 was the most widely used variety in the 1997 WS (two-thirds of all farmers), but IR64 was the preferred choice in all subsequently grown crops. Farmers used both transplanting and direct seeding (pregerminated seeds at 150 to 200 kg ha^{-1} broadcast on wet soil), but we observed distinct seasonal and location differences (Table 8.4). Transplanting dominated in the WS (60%), but direct seeding in the DS (60–70%). Most farmers on heavy clay soil in Bantug and Lagare direct-seeded in the DS and some also in WS crops, whereas almost all farmers on silty soil in Burgos preferred transplanting in the WS and in the DS.

Pest management was done by the farmers, but with occasional guidance from the researchers. Prophylactic fungicide application for sheath blight control was done in the SSNM plots in the 1998 DS, but sheath blight incidence was low in both SSNM and the FFP. Handweeding was occasionally done in SSNM plots, but did not affect the comparison between SSNM and FFP treatments because sampling areas did not include weedy patches that occurred in a few direct-seeded fields. In general, no complete pest control was achieved and pests caused yield losses on many farms in both SSNM and FFP plots, particularly in WS crops. Other problems encountered in 1997 to 1999 were related to climate and water management, but affected SSNM and FFP plots similarly.

Fertilizer applications for SSNM were prescribed on a field- and crop-specific basis following the approach described in Chapter 5 (Witt et al 1999). In the 1997 WS, average values of INS measured in the 1995 and 1996 DS crops were used as model input. Similarly, the average IPS and IKS were estimated from the plant P and K accumulation in the FFP sampled during this period, assuming recovery fractions of 0.2 kg kg^{-1} for fertilizer P and 0.4 kg kg^{-1} for fertilizer K. Note that many farmers applied little K so that FFP treatments were essentially NP treatments and plant K accumulation measured in the FFP was roughly equivalent to IKS. Beginning with the 1998 DS crop, estimates of INS, IPS, and IKS were continuously improved by incorporating the values measured in 1997 and 1998 and by adjusting IPS and IKS according to the actual P and K input-output balance. For example, INS used for predicting fertilizer rates in the 1999 DS were averages of the INS measured on each farm in the 1995, 1996, 1997, and 1998 DS crops and the 1997 WS crop. IPS used for predicting fertilizers rates in the 1999 DS were based on the original IPS estimate (see above), the IPS measured in the 1997 WS and 1998 DS crops, and the P balance of the 1998 DS crop, assuming that 20% of residual P is plant-available for succeeding crops. The adjusted values were then used as model input for the succeeding crop. Target yields for working out field-specific fertilizer rates (see Chapter 5) were set to 6 t ha^{-1} in the WS and 8 t ha^{-1} in the DS. First-crop recovery fractions of 0.5 (WS)–0.6 (DS), 0.2–0.25, and 0.50 kg kg^{-1} were assumed for fertilizer N, P, and K, respectively. The climatic yield potential was set to 7.5 t ha^{-1} for the WS and 10.5 t ha^{-1} for the DS, based on crop simulations using ORYZA1 (Kropff et al 1993) and maximum-yield field experiments conducted at the PhilRice site.

In the first season, the 1997 WS, N was applied based on INS and SPAD readings. Basal N at 20 kg N ha^{-1} was applied if INS was less than 40 kg N ha^{-1} (on 5 farms only). After 15 DAT (20 DAS in direct-seeded rice) until about 60 DAT, SPAD readings were taken weekly and 30–50 kg N ha^{-1} were applied if SPAD values had dropped below 33. This regime resulted in too-high late-N applications, causing increased pest incidence, and the use of a single SPAD threshold value risks missing applications at critical growth stages such as early tillering (ET), midtillering (MT), panicle initiation (PI), and first flowering (FF). Therefore, N management was gradually improved to take into account the observations of the previous seasons and to account better for critical growth stages that also depend on the crop establishment method. Different ranges of SPAD values were first used in the 1998 WS and the N management strategy used in the 1999 DS was as follows:

FN 1	Basal	If INS >40	0 kg N ha^{-1}
		If INS <40	20 kg N ha^{-1}
FN 2	14–21 DAT/DAS		30 kg N ha^{-1}
FN 3	28–35 DAT/DAS (MT)	If SPAD >37	0 kg N ha^{-1}
		If SPAD 35–37	30 kg N ha^{-1}
		If SPAD 33–35	40 kg N ha^{-1}
		If SPAD <33	50 kg N ha^{-1}
FN 4	40–50 DAT/DAS (PI)	If SPAD >35	40 kg N ha^{-1}
		If SPAD 33–35	50 kg N ha^{-1}
		If SPAD <33	60 kg N ha^{-1}
FN 5	65–70 DAT/DAS (FF)	If SPAD >35	0 kg N ha^{-1}
		If SPAD 33–35	20 kg N ha^{-1}
		If SPAD <33	30 kg N ha^{-1}

Note that in this approach the SPAD meter was not used to make a decision about when to apply, but only about how much N to apply at a certain growth stage. This reduced the number of SPAD measurements and allowed a more gradual, real-time

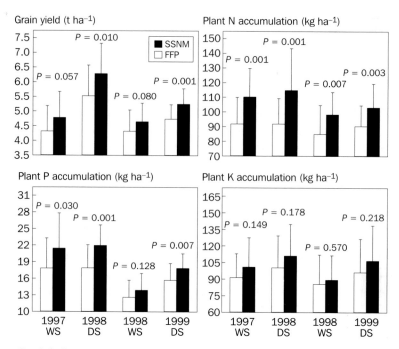

Fig. 8.4. Grain yield and plant N, P, and K accumulation in the farmers' fertilizer practice (FFP) and site-specific nutrient management (SSNM) plots (n = 27) in Nueva Ecija, Philippines (1997-99; bars: mean; error bars: standard deviation).

adjustment of N depending on the climatic conditions affecting growth and need for N. Moreover, basal N was applied only on two farms and late-N application at first flowering was done on nine farms only so that most fields received N in 3–4 split applications. All P and 50% of the K fertilizer were incorporated in the soil before sowing or planting. Another 50% of the K rate was topdressed at PI. Fertilizer sources used in the SSNM were urea (46% N), triple superphosphate (20% P), and muriate of potash (50% K). For comparison, farmers used urea, ammonium phosphate (AP), diammonium phosphate (DAP), MOP, and compound fertilizers such as "14-14-14" or "17-0-17" as N, P, and K sources in the FFP. They usually applied fertilizers in two to three splits, mostly at ET and PI or at ET, MT, and PI, but there was little consistency among farmers and among seasons for the same farmer. No organic manure was applied.

Effect of SSNM on grain yield and nutrient uptake

Compared with the FFP, SSNM significantly increased grain yield, plant N, and plant P uptake in each of the four crops grown in 1998 and 1999 (Fig. 8.4). The average yield difference between SSNM and the FFP across all four was 0.52 t ha^{-1} (11%, P = 0.000) and significantly larger in the DS (0.63 t ha^{-1}, 12%) than in the WS (0.40 t

Table 8.6. Effect of site-specific nutrient management on agronomic characteristics at Nueva Ecija, Philippines (1997-99).

Parameters	Levels[a]	Treatment[b]		Δ^c	$P>|T|^c$	Effects[d]	$P>|F|^d$
		SSNM	FFP				
Grain yield (GY)	All	5.24	4.73	0.52	0.000	Village	0.580
(t ha^{-1})	Year 1	5.53	4.92	0.61	0.008	Year[e]	0.040
	Year 2	4.95	4.53	0.42	0.001	Season[e]	0.012
	DS	5.77	5.13	0.63	0.001	Year × season[e]	0.540
	WS	4.72	4.32	0.40	0.009	Village × crop	0.468
Plant N uptake (UN)	All	106.7	89.7	16.9	0.000	Village	0.397
(kg ha^{-1})	Year 1	112.6	91.8	20.7	0.000	Year	0.026
	Year 2	100.7	87.6	13.1	0.000	Season	0.559
	DS	109.0	91.1	17.9	0.000	Year × season	0.447
	WS	104.3	88.4	15.9	0.000	Village × crop	0.220
Plant P uptake (UP)	All	18.8	16.0	2.8	0.000	Village	0.855
(kg ha^{-1})	Year 1	21.7	17.9	3.8	0.000	Year	0.591
	Year 2	15.9	14.2	1.7	0.010	Season	0.866
	DS	19.9	16.8	3.1	0.000	Year × season	0.747
	WS	17.7	15.2	2.4	0.028	Village × crop	0.417
Plant K uptake (UK)	All	102.1	93.4	8.6	0.023	Village	0.115
(kg ha^{-1})	Year 1	106.0	95.9	10.1	0.051	Year	0.456
	Year 2	98.1	90.9	7.1	0.196	Season	0.328
	DS	108.9	98.3	10.6	0.066	Year × season	0.482
	WS	95.2	88.5	6.7	0.158	Village × crop	0.416
Agronomic efficiency	All	14.9	12.2	2.8	0.020	Village	0.002
of N (AEN)	Year 1	15.8	13.0	2.8	0.154	Year	0.733
(kg grain kg^{-1} N)	Year 2	14.1	11.4	2.7	0.042	Season	0.156
	DS	16.6	13.1	3.5	0.012	Year × season	0.089
	WS	13.3	11.3	2.0	0.297	Village × crop	0.238
Recovery efficiency	All	0.46	0.33	0.13	0.000	Village	0.008
of N (REN)	Year 1	0.49	0.34	0.15	0.000	Year	0.445
(kg plant N kg^{-1} N)	Year 2	0.43	0.32	0.11	0.003	Season	0.240
	DS	0.41	0.32	0.09	0.009	Year × season	0.421
	WS	0.50	0.33	0.17	0.000	Village × crop	0.746
Partial productivity	All	51.7	52.6	−1.0	0.775	Village	0.001
of N (PFPN)	Year 1	53.9	56.6	−2.7	0.633	Year	0.554
(kg grain kg^{-1} N)	Year 2	49.5	48.6	0.9	0.808	Season	0.126
	DS	45.0	42.4	2.6	0.347	Year × season	0.469
	WS	58.4	63.1	−4.7	0.408	Village × crop	0.317

[a]All = all four crops grown from 1997 WS to 1999 DS; year 1 = 1997 WS and 1998 DS; year 2 = 1998 WS and 1999 DS; WS = 1997 WS and 1998 WS; DS = 1998 DS and 1999 DS. [b]FFP = farmers' fertilizer practice; SSNM = site-specific nutrient management. $^c\Delta$ = SSNM – FFP. $P>|T|$ = probability of a significant mean difference between SSNM and FFP. [d]Source of variation of analysis of variance of the difference between SSNM and FFP by farm; $P>|F|$ = probability of a significant F-value. [e]Year refers to two consecutive cropping seasons.

ha^{-1}, 9%, Table 8.6). This season effect was significant ($P = 0.012$), suggesting better performance of SSNM in the DS because of greater yield potential and fewer biotic stresses. On average and across all seasons, plant N accumulation was 16.9 kg ha^{-1} (19%, $P = 0.000$), plant P 2.8 kg ha^{-1} (18%, $P = 0.000$), and plant K 8.6 kg ha^{-1} (9%, $P = 0.023$) greater in SSNM than in the FFP. Note that the increase in plant K because of SSNM became significant only when data across all seasons were analyzed because of the greater degree of freedom in the ANOVA model. Crop-season effects were not significant, that is, similar increases in nutrient uptake were achieved in DS and WS crops (Table 8.6).

Poor weather conditions in the 1998 WS and 1999 DS reduced yields and nutrient uptake in all treatments compared with previous years (Fig. 8.4). As a result, the average yield increase from SSNM was greater in 1997-98 (0.61 t ha^{-1}) than in 1998-99 (0.42 t ha^{-1}) and increases in plant N, P, and K accumulation over the FFP were also much less in the second year (Table 8.6). Village and village-crop effects on grain yield and nutrient uptake were not statistically significant (Table 8.6), indicating that SSNM increased yields and nutrient uptake similarly in all three villages, irrespective of differences in soil type and crop management practices.

Yield increases by SSNM were largest in the 1998 DS (on average 0.8 t ha^{-1} or 14%; = 1 t ha^{-1} on 7 farms), the crop with the highest climatic yield potential in our study. Note, however, that even the 1998 DS crop had a reduced yield potential because of unusual temperatures (see below), so SSNM will likely perform even better in years with more normal climate.

Effect of SSNM on nitrogen-use efficiency

Significant increases in the recovery efficiency of fertilizer N were achieved through the field- and season-specific N management practiced in SSNM (Fig. 8.5). Average REN was 0.46 kg kg^{-1} in SSNM, which represents a 0.13 kg kg^{-1} (39%, $P = 0.000$) increase in REN over that achieved in the FFP in the same years (Table 8.6). Note that crop-year and crop-season effects on REN were not statistically significant, suggesting that this level of increase in REN was consistently achieved in all four crops grown (Fig. 8.5) with little change over time. On many farms, REN exceeded 0.50 kg kg^{-1}, a level that is often achieved only in well-managed field experiments at research stations. The modifications in N management introduced in 1998 and 1999 did not result in a further increase in REN, mainly because of the unfavorable climatic conditions affecting growth of the 1998 WS and 1998 DS crop.

Although AEN increased on average by 2.8 kg kg^{-1} (22%, $P = 0.020$), this increase was not statistically significant in two out of four crops grown because of the large variability among farms (Fig. 8.5). The PEN was the same in the FFP and SSNM (Fig. 8.5) in all four crops grown and remained below 50 kg kg^{-1} on most farms. SSNM did not change the average PFPN in any of the four cropping seasons.

The different parameters characterizing N-use efficiency suggest that in SSNM we were able to get much more fertilizer N into the plant and reduce N losses, but yield did not increase correspondingly because other biotic and abiotic stresses occurred. However, the large and consistent gains in REN indicate that the plant-based

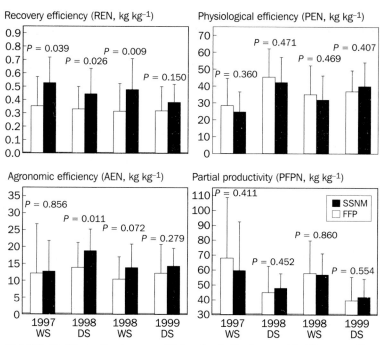

Fig. 8.5. Fertilizer nitrogen-use efficiencies in the farmers' fertilizer practice (FFP) and site-specific nutrient management (SSNM) plots (n = 27) in Nueva Ecija, Philippines (1997-99; bars: mean; error bars: standard deviation).

N management strategy applied in the SSNM was dynamic and flexible enough to account for widely differing INS (Fig. 8.3) and climatic effects among sites and seasons. Similar results were obtained in other on-farm studies in Nueva Ecija (Cassman et al 1996, Peng et al 1996a,b). Thus, it is possible to manage the spatial and temporal variation in optimal N rates by applying a real-time N management strategy (Dawe and Moya 1999).

Effect of SSNM on fertilizer use and profit

Site-specific nutrient management increased profit in the DS, but not in the WS (Table 8.7; Fig. 8.6). The average profit increase was $69 ha^{-1} crop^{-1} ($P = 0.012$), but significantly larger ($P = 0.043$) in DS crops ($90 ha^{-1} crop^{-1}) than in WS crops ($48 ha^{-1} crop^{-1}). Profit increases due to SSNM varied among villages ($P = 0.032$) and also widely among farms. Large potential for increasing yields and profit exists if SSNM is associated with an overall improvement in crop management. For example, in the 1998 DS, average profit increase was $119 ha^{-1}, with a range of –$13 to $271 ha^{-1}.

This profit increase was achieved despite a significantly larger fertilizer cost in SSNM than in the FFP. Average N rates in SSNM and FFP were the same (about 110 kg N ha^{-1}), but more P and K were applied in SSNM. On average, fertilizer cost in

Table 8.7. Effect of site-specific nutrient management on fertilizer use, fertilizer cost, and gross returns above fertilizer costs from rice production at Nueva Ecija, Philippines (1997-99).

Parameters	Levels[a]	Treatment[b] SSNM	Treatment[b] FFP	Δ[c]	P>|T|[c]	Effects[d]	P>|F|[d]
N fertilizer (FN)	All	111	109	2.3	0.675	Village	0.001
(kg ha⁻¹)	Year 1	113	109	4.4	0.562	Year[e]	0.671
	Year 2	109	109	0.1	0.986	Season[e]	0.400
	DS	133	133	−0.3	0.967	Year × season[e]	0.616
	WS	89	84	5.2	0.367	Village × crop	0.422
P fertilizer (FP)	All	19.2	14.4	4.8	0.000	Village	0.000
(kg ha⁻¹)	Year 1	20.9	13.5	7.4	0.000	Year	0.008
	Year 2	17.5	15.3	2.1	0.197	Season	0.606
	DS	20.9	15.6	5.3	0.002	Year × season	0.553
	WS	17.5	13.2	4.3	0.006	Village × crop	0.110
K fertilizer (FK)	All	49.4	21.3	28.1	0.000	Village	0.134
(kg ha⁻¹)	Year 1	55.0	19.2	35.8	0.000	Year	0.002
	Year 2	43.8	23.4	20.4	0.000	Season	0.021
	DS	56.9	23.0	33.8	0.000	Year × season	0.314
	WS	41.9	19.4	22.5	0.000	Village × crop	0.084
Fertilizer cost	All	118.6	81.5	37.1	0.000	Village	0.001
(US$ ha⁻¹)	Year 1	127.5	77.9	49.5	0.000	Year	0.002
	Year 2	109.7	85.1	24.6	0.001	Season	0.160
	DS	136.8	93.9	42.9	0.000	Year × season	0.523
	WS	100.3	68.8	31.5	0.000	Village × crop	0.250
Gross returns above	All	982	914	69	0.012	Village	0.032
fertilizer costs	Year 1	1,035	956	79	0.094	Year	0.337
(US$ ha⁻¹)	Year 2	930	871	59	0.021	Season	0.043
	DS	1,075	984	90	0.019	Year × season	0.526
	WS	890	842	48	0.138	Village × crop	0.370

[a]All = all four crops grown from 1997 WS to 1999 DS; year 1 = 1997 WS and 1998 DS; year 2 = 1998 WS and 1999 DS; WS = 1997 WS and 1998 WS; DS = 1998 DS and 1999 DS. [b]FFP = farmers' fertilizer practice; SSNM = site-specific nutrient management. [c]D = SSNM − FFP. P>\|T\| = probability of a significant mean difference between SSNM and FFP. [d]Source of variation of analysis of variance of the difference between SSNM and FFP by farm; P>\|F\| = probability of a significant F-value. [e]Year refers to two consecutive cropping seasons.

the SSNM was $119 ha⁻¹ crop⁻¹ vis-à-vis $82 ha⁻¹ crop⁻¹ in the FFP. Note, however, that, following the SSNM concept (Chapter 5), P and K rates declined after the first year, leading to a significant decrease in the additional fertilizer cost from $50 ha⁻¹ in year 1 to $25 ha⁻¹ in year 2. Although some farmers tended to increase their P and K use over time (Fig. 8.6), farmers' fertilizer use remained unbalanced. On average, farmers applied about 30% less P and 130% less K than in SSNM. Summarizing, additional profit produced by using SSNM was large in comparison to the additional fertilizer cost involved and the higher P and K rates in SSNM are probably more appropriate for sustaining productivity.

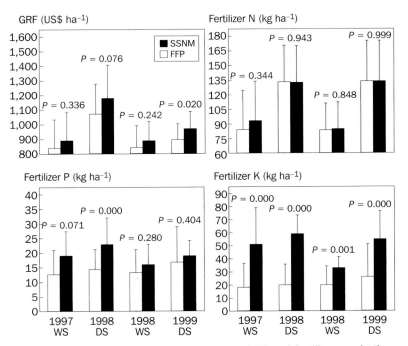

GRF (US$ ha⁻¹)

Fertilizer N (kg ha⁻¹)

Fertilizer P (kg ha⁻¹)

Fertilizer K (kg ha⁻¹)

Fig. 8.6. Gross returns above fertilizer costs (GRF) and fertilizer use in the farmers' fertilizer practice (FFP) and site-specific nutrient management (SSNM) plots (n = 27) in Nueva Ecija, Philippines (1997-99; bars: mean; error bars: standard deviation).

Examples of the performance of SSNM over time

To illustrate some of the principles of SSNM applied in our study, Table 8.8 shows the performance of SSNM and FFP in two farmers' fields at Bantug for five successive crops, including the first SSNM "test crop" in the 1997 DS. Note that both fields are located within about 300 m distance on the same soil type (Bantug clay), but they differed significantly in soil fertility, yield, and N-use efficiency. These differences are probably the result of historical differences in land management during the past 20 to 30 years of intensive rice cropping.

Mr. de Regla (farm no. 110) never applied much P and K in the past and his soil became depleted in these nutrients. Resin adsorption quantity (RAQ) measured in the 1998 DS during a 14-d *in situ* placement of mixed-bed ion exchange resin capsules (Dobermann et al 1997) was 0.07 µmol P cm⁻² (in the 0-P plot) and 0.69 µmol K cm⁻² (in the 0-K plot). Measured in the same omission plots, IPS was 17 kg P ha⁻¹ and IKS was 62 kg K ha⁻¹. In contrast, Mr. Villanueva applied more P and K than Mr. de Regla (Table 8.8) and maintained higher soil fertility in his field.[1] RAQ measured

---

[1]During 1995 to 1996, before the introduction of an SSNM plot, average fertilizer rates in four crops monitored were 12 kg P ha⁻¹ and 23 kg K ha⁻¹ (Villanueva farm) and 11 kg P ha⁻¹ and 20 kg K ha⁻¹. Note that Mr. Villanueva increased his P use significantly after the 1997 DS, whereas Mr. de Regla did not.

Table 8.8. Performance of site-specific management (SSM) and farmers' fertilizer practice (FFP) in two farmers' fields at Bantug, Nueva Ecija.[a]

Season	Grain yield (t ha^{-1})		FN (kg ha^{-1})		FP (kg ha^{-1})		FK (kg ha^{-1})		AEN (kg kg^{-1})		REN (kg kg^{-1})	
	FFP	SSM	FFP	SSM	FFP	SSM	FFP	SSM	FFP	SSM	FFP	SSM
Gerardo de Regla (farm no. 110), Bantug												
1997 DS	5.89	6.99	85	193	9	38	18	184	23	16	0.32	0.22
1997 WS	4.02	4.90	109	140	5	28	10	99	10	14	0.22	0.37
1998 DS	6.13	6.96	118	158	13	14	25	49	18	19	0.30	0.73
1998 WS	2.37	4.30	68	109	9	19	17	30	12	25	0.19	0.67
1999 DS	4.82	6.10	154	140	10	24	19	60	12	22	0.23	0.54
Mean	4.65	5.85	107	148	9	25	18	85	15	19	0.25	0.50
Romeo Villanueva (farm no. 103), Bantug												
1997 DS	7.96	8.96	127	172	11	31	43	181	25	24	0.45	0.54
1997 WS	5.23	6.02	123	120	34	9	68	66	17	24	0.55	0.65
1998 DS	6.57	7.08	133	140	24	19	40	49	26	28	0.56	0.72
1998 WS	4.22	4.27	79	69	32	18	23	29	15	18	0.39	0.57
1999 DS	4.96	5.54	156	170	31	14	24	68	9	12	0.28	0.40
Mean	5.78	6.37	124	134	26	18	40	79	18	21	0.45	0.58

[a]FN = fertilizer N; FP = fertilizer P; FK = fertilizer K; AEN = agronomic N-use efficiency; REN = recovery efficiency of applied N.

in the 1998 DS was 0.74 µmol P cm^{-2} and 2.66 µmol K cm^{-2}. IPS was 25 kg P ha^{-1} and IKS was 105 kg K ha^{-1}. On average, Mr. Villanueva's yield was 5.8 t ha^{-1} in 1997-99 vis-à-vis only 4.7 t ha^{-1} achieved by Mr. de Regla. Visual observations also indicated a better overall quality of crop management in the field of Mr. Villanueva, particularly for land leveling, crop establishment, and weed and N management. His average REN of 0.45 kg kg^{-1} was almost twice as high as that of Mr. de Regla.

Therefore, our strategy was to build up soil fertility in Mr. de Regla's field in combination with improving crop and N management. The first two SSNM crops received relatively large P and K doses, but rates declined over time to levels that were close to crop removal. In the first two crops, N management was not optimal yet. Rates were too high and the amount and timing of split applications was not optimal, resulting in no significant increases in AEN or REN. However, with soil fertility improving and by gradually modifying the N management strategy, significant increases in yield and N-use efficiency were achieved in the three successive crops grown since the 1998 DS. Both AEN and REN were maintained at high levels and, compared to the first two SSNM crops, fertilizer N rates declined. The average yield increase over the FFP across all five crops was 1.2 t ha^{-1} crop^{-1} or a total production increase of 6 t ha^{-1} during 1997 to 1999. Accounting for the extra fertilizer cost in the SSNM, the average profit increase over FFP was $148 ha^{-1} crop^{-1} or a total of $742 ha^{-1} for all five crops grown.

Despite the higher soil fertility and yield in Mr. Villanueva's field, SSNM increased yield by an average of 0.6 t ha^{-1} on his farm. Nitrogen rates in the FFP and SSNM were similar, but the timing of N applications differed and about twice as much K was applied in SSNM. Our data also suggest that Mr. Villanueva applied too much P, on average 26 kg P ha^{-1} crop^{-1} versus 18 kg P ha^{-1} crop^{-1} in SSNM. High AEN and REN were achieved in the SSNM plot. Note that the lower AEN and REN in the 1999 DS resulted from unfavorable climatic conditions, but, in all four crops grown from 1997 to 1998, REN was maintained at 0.54 kg kg^{-1} and AEN at 18 kg kg^{-1}. Following SSNM, average profit increased by $90 ha^{-1} crop^{-1} or a total of $450 ha^{-1} for all five crops grown. These numbers illustrate what can be achieved in farmers' fields with location-specific crop management. They are about double that of current average N-use efficiencies.

Factors affecting the performance of SSNM

Average actual yields in SSNM were 78% of the model-predicted target yield in the 1997 WS, 75% in the 1998 DS, 77% in the 1998 WS, and 65% in the 1999 DS. In SSNM, six out of 27 farms were within 10% of the 6 t ha^{-1} yield target in the 1997 WS, but only two in the 1998 WS. Only two out of 27 farms were within 10% of the 8 t ha^{-1} yield target in the 1998 DS and none in the 1999 DS. The yield targets set for the SSNM treatment assumed average climatic conditions and yield potential, good water supply, and no pest losses, conditions that were achieved only on a few farms. Some of the yield losses were due to uncontrollable factors such as climate, others were due to variation in the quality of crop management.

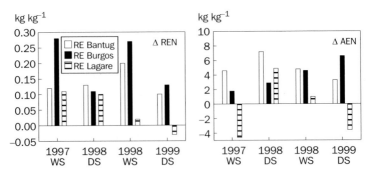

Fig. 8.7. Average increase in the recovery efficiency of N (Δ REN) and the agronomic N efficiency (Δ AEN) with site-specific nutrient management compared with the farmers' fertilizer treatment in the three survey villages, Bantug (n = 9), Burgos (n = 9), and Lagare (n = 9), in Nueva Ecija, Philippines, 1997-99.

The experiments were conducted during the El Niño–La Niña climatic cycle (1997-99), which resulted in lower than normal climatic yield potential, occasional problems with water supply (early or late water release), and nonoptimal planting dates. Crop simulation modeling indicated that potential yields in most crops grown in 1998 to 1999 were about 1 t ha^{-1} lower than in normal years. For Nueva Ecija, the mean climatic yield potential for a wet-season crop established in July is 9.4 t ha^{-1}. However, the average yield potential of the varieties used by the participating farmers was only 8.4 t ha^{-1} during the 1997 WS and 8.2 t ha^{-1} during the 1998 WS. The mean climatic yield potential for a dry-season crop established in December is around 11.0 t ha^{-1}. The average yield potential of the farmers' varieties was 9.8 t ha^{-1} during the 1998 DS and 8.7 t ha^{-1} during the 1999 DS. Yields declined significantly in the 1998 DS because of high temperatures at flowering causing increased spikelet sterility (only 73% filled spikelets), in the 1998 WS because of high rainfall during flowering and grain filling, and in the 1999 DS because of high rainfall and less than normal solar radiation. Major crop management problems observed included poor seed quality, heterogeneous crop establishment in direct-seeded rice, snail damage, weeds, diseases, and poor water management. In the 1997 WS, the coincidence of high temperature and humidity led to a late bacterial leaf blight (BLB) outbreak, which reduced grain filling to 70% on average. Severe yield losses because of BLB and weeds occurred on 12 farms. Yields in damaged SSNM plots were on average 1.3 t ha^{-1} lower than on the 15 farms with no or less severe problems. In 1998, water supply was limited because of El Niño phenomenon and fields often remained nonflooded during early growth stages, with more severe drought on at least five farms. At later stages, waterlogging and lodging caused by a typhoon during grain filling reduced yields on many farms. In the 1999 DS, unusual rat damage occurred at some sites.

An unresolved issue are differences among villages in crop management practices. Significant village differences in the effect of SSNM on AEN, REN, and PFPN

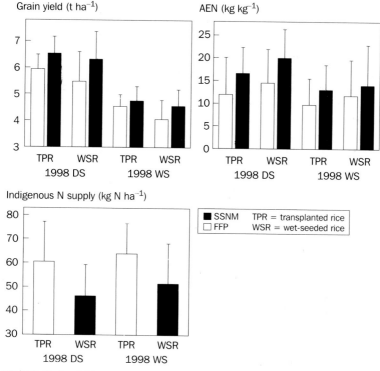

Fig. 8.8. Grain yield, agronomic N-use efficiency (AEN), and indigenous N supply (INS) in the farmers' fertilizer practice (FFP) and site-specific nutrient management (SSNM) plots (n = 27) as affected by crop establishment method. Bars: mean; error bars: standard deviation. TPR = transplanted rice, WSR = wet-seeded rice (broadcast).

were observed (Table 8.6). In general, increases in N-use efficiency because of SSNM were relatively consistent in Bantug and Burgos, but not in Lagare (Fig. 8.7). Reasons for the more erratic performance in Lagare remain uncertain. We observed, however, that farmers in Lagare frequently faced water shortage because of the ongoing repair of the local irrigation system in the last years. This may have contributed to lower rice yields than in the other two villages. Further, weed problems are more serious in Lagare than in Bantug and Burgos.

Interactions among crop establishment method, rooting depth, and nutrient uptake need further study because our data suggest differences in the performance of SSNM between transplanted (TPR) and wet-seeded rice (WSR). Average grain yields in TPR tended to be slightly larger than those in WSR, but this difference was not statistically significant (Fig. 8.8). However, the relative increases in grain yield with SSNM were always greater in WSR than in TPR. For example, in the 1998 DS, six farmers transplanted rice while the remaining 21 farmers wet-direct seeded. In TPR, average increases in grain yield vis-à-vis the control were 31% in the FFP and 45% in SSNM. In WSR, average increases in grain yields over the control were 51% in the

FFP and 74% in SSNM. Comparing SSNM with FFP, grain yield in SSNM exceeded that of the FFP by 10% in TPR and by 15% in WSR. Nitrogen-use efficiency tended to be higher in WSR than in TPR (Fig. 8.8), which was mainly an artifact of lower grain yield and N uptake without applied N (INS) in the WSR plots (Peng et al 1996b). However, fast and dense early plant growth in WSR favors recovery of topdressed applied N fertilizer. Visual observations in DS crops also indicated that P and K deficiency symptoms were much more pronounced in the FFP of WSR fields, perhaps because of the rapid early P and K uptake in WSR causing a fast depletion of soil resources in the upper few centimeters. It is likely that shallow soil tillage and the high plant density in WSR plots result in a shallower root system and a smaller soil volume from which nutrients are extracted by the plants.

8.3 Future improvements of SSNM

The SSNM approach increased yields and profit on both "poor" and "good" farms and, at the same time, should have positive effects on soil fertility and sustainability of the production system over the longer term. However, extrapolation to other domains and more widespread adoption will require significant improvements and simplifications (see Chapter 17). In Central Luzon, we will need an approach that is fine-tuned to the differences in crop establishment methods because both WSR and TPR are common.

Different plant-based N management strategies are required because of the differences in growth dynamics between TPR and WSR (Dingkuhn et al 1992, Peng et al 1996b, Schnier et al 1990a,b). Work is in progress to specify generalized strategies for applying N (see Chapter 6). However, little is known about the best methods for applying P and K in WSR. Many farmers do not incorporate P and K fertilizer in WSR fields, but surface-apply it within about 2 weeks after sowing. First-crop recovery efficiencies of surface-applied P and K in WSR are probably larger than commonly assumed recovery efficiencies of about 20% for P and 50% for K, but we need better measurements of this to improve our model-based fertilizer recommendations.

Developing objective criteria for setting a realistic yield goal in an SSNM approach remains a key issue for further research. The currently used assumptions of average climatic conditions and no yield reductions because of stresses other than nutrients are probably too optimistic for what can be achieved in farmers' fields. They may lead to excessive application of P and K fertilizer in some cases, although these nutrients will not be lost. Therefore, because there is no yield penalty associated with applying too much P or K, we may also regard this as a "crop insurance" or capital investment into soil fertility. This is of particular interest on soils with low to moderate indigenous supplies of P and K and on land owned by a farmer or rented for an extended period of time (+10 years). Extending this thought, it is probably sufficient to simplify P and K management by specifying field-specific rates valid for a period of several years instead of adjusting them after each crop as was done in our study.

8.4. Conclusions

Rice yield growth rates in Central Luzon have slowed down in recent years. There are signs that intensive, irrigated rice monoculture has depleted some soil nutrients such as P, K, and Zn in many farmers' fields because of the historical unbalanced fertilizer use. Potassium deficiency has probably become a serious constraint to increasing DS rice yields beyond about 6 t ha^{-1}, particularly in direct-seeded areas with high plant population densities. In addition, productivity, yield, N-use efficiency, and soil fertility vary widely among individual farmers and even within small domains with similar overall environmental characteristics.

We tested a site-specific nutrient management approach in four successive rice crops to manage the variability in indigenous soil nutrient supply on a field- and season-specific basis. In comparison to the farmers' fertilizer practice, average plant N and P uptake increased by almost 20% and plant K uptake by about 10% because of this dynamic nutrient management. Average rice yields increased by 11%. Average profit increased by $87 ha^{-1} crop^{-1}, even though more P and K fertilizer were applied in SSNM to achieve balanced nutrition and sustain soil fertility. Average REN increased by 39% compared to the farmers' practice in the same cropping season, suggesting enormous potential for improving N-use efficiency in farmers' fields and reducing N losses into the environment.

The true potential gain in yield, profit, and nutrient efficiency through SSNM is likely even larger than what was measured in our study. We conducted our experiments in a 2-year period with unfavorable climatic conditions related to the El Niño–La Niña climatic cycle so that the yield goals were not achieved.

Considering the good performance of SSNM in our study involving 27 farms and four consecutive rice crops, we recommend that this approach be adopted for larger irrigated rice areas in the Philippines. Key areas to study in further extrapolation and extension work include the development of (1) feasible methods for estimating INS, IPS, and IKS, (2) simplified guidelines for N, P, and K use, and (3) guidelines for timing and mode of applying N, P, and K that are specific for TPR and WSR grown in the DS and WS in the Philippines.

References

Alicante MM, Rosell DZ, Marfori RT, Hernandez S. 1948. Soil survey report of the Nueva Ecija province. Soil report 8. Manila (Philippines): Department of Agriculture and Commerce.

Cassman KG, Gines HC, Dizon M, Samson MI, Alcantara JM. 1996. Nitrogen-use efficiency in tropical lowland rice systems: contributions from indigenous and applied nitrogen. Field Crops Res. 47:1-12.

Cassman KG, Peng S, Dobermann A. 1997. Nutritional physiology of the rice plant and productivity decline of irrigated lowland rice systems in the tropics. Soil Sci. Plant Nutr. 43:1111-1116.

Cassman KG, Pingali PL. 1995. Extrapolating trends from long-term experiments to farmers' fields: the case of irrigated rice systems in Asia. In: Barnett V, Payne R, Steiner R, editors. Agricultural sustainability in economic, environmental, and statistical terms. London (UK): J. Wiley & Sons. p 64-84.

Dawe D, Moya P. 1999. Variability of optimal nitrogen applications for rice. In: Program report for 1998. Makati City (Philippines): International Rice Research Institute. p 15-17.

De Datta SK, Gomez KA, Descalsota JP. 1988. Changes in yield response to major nutrients and in soil fertility under intensive rice cropping. Soil Sci. 146:350-358.

Dingkuhn M, De Datta SK, Javellana C, Pamplona R, Schnier HF. 1992. Effect of late-season N fertilization on photosynthesis and yield of transplanted and direct-seeded tropical flooded rice. I. Growth dynamics. Field Crops Res. 28:223-234.

Dobermann A, Dawe D, Roetter R, Cassman KG. 2000. Reversal of rice yield decline in a long-term continuous cropping experiment. Agron. J. 92:633-643.

Dobermann A, Oberthür T. 1997. Fuzzy mapping of soil fertility: a case study on irrigated riceland in the Philippines. Geoderma 77:317-339.

Dobermann A, Pampolino MF, Adviento MAA. 1997. Resin capsules for on-site assessment of soil nutrient supply in lowland rice fields. Soil Sci. Soc. Am. J. 61:1202-1213.

Frère M, Popov GF. 1979. Agrometeorological crop monitoring and forecasting. FAO Plant Production and Protection Paper 17. Rome (Italy): FAO.

Fujisaka S. 1993. Were farmers wrong in rejecting a recommendation? The case of nitrogen at transplanting for irrigated rice. Agric. Syst. 43:271-286.

Gines HC. 1982. Paddy land suitability classification in relation to its potential for multiple cropping systems: a case study of the central plain of Luzon. Discussion Paper No. 115. Kyoto (Japan): The Center for Southeast Asian Studies, Kyoto University.

Kawaguchi K, Kyuma K. 1977. Paddy soils in tropical Asia: their material nature and fertility. Honolulu, Haw. (USA): The University Press of Hawaii. 258 p.

Kropff MJ, van Laar HH, ten Berge HFM. 1993. ORYZA1: a basic model for irrigated lowland rice production. Simulation and Systems Analysis for Rice Production (SARP) publication. Los Baños (Philippines): International Rice Research Institute.

Olk DC, Cassman KG, Mahieu N, Randall EW. 1998. Conserved chemical properties of young humic acid fractions in tropical lowland soil under intensive irrigated rice cropping. Eur. J. Soil Sci. 49:337-349.

Olk DC, Cassman KG, Simbahan GC, Sta. Cruz PC, Abdulrachman S, Nagarajan R, Tan PS, Satawathananont S. 1999. Interpreting fertilizer-use efficiency in relation to soil nutrient-supplying capacity, factor productivity, and agronomic efficiency. Nutr. Cycl. Agroecosyst. 53:35-41.

Peng S, Garcia FV, Gines HC, Laza RC, Samson MI, Sanico AL, Visperas RM, Cassman KG. 1996b. Nitrogen use efficiency of irrigated tropical rice established by broadcast wet-seeding and transplanting. Fert. Res. 45:123-134.

Peng S, Garcia FV, Laza RC, Sanico AL, Visperas RM, Cassman KG. 1996a. Increased N-use efficiency using a chlorophyll meter on high-yielding irrigated rice. Field Crops Res. 47:243-252.

Schnier HF, Dingkuhn M, De Datta SK, Mengel K, Faronilo JE. 1990a. Nitrogen fertilization of direct-seeded flooded vs. transplanted rice. I. Nitrogen uptake, photosynthesis, growth, and yield. Crop Sci. 30:1276-1284.

Schnier HF, Dingkuhn M, De Datta SK, Mengel K, Wijangco E, Javellana C. 1990b. Nitrogen economy and canopy carbon dioxide assimilation of tropical lowland rice. Agron. J. 82:451-459.

Tiongco M, Dawe D. 2000. Long-term productivity trends in a sample of Philippine rice farms. Poster presented at XXIVth International Conference of Agricultural Economists, Berlin, 13-18 Aug. 2000.

Witt C, Dobermann A, Abdulrachman S, Gines HC, Wang GH, Nagarajan R, Satawathananont S, Son TT, Tan PS, Tiem LV, Simbahan GC, Olk DC. 1999. Internal nutrient efficiencies of irrigated lowland rice in tropical and subtropical Asia. Field Crops Res. 63:113-138.

Notes

Authors' addresses: H.C. Gines, G.O. Redondo, A.P. Estigoy, Philippine Rice Research Institute (PhilRice), Maligaya, Muñoz, 3119 Nueva Ecija, Philippines, e-mail: asd@mozcom.com; A. Dobermann, International Rice Research Institute, Los Baños, Philippines, and University of Nebraska, Lincoln, Nebraska, USA.

Acknowledgments: We are particularly grateful to the farmers in Bantug, Burgos, and Lagare for their patience and excellent cooperation in conducting the on-farm experiments since 1994. We wish to thank Piedad Moya, Don Pabale, and Marites Tiongco (IRRI) for help with the socioeconomic farm monitoring, Ma. Arlene Adviento (IRRI) for help with the soil and plant analysis work, and Gregorio C. Simbahan, Rico Pamplona, Olivyn Angeles, Julie Mae Criste Cabrera-Pasuquin, and Edsel Moscoso (IRRI) for help with data management and statistical analysis. The authors are much obliged to Dr. Santiago Obien, Director of PhilRice and Irrigated Rice Research Consortium steering committee member, for his steady support of our research. We acknowledge contributions made by various other scientists involved in this research, including Segfredo R. Serrano, Edna M. Punzalan, Jocelyn Bajita, Fernando. D. Garcia, and Rolando T. Cruz of PhilRice and Pompe Sta. Cruz, Christian Witt, Kenneth G. Cassman, David Dawe, and Daniel C. Olk of IRRI. Last but not least, we are grateful to our field workers, Shiela Pablo, Mario Elliot, Cenen España, and Renato Dawang.

Citation: Dobermann A, Witt C, Dawe D, editors. 2004. Increasing productivity of intensive rice systems through site-specific nutrient management. Enfield, N.H. (USA) and Los Baños (Philippines): Science Publishers, Inc., and International Rice Research Institute (IRRI). 410 p.

9 | Site-specific nutrient management in intensive irrigated rice systems of West Java, Indonesia

S. Abdulrachman, Z. Susanti, Pahim, A. Djatiharti, A. Dobermann, and C. Witt

9.1 Characteristics of rice production in West Java

Trends

West Java (Fig. 9.1) is one of the most important rice-growing areas of Indonesia and represents a typical region with early adoption of intensification because of Green Revolution technologies. The annual irrigated rice harvest area is currently about 2 million ha, with an annual production of about 10 million t (Table 9.1). Growing conditions are generally favorable. Rice is mostly grown on soils with heavy texture, such as Ultisols, Inceptisols, and Alfisols, with varying degrees of aquic conditions and soil fertility. The humid, semihot equatorial climate with around 1,400–1,800

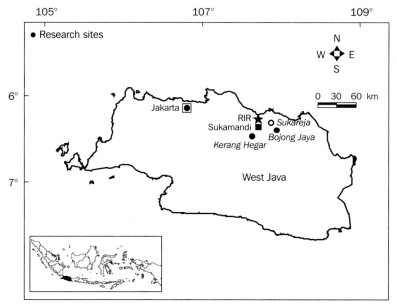

Fig. 9.1. Location of the experimental sites at Sukamandi, West Java, Indonesia.

Table 9.1. Changes in rice production in West Java, Indonesia.

Item	1976-80	1981-85	1986-90	1991-95	1996-2000
Total rice area (million ha)	1.80	1.93	2.08	2.08	2.14
Irrigated rice (%)	93.8	93.3	92.8	92.5	92.2
Total rice production (million t)	5.76	8.01	9.72	10.43	10.34
Irrigated rice yield (t ha $^{-1}$)	3.2	4.1	4.7	5.0	4.8
Fertilizer consumption (1,000 t)					
Fertilizers (N, P_2O_5, K_2O)	–	–	427.4	376.4	308.2[a]
N fertilizers (N)	–	–	266.9	260.0	238.4[a]
P fertilizers (P_2O_5)	–	–	134.4	102.4	60.9[a]
K fertilizers (K_2O)	–	–	26.1	14.0	8.9[a]

[a]Until 1998.
Sources: Statistical Year Book of Indonesia, Central Bureau of Statistics (1977, 1980-1981, and 1986), and Statistical Year Book of Indonesia, BPS Statistics Indonesia (1990 and 2000).

mm rainfall per year (Fig. 9.2) allows growing two to three crops per year. The most important cropping systems are either rice-rice or rice-rice-secondary crops. Two main rice-growing seasons are distinguished, a wet season (November-March) and a dry season (May-August). Most rainfall usually occurs during the WS, but rainfall in the DS can be substantial in certain years. In addition to rainfall, surface water from an extensive canal system is used for irrigation. In 1996-2000, solar radiation was generally lower in the WS than in the DS (20.9 vs. 18.9 MJ m^{-2} d^{-1}, Fig. 9.2) and the model ORYZA (Bouman et al 2001) predicted, on average, a 1.7 t ha^{-1} lower yield potential for IR72 in the WS (8.7 and 10.4 t ha^{-1}). Great variation in climatic conditions occurred among seasons and years, however, sometimes resulting in greater yield potential in the WS than in the DS, as in 1998 (9.8 vs. 8.5 t ha^{-1}). Note that solar radiation was obtained from sunshine hours using the standard coefficients for dry tropical conditions (Frère and Popov 1979, Supit et al 1994) and that some uncertainties are associated with the use of standard coefficients. Simulations showed that potential production would differ by up to 7% depending on the choice of the coefficients (Cabrera et al 1998).

Crop management practices include one plowing and two harrowing operations for common land preparation and crop establishment is mainly done by transplanting. Farmers harvest their rice by hand (sickle), which leaves medium-long stubble residues. All loose straw is burned in heaps after threshing is done on-site.

Irrigated rice area expanded slightly until the late 1980s, but both area and production have stagnated since then (Table 9.1). From 1975 to 1990, irrigated rice yields in West Java increased annually at 154 kg ha^{-1}, but the growth rate has declined markedly and yields have remained in the 4.5 to 5 t ha^{-1} range since then (Cassmann and Dobermann 2001, Abdulrachman et al 2002). The initial yield increase period was associated with the fast adoption of modern rice varieties, expansion of irrigated area, and steep rise in NPK fertilizer consumption. During the 1980s,

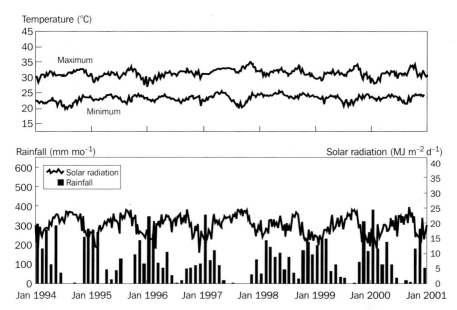

Fig. 9.2. Climatic conditions at Sukamandi, West Java, Indonesia, during 1994 to 2000. Solar radiation and temperature data are 7-day moving averages; rainfall is monthly total. Solar radiation was obtained from sunshine hours using the standard Ångström coefficients for humid tropical conditions of 0.29 for a and 0.45 for b (Frère and Popov 1979) for Sukamandi (6°20′S, 107°39′E).

government subsidies on fertilizers led to blanket recommendations for fertilizer use and their relatively high use. Soil nutrient surveys conducted thereafter suggested that soil P and K levels had increased substantially and it was recommended that on such soils P fertilizers could be applied only once in two to three cropping seasons (Sri Adiningsih et al 1991). These recommendations may have led to the decline in P and K fertilizer use from 1986-90 to 1991-95 (Table 9.1). Subsequently, the financial crisis that began in 1997 led to even sharper declines in fertilizer use, especially for P and K (see Chapter 3 by Moya et al for more information). Despite the worrisome yield stagnation, rice yields in West Java remain higher than in other parts of Indonesia and rival an average rice yield of 4.3 t ha^{-1} for the whole country (Anonymous 1998).

Earlier research conducted in Indonesia hypothesized that deteriorating soil quality, indiscriminate use of N fertilizer, and virtual abandonment of organic manuring may have caused stagnating or even declining productivity in intensive rice cropping. Ismunadji and Sudjadi (1983) reported that intensive fertilization for irrigated rice could stimulate iron toxicity. Isgiyanto et al (1992) also reported that soil organic matter deterioration, disturbance of soil physical condition, and micronutrient deficiency may be the factors causing stagnant rice production growth. However, no clear cause-effect relationships have been established to explain the productivity trends observed. "Technology packages" that promote fertilizer in the form of blanket doses

Table 9.2. General soil properties on 20 rice farms at Sukamandi, Indonesia.[a]

Soil properties	Min.	25%	Median	75%	Max.
Clay content (%)	38.0	44.0	67.0	74.0	80.0
Silt content (%)	19.0	25.0	31.0	41.0	53.0
Sand content (%)	1.0	1.0	2.0	9.0	19.0
Soil organic C (g kg^{-1})	7.9	9.3	14.0	15.4	18.5
Total soil N (g kg^{-1})	0.78	0.99	1.38	1.50	1.60
Soil pH (1:1 H$_2$O)	4.7	4.8	5.2	5.5	6.5
Cation exchange capacity (cmol$_c$ kg^{-1})	12.9	16.0	34.5	36.4	39.2
Exchangeable K (cmol$_c$ kg^{-1})	0.08	0.20	0.46	0.58	0.74
Exchangeable Na (cmol$_c$ kg^{-1})	0.27	0.34	1.23	1.67	2.96
Exchangeable Ca (cmol$_c$ kg^{-1})	6.59	9.46	14.40	20.30	22.30
Exchangeable Mg (cmol$_c$ kg^{-1})	1.89	2.79	10.00	10.90	12.30
Extractable P (Olsen-P, mg kg^{-1})	2.00	3.30	6.00	8.90	13.00
Extractable Zn (0.05N HCl, mg kg^{-1})	0.96	1.20	1.60	2.00	3.00

[a]Measured on soil samples collected before the 1998 dry season.

across large areas appear to have limitations for regaining momentum in yield and production growth in West Java. On-farm research conducted from 1994 to 1996 confirmed the existence of substantial spatial variability in soil nutrient supply, fertilizer use, and nutrient-use efficiency (Olk et al 1999). Therefore, sustaining high rice productivity may require more site-specific nutrient management (SSNM) approaches to replace the regional blanket recommendations. The objectives of this study were to (1) quantify the variability of rice productivity and soil fertility and (2) develop and test a new SSNM approach under the specific conditions of West Java.

Current biophysical and socioeconomic farm characteristics

The experimental domain of the RTDP project is located around Sukamandi (6°21'S, 107°40'E) in Subang, West Java (Fig. 9.1). To conduct a more detailed analysis of current farm productivity and its variation among farms, on-farm monitoring experiments began in 1995. The experimental approaches were the same as described in Chapter 2. Twenty farms were used in this study, located in three villages (Karang Hegar, Sukareja, and Bojongjaya) about 15 to 20 km away from the Research Institute for Rice at Sukamandi, West Java (Fig. 9.1). Soil types differed among the three sites and ranged from acid, low-CEC Ultisols (Latosols) at Karang Hegar to more fertile Inceptisols (gley humic soil) at Sukareja and Entisols (alluvial soil) at Bojong Jaya. All soils had high clay content of typically 40% to 75% (Table 9.2). Although the median exchangeable K content was 0.46 cmol kg^{-1}, it was low (<0.2 cmol kg^{-1}) at about 25% of all sites. Contrary to the earlier soil fertility surveys (Sri Adiningsih et al 1991), relatively low Olsen-P content of less than 10 mg kg^{-1} was measured at about 80% of all experimental sites. Available zinc was above the critical level of 1 mg kg^{-1} on all farms (Table 9.2).

Table 9.3. Demographic and economic characteristics of rice production on 20 farms at Sukamandi, West Java. Values shown are based on socioeconomic farm surveys conducted for whole farms.

Production characteristics	Min.	25%	Median	75%	Max.
Total cultivated area (ha)	0.16	0.46	0.62	0.90	2.84
Age of household head (y)	27	32	37	54	81
Education of household head (y)	1	6	6	9	12
Household size (persons)	2	3	3	5	8
1998 dry-season crop					
Rice area (ha)[a]	0.15	0.33	0.52	0.78	2.84
Yield (t ha^{-1})	1.11	1.46	2.54	4.03	5.18
N fertilizer (kg ha^{-1})	46	79	103	151	234
P fertilizer (kg ha^{-1})	0	0	0	5	32
K fertilizer (kg ha^{-1})	0	0	0	0	31
Insecticide (kg ai ha^{-1})	0.01	0.54	1.22	1.87	2.73
Herbicide (kg ai ha^{-1})	0.0	0.0	0.17	0.83	1.61
Other pesticides (kg ai ha^{-1})[b]	0.0	0.008	0.01	1.37	3.22
Total labor (8-h d ha^{-1})	80	103	142	202	295
Net return from rice (US$ ha^{-1} crop^{-1})	−76	40	165	371	491
1998 wet-season crop					
Rice area (ha)[a]	0.15	0.24	0.52	0.78	2.84
Yield (t ha^{-1})	4.17	4.96	5.67	6.17	8.59
N fertilizer (kg ha^{-1})	44	80	99	122	147
P fertilizer (kg ha^{-1})	0	0	10	15	28
K fertilizer (kg ha^{-1})	0	0	0	0	25
Insecticide (kg ai ha^{-1})	0	0.02	0.27	0.61	2.70
Herbicide (kg ai ha^{-1})	0	0.30	0.79	1.33	1.64
Other pesticides (kg ai ha^{-1})[b]	0	0	0	0	0
Total labor (8-h d ha^{-1})	96	119	159	208	317
Net return from rice (US$ ha^{-1} crop^{-1})	164	216	226	282	455

[a]Rice area in which the treatment plots were embedded in subsequent years. The total rice area may even be larger. [b]Includes fungicide, molluscicide, rodenticide, and crabicide.

Table 9.3 shows the variation in demographic and economic characteristics of rice production among the 20 farms in 1998. Farm sizes ranged from 0.16 to 2.84 ha (median 0.62 ha), with 75% landowners, 12.5% leaseholders, and 12.5% share tenants. More than 80% of all farmers manage less then 1 ha of land and both education level and household size cover a wide range. The highest level of farmers' education is senior high school, but many farmers have only an elementary school education. The median age of household heads was 37. Labor input in West Java is generally high and typically ranges from about 100 to 200 8-h person-days ha^{-1}. Much of this is associated with labor-intensive activities such as transplanting, hand-weeding, and manual harvest.

Table 9.4. Baseline agronomic characteristics of rice production on 20 farms at Sukamandi, West Java. Values shown are means and standard deviations (SD) of the farmers' fertilizer practice for two consecutive rice crops monitored before SSNM plots were established.

Agronomic characteristics[a]	1995 DS crop		1996 WS crop	
	Mean	SD	Mean	SD
Grain yield (t ha [1])	3.92	0.92	4.95	0.49
Harvest index	0.48	0.04	0.47	0.05
No. of panicles m [2]	207	36	208	25
Total no. of spikelets m [2]	21,354	3,614	20,262	3,637
Total no. of spikelets panicle [1]	104	10	98	17
Filled spikelets (%)	84	4	87	6
Fertilizer N use (kg ha [1])	135.8	31.2	131.8	52.9
Fertilizer P use (kg ha [1])	15.5	15.6	10.0	11.7
Fertilizer K use (kg ha [1])	18.7	24.3	14.1	28.7
N uptake (kg ha [1])	87.2	23.7	99.5	15.5
P uptake (kg ha [1])	7.8	1.8	14.1	2.5
K uptake (kg ha [1])	86.6	20.1	81.7	21.0
Input-output N balance (kg ha [1] crop [1])	−10.8	24.1	−15.7	23.9
Input-output P balance (kg ha [1] crop [1])	7.9	15.9	−3.8	12.2
Input-output K balance (kg ha [1] crop [1])	−33.3	24.9	−36.2	33.4
Partial productivity of N (kg kg [1])	30.6	10.8	45.2	21.6
Agronomic efficiency of N (kg kg [1])	5.2	4.8	7.8	7.3
Recovery efficiency of N (kg kg [1])	0.14	0.09	0.19	0.14
Physiological efficiency of N (kg kg [1])	32.2	18.2	33.2	21.9

Farmers prefer rice varieties such as IR64, IR42, Muncul, Way Apoburu, and Cisadane. Urea and ammonium sulfate are common sources for N, single superphosphate for P, and muriate of potash for K. Baseline agronomic characteristics of rice production are shown for the two rice crops grown in 1995-96, before the onset of the economic crisis in Indonesia (Table 9.4). At that time, farmers applied significantly larger amounts of fertilizer than after 1997 (Table 9.3), but wide variation in NPK rates and rice yields was observed. Crop N and P uptake were generally positively correlated with grain yield, but the estimated average K input-output nutrient balance was highly negative. Fertilizer-N-use efficiency was usually low. Agronomic efficiency of N (AEN) averaged 5.2 kg kg^{-1} in the 1995 DS and 7.8 kg kg^{-1} in the 1996 WS. The average recovery efficiency of fertilizer N (REN) was less than 20% in both cropping seasons; the average physiological N-use efficiency (PEN) was 33 kg kg^{-1}. These values are well below achievable levels and they suggest a lack of congruence of N supply and crop N demand as well as the presence of other factors that reduce the yield response to fertilizer N (Dobermann and Fairhurst 2000). Because of the increase in fertilizer prices in 1997-99, farmers applied much smaller amounts of fertilizer than before. Nitrogen rates averaged about 100 kg N ha^{-1} per rice crop and more than 75% of all farmers did not apply any K. The median P rate decreased to 0 in the dry season and 10 kg P ha^{-1} in the wet season (Table 9.3). Because no organic manure is applied to rice and surface water with low nutrient content is used for irrigation, negative P and K input-output balances have probably become common in recent years. The canal water used for irrigation was analyzed at the three project sites in the 1999 WS and the 2000 DS. The phosphate content of the canal water was negligible and the K content averaged 1.84 mg K L^{-1} (median, n = 24). Even at high irrigation water use of 1,000 to 1,500 L m^{-2} crop^{-1}, for instance in a DS on soil with higher percolation rates, K inputs would range from only 20 to 30 kg K ha^{-1} crop^{-1}, which may not be sufficient to balance K removed with grain and straw.

Pest surveys conducted in 1998 indicated relatively high levels of pest incidence at the on-farm sites around Sukamandi. Rats, stem borer, grain discoloration, and various diseases (stem rot, red stripe, narrow brown spot, sheath blight) were the most important pest problems in this area (Sta. Cruz et al 2001). In other years, yield losses were also caused by brown planthopper (BPH), leaffolder, rice bug, and golden apple snail. Rat damage has become particularly severe in recent years. Grain yield was higher in the WS than in the DS in 1997-99 although the model ORYZA (Bouman et al 2001) predicted a higher yield potential for the DS (10.1 vs. 8.8 t ha^{-1}) based on IR72. The actual average number of days to maturity was 95 in the DS and 100 in the WS, whereas ORYZA predicted 97 d for the DS and 95 d for the WS based on IR72. This discrepancy is not fully understood since farmers used the same varieties in the WS and DS. There may have been positive effects of the longer, dry fallow periods prior to the WS crops on nutrient availability, nutrient uptake, and plant growth. Differences in soil aeration status are further enhanced because of differences in management practices before the two seasons. Farmers plow their fields (0 to 15-cm soil layer) prior to the WS but not the DS, whereas other tillage practices are the same (use of rotary tiller, harrowing). We can only speculate that this may increase the availability of nutrients, stimulate soil N mineralization, and create more favorable conditions for root growth. There is also evidence that the rainwater contributes considerably to plant N nutrition in the WS but not in the DS (S. Abdulrachman, personal communication). In 1998, DS yields and net return from rice production were particularly low because of unusual rainy and hot weather, delayed planting of the DS crop by 2 to 4 weeks, severe pest problems, and increasing costs for fertilizers (by 150%) and pesticides (by 400%). The average DS grain yield was only 2.54 t ha^{-1}.

Figure 9.3. shows the variability of the indigenous N (INS), P (IPS), and K (IKS) supplies among 20 farmers' fields as measured from 1997 to 1999. Differences among the three to four rice crops are mainly attributed to seasonal fluctuations in climate and pest incidence, which were strongly affected by El Niño phenomenon. However, within each season, ranges of indigenous N, P, and K supplies were large and differences also occurred among the three villages. The highest

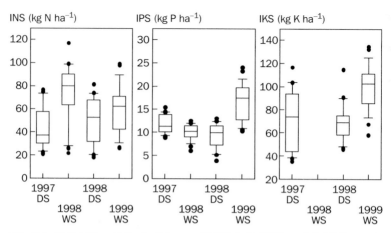

Fig. 9.3. Variability of the indigenous N (INS), P (IPS), and K (IKS) supply among 20 farmers' fields at Sukamandi, Indonesia (1997-99). Median with 10th, 25th, 75th, and 90th percentiles as vertical boxes with error bars; outliers as bullets.

values were generally found on the gley humic and alluvial soil at Sukareja and Bojong Jaya. Delayed water availability, weeds, and rats caused poor growth of the 1997 DS crop, which probably explains the lower than normal estimates of INS (average of 44 kg N ha^{-1}) in this season. This crop was followed by an abnormally long and dry fallow period (Fig. 9.2), causing an increase in the average INS to 73 kg N ha^{-1} in the 1997-98 WS crop, similar to values observed in 1995 and 1996 (Olk et al 1999). The average INS measured in the 1998 DS and 1998-99 WS was 51 and 59 kg N ha^{-1}, respectively.

Although it was commonly thought that lowland rice soils on Java have been overfertilized with P (Sri Adiningsih et al 1991), crop-based measurements of IPS suggested only low to moderate levels. Average IPS ranged from about 10 to 16 kg P ha^{-1} in the four crops sampled (Fig. 9.3). Coefficients of variation (CV) in IPS among farms were lower than for INS and IKS (typically about 15% to 20%), which may result from a history of similar P application because of subsidies on P in Indonesia practiced until a few years ago. Values of IKS varied widely among farms as well as seasons (Fig. 9.3) for reasons that are not clearly understood. The average levels of IPS and IKS found on most farms would be sufficient to support rice yields of only 5 to 6 t ha^{-1} without P or K application and these nutrients may represent constraints to regaining momentum in yield growth rates beyond such levels. This is further aggravated by the fact that nutrient inputs from sources such as irrigation, rainfall, or sediments are probably small in this domain.

9.2 Effect of SSNM on productivity and nutrient-use efficiency

Management of the SSNM plots

Beginning with the 1997 DS crop, a site-specific nutrient management (SSNM) plot was established in each of the 20 farm fields as a comparison with the farmers' fertilizer practice (FFP, see Chapter 2). The initial SSNM approach was tested over a period of four consecutive rice crops as described for other sites (see Chapter 5).

The size of the SSNM plots ranged from 270 to 520 m^2, depending on the size of the farmers' fields in which the SSNM and FFP plots were located. Rice varieties were chosen by the farmers. In general, farmers planted conventional modern varieties, mostly IR64 and some IR42, Way Apoburu, Ketan, or Muncul. Average plant density was only 13 to 14 hills m^{-2} (27 × 27-cm spacing of hills) using 25–30-d-old seedlings and 3–4 seedlings hill^{-1}. Farmers did all water management and pest control in both FFP and SSNM plots following the commonly adopted methods. However, severe pest damage was observed in the first crop (1997 DS). Since then, on each farm, a plastic barrier was installed to surround all treatments, including the FFP, to reduce rat damage. Carbofuran, bensultrap, deltametrin, fenpropatrin, fipronil, and chlorpyrifos were used as pesticides for controlling pests such as BPH, stem borer, leaffolder, golden snail, black bug, and leaf blights.

Fertilizer applications for SSNM were prescribed on a field- and crop-specific basis following the approach described in Chapter 5. In the 1997 DS, average values of the INS measured in the 1995 and 1996 crops were used as model input. Similarly, the average IPS and IKS were estimated from the plant P and K accumulation in the FFP sampled during this period, assuming recovery fractions of 0.2 kg kg^{-1} for fertilizer P and 0.4 kg kg^{-1} for fertilizer K. Beginning with the 1998 WS crop, estimates of INS, IPS, and IKS were continuously improved by incorporating the values measured in omission plots in each year and by adjusting IPS and IKS according to the actual P and K input-output balance. Target yields for working out field-specific fertilizer rates ranged from 4.5 to 6.5 t ha^{-1} in the DS and 5.5 to 7 t ha^{-1} in the WS, but varied among villages depending on soil differences and the water release and planting schedule in the irrigation system. Yield targets were generally set lowest for the poorer soils with early planting at Karang Hegar, where other constraints such as weeds, rats, and insects were also severe. The climatic yield potential was set to 8 to 9 t ha^{-1} for the WS and 6.5 to 7.5 t ha^{-1} for the DS, based on earlier crop simulations and maximum yields achieved in other studies. Selected yield potentials were probably suitable for the period of measurements, although recent yield data obtained from farmers' fields indicate that WS and DS yield trends are not as consistent as expected. Thus, the above given yield potentials may not hold for this site. First-crop recovery fractions of 0.4 to 0.5, 0.2, and 0.4 to 0.5 kg kg^{-1} were assumed for fertilizer N, P, and K, respectively. Urea and ammonium sulfate were used as N sources, single superphosphate for P, and muriate of potash for K. None of the farmers applied organic manure. All P and 50% of the K fertilizer were incorporated in the soil before sowing or planting. Another 50% of the K rate was topdressed at panicle initiation (PI).

In the first season, the 1997 DS, N was applied based on INS and chlorophyll meter (SPAD) readings. Basal N at 20–30 kg N ha^{-1} was applied if INS was less than 70 kg N ha^{-1}, that is, on all nine farms at Karang Hegar. At 14 days after transplanting (DAT), all SSNM plots received 40–50 kg N ha^{-1}. After 14 DAT, SPAD readings were taken weekly, but not used to adjust N rates. At Karang Hegar, all SSNM plots received 80–90 kg N ha^{-1} at PI, and at Bojong Jaya and Sukareja a standard dose of 106 kg N ha^{-1} was applied at this stage. This management regime resulted in high total N application of 150 to 160 kg N ha^{-1}, causing low N-use efficiency and increased pest incidence. Therefore, the N management scheme was gradually improved over time to incorporate SPAD readings into the decision making to improve the congruence between N supply and crop N demand. In the fourth crop (1999 WS), no preplant N was applied and the scheme used was

FN 1	14–20 DAT		30 kg N ha^{-1}
FN 2	21–35 DAT	If SPAD >37	0 kg N ha^{-1}
		If SPAD <37	30 kg N ha^{-1}
FN 3	42–49 DAT (PI)	If SPAD >37	0 kg N ha^{-1}
		If SPAD 35–37	40 kg N ha^{-1}
		If SPAD <35	50 kg N ha^{-1}
FN 5	56–65 DAT (FF)	If SPAD >37	0 kg N ha^{-1}
		If SPAD 35–37	20 kg N ha^{-1}
		If SPAD <35	30 kg N ha^{-1}

In this approach, the SPAD meter was not used to make a decision about when to apply, but only about how much N to apply at a certain growth stage. Nitrogen management was done similarly in the 2000 DS and 2001 WS crop, but with slightly different SPAD thresholds (34 instead of 35 and 36 instead of 37) and following a typical N schedule proposed for tropical transplanted rice (Dobermann and Fairhurst 2000).

Effect of SSNM on grain yield and nutrient uptake

Compared with the FFP plots sampled in the same season, SSNM did not significantly increase grain yield from 1997 to 1999 (Fig. 9.4, Table 9.5). Although average WS yield (5.13 t ha^{-1}) was higher than DS yield (3.85 t ha^{-1}), there were no significant differences in the performance of SSNM among DS and WS crops (nonsignificant crop-season effect). Average yield differences between SSNM and FFP increased from 0.14 t ha^{-1} in year 1 to 0.29 t ha^{-1} in year 2, but the crop-year effect was not significant ($P = 0.234$). Yield increases larger than 0.4 t ha^{-1} were observed in only about 25% of all cases. Yields were particularly low in the 1998 DS (3.1 t ha^{-1} in SSNM, Fig. 9.4).

On average, SSNM increased plant N uptake by 8.9 kg N ha^{-1} (10%, $P = 0.03$, Table 9.5) vis-à-vis the farmers' practice. Increases in plant N of more than 15 kg N ha^{-1} were observed in about 30% of all cases, illustrating the potential for using a more crop-based approach of N management. Increases in N uptake were larger in

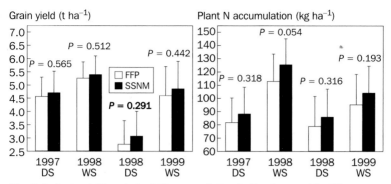

Fig. 9.4. Grain yield and plant N accumulation in the farmers' fertilizer practice (FFP) and site-specific nutrient management (SSNM) plots at Sukamandi, Indonesia (1997-99; bars: mean; error bars: standard deviation).

WS crops (10.9 kg N ha^{-1}) than in DS crops (6.8 kg N ha^{-1}), but the crop-season effect was not statistically significant ($P = 0.255$) because of large variability among sites (Fig. 9.4). Site-specific management in the West Java domain did not significantly increase crop P and K uptake (data not shown). The average difference in P uptake between SSNM and FFP was 1.5 kg P ha^{-1} crop^{-1} and that for K was 2.5 kg K ha^{-1}. However, larger differences of more than 2 kg P ha^{-1} or 10 kg K ha^{-1} crop^{-1} were observed in about 30% of all cases.

In summary, results obtained for the best 25% to 30% of all cases indicate the potential for increasing yields and nutrient uptake by rice in West Java, but crop growth was mostly constrained by factors other than those managed through SSNM. Poor weather, insufficient plant density, multiple pest problems, El Niño, and socioeconomic problems caused a poor average performance of SSNM (see below).

Effect of SSNM on nitrogen-use efficiency

Significant increases in N-use efficiency were achieved through the field- and season-specific N management practiced in SSNM (Fig. 9.5). Average REN was 0.46 kg kg^{-1} in the SSNM treatment, which represents a 0.15 kg kg^{-1} (48%, $P = 0.000$) increase in REN over that achieved in the FFP in the same years (Table 9.5). In about 20% of all cases, that increase was even larger than 0.25 kg kg^{-1}. Similarly, average AEN increased from 9.2 kg kg^{-1} in the FFP to 13.0 kg kg^{-1} in SSNM plots (41%, $P = 0.001$). No significant increase in PEN was observed (Fig. 9.5), but the partial factor productivity of N (PFPN) increased by 21% ($P = 0.012$). Average PFPN in SSNM was about 48 kg kg^{-1}, a level that comes close to what should be achieved in irrigated rice (Dobermann and Fairhurst 2000).

However, significant crop-year-season effects suggest that the performance of SSNM with regard to N-use efficiency improved greatly after the first crop grown. Because of excessive N rates applied and a splitting scheme that was not based on actual SPAD readings, N-use efficiency in the first crop (1997 DS) was generally low and the same in both SSNM and FFP. As described above, an improved N manage-

Table 9.5. Effect of site-specific nutrient management on agronomic characteristics at Sukamandi, Indonesia (1997-99).

Parameters	Levels[a]	Treatment[b] SSNM	Treatment[b] FFP	Δ[c]	P>|T|[c]	Effects[d]	P>|F|[d]
Grain yield (GY)	All	4.51	4.29	0.22	0.282	Village	0.307
(t ha^{-1})	Year 1	5.07	4.93	0.14	0.430	Year[e]	0.234
	Year 2	3.97	3.68	0.29	0.345	Season[e]	0.550
	DS	3.85	3.62	0.23	0.402	Year × season[e]	0.992
	WS	5.13	4.93	0.20	0.337	Village × crop	0.956
Plant N uptake (UN)	All	101.3	92.4	8.9	0.030	Village	0.669
(kg ha^{-1})	Year 1	107.9	98.1	9.7	0.109	Year	0.669
	Year 2	95.1	87.0	8.1	0.125	Season	0.255
	DS	87.0	80.2	6.8	0.153	Year × season	0.618
	WS	114.9	104.0	10.9	0.037	Village × crop	0.873
Agronomic efficiency	All	13.0	9.2	3.8	0.001	Village	0.947
of N (AEN)	Year 1	13.0	10.9	2.1	0.143	Year	0.022
(kg grain kg^{-1} N)	Year 2	12.9	7.6	5.3	0.001	Season	0.128
	DS	10.5	8.0	2.6	0.031	Year × season	0.199
	WS	15.2	10.3	4.9	0.005	Village × crop	0.437
Recovery efficiency	All	0.46	0.31	0.15	0.000	Village	0.850
of N (REN)	Year 1	0.44	0.32	0.12	0.002	Year	0.258
(kg plant N kg^{-1} N)	Year 2	0.48	0.30	0.18	0.000	Season	0.011
	DS	0.35	0.26	0.10	0.000	Year × season	0.014
	WS	0.56	0.36	0.20	0.000	Village × crop	0.437
Physiological	All	28.8	28.7	–0.1	0.961	Village	0.680
efficiency of N	Year 1	31.4	33.6	–3.2	0.532	Year	0.003
(PEN) (kg grain	Year 2	26.4	24.0	2.8	0.325	Season	0.071
kg^{-1} N)	DS	30.8	29.8	1.6	0.767	Year × season	0.881
	WS	26.9	27.7	–1.6	0.759	Village × crop	0.854
Partial productivity	All	47.9	39.7	8.1	0.012	Village	0.170
of N (PFPN)	Year 1	48.1	42.6	5.5	0.239	Year	0.264
(kg grain kg^{-1} N)	Year 2	47.7	37.0	10.6	0.018	Season	0.060
	DS	34.2	29.8	4.3	0.193	Year × season	0.011
	WS	60.9	49.1	11.7	0.004	Village × crop	0.008

[a]All = all four crops grown from 1997 DS to 1999 WS; year 1 = 1997 DS and 1998 WS; year 2 = 1998 DS and 1999 WS; WS = 1998 WS and 1999 WS; DS = 1997 DS and 1998 DS. [b]FFP = farmers' fertilizer practice; SSNM = site-specific nutrient management. [c]Δ = SSNM – FFP. P>\|T\| = probability of a significant mean difference between SSNM and FFP. [d]Source of variation of analysis of variance of the difference between SSNM and FFP by farm; P>\|F\| = probability of a significant F-value. [e]Year refers to two consecutive cropping seasons.

ment scheme was introduced thereafter, resulting in a large decrease in N rates (Fig. 9.6) and significant increases in N-use efficiency (Fig. 9.5). Average REN in SSNM was high in both WS crops (0.56 kg kg^{-1}) vis-à-vis 0.36 kg kg^{-1} in the FFP (56% difference, $P = 0.000$). Differences in REN were smaller in the low-yielding DS crops, but still about 35% greater with SSNM, and this difference was statistically significant ($P = 0.000$). However, PEN (increase in grain yield per increase in plant N) was usually low for both SSNM and FFP, ranging from 24 to 31 kg grain kg^{-1} N

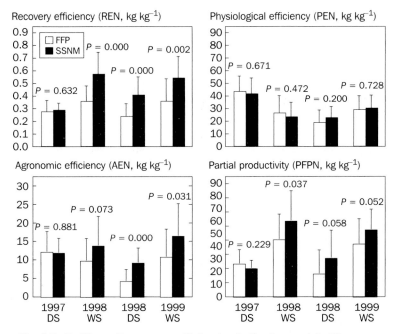

Fig. 9.5. Fertilizer nitrogen-use efficiencies in the farmers' fertilizer practice (FFP) and site-specific nutrient management (SSNM) plots at Sukamandi, Indonesia (1997-99; bars: mean; error bars: standard deviation).

(Table 9.5, Fig. 9.5). In healthy rice crops, PEN should be close to 50 kg grain kg^{-1} taken up from fertilizer, indicating that the plant's ability to transform nutrient uptake into grain yield was negatively affected during the period of observation, which was further supported by low internal N-use efficiencies (IEN). The average IEN was 44.6 kg grain kg^{-1} plant N in SSNM and 46.4 kg kg^{-1} for the FFP, which is substantially lower than the optimal IEN of about 68 kg kg^{-1} that was achieved in the absence of stress (Witt et al 1999). Agronomic efficiencies were also low in all treatments (Table 9.5) as compared with optimal values of $\geq$ 20 kg grain kg^{-1} fertilizer N applied (see Chapter 2).

In summary, N-use efficiency was significantly higher in SSNM than in the FFP because of a combination of better adjustment of N rates according to INS levels as well as better timing of split applications. The average number of N applications in the FFP was 1.6 for DS crops and 2.0 for WS crops. This compares with 2.3 (DS) and 2.4 (WS) N applications in SSNM plots. A more balanced N:P:K nutrition may have contributed to increased N-use efficiency as well, but the effect cannot be clearly separated from the N timing effect. Nutrient uptake appeared to be sufficient to support higher yields than the observed yields, but transformation of plant nutrient uptake into grain yield was inefficient in both SSNM and FFP because of crop, pest, and water management problems (see below).

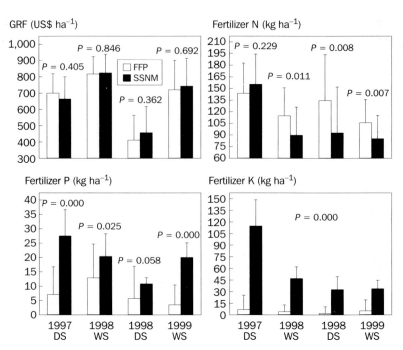

GRF (US$ ha⁻¹)

Fertilizer N (kg ha⁻¹)

Fertilizer P (kg ha⁻¹)

Fertilizer K (kg ha⁻¹)

Fig. 9.6. Gross returns above fertilizer costs and fertilizer use in the farmers' fertilizer practice (FFP) and site-specific nutrient management (SSNM) plots at Sukamandi, Indonesia (1997-99; bars: mean; error bars: standard deviation).

Effect of SSNM on fertilizer use and profit

Site-specific nutrient management decreased N use but increased P and K rates vis-à-vis the farmers' practice. Only small yield increases resulted in nonsignificant average profit increases over the FFP (Table 9.6, Fig. 9.6). On average, about 20 kg N ha⁻¹ less was applied in SSNM than in the FFP. However, this difference changed from +14 kg ha⁻¹ in the first crop to –32 kg ha⁻¹ for all three crops grown thereafter, when crop-based N management was practiced (Fig. 9.6). Despite much lower yields, farmers applied higher N rates in the DS (139 kg ha⁻¹) than in the WS (110 kg ha⁻¹), probably in attempts to recover or improve poor-looking rice crops. Clearly, for current yield levels and standard cropping practices, N rates at Sukamandi can be decreased significantly without a yield penalty provided that a better timing of N applications occurs.

Both fertilizer-P and -K rates in SSNM were much higher than in the FFP, but decreased over time (Fig. 9.6). On average across all four crops, application differences between SSNM and FFP were about 12 kg P ha⁻¹ and 51 kg K ha⁻¹ crop⁻¹ and, despite less N use, the total fertilizer cost increased by about US$23 ha⁻¹ crop⁻¹ (Table 9.6). Note, however, that average P and K rates applied by the farmers reached very low levels during the experimental period (only 4.7 kg P ha⁻¹ and 3.6 kg K ha⁻¹ in year 2), mainly because of sharply increasing fertilizer prices. Such rates are

Table 9.6. Effect of site-specific nutrient management on fertilizer use, fertilizer cost, and gross returns above fertilizer costs from rice production at Sukamandi, Indonesia (1997-99).

Parameters	Levels[a]	Treatment[b] SSNM	Treatment[b] FFP	Δ[c]	$P > \|T\|$[c]	Effects[d]	$P > \|F\|$[d]
N fertilizer (FN)	All	105	124	−19.7	0.002	Village	0.066
(kg ha^{-1})	Year 1	121	129	−7.6	0.388	Year[e]	0.005
	Year 2	89	120	−31.1	0.000	Season[e]	0.362
	DS	122	139	−16.5	0.107	Year × season[e]	0.000
	WS	87	110	−22.7	0.000	Village × crop	0.076
P fertilizer (FP)	All	19.5	7.4	12.2	0.000	Village	0.130
(kg ha^{-1})	Year 1	23.8	10.2	13.6	0.000	Year	0.080
	Year 2	15.4	4.7	10.8	0.000	Season	0.542
	DS	18.8	6.4	12.4	0.000	Year × season	0.000
	WS	20.2	8.2	11.9	0.000	Village × crop	0.363
K fertilizer (FK)	All	55.6	4.5	51.1	0.000	Village	0.506
(kg ha^{-1})	Year 1	79.1	5.5	73.7	0.000	Year	0.000
	Year 2	33.3	3.6	29.7	0.000	Season	0.000
	DS	71.6	4.3	67.3	0.000	Year × season	0.000
	WS	40.4	4.7	35.8	0.000	Village × crop	0.005
Fertilizer cost	All	63.4	40.7	22.7	0.000	Village	0.093
(US$ ha^{-1})	Year 1	78.9	45.2	33.7	0.000	Year	0.000
	Year 2	49.5	36.7	12.9	0.000	Season	0.003
	DS	71.4	43.7	27.7	0.000	Year × season	0.000
	WS	56.2	38.0	18.2	0.000	Village × crop	0.025
Gross returns above	All	676	662	13.9	0.680	Village	0.347
fertilizer costs	Year 1	760	768	−8.4	0.791	Year	0.012
(US$ ha^{-1})	Year 2	601	567	33.9	0.498	Season	0.886
	DS	556	542	13.2	0.775	Year × season	0.083
	WS	784	770	14.5	0.670	Village × crop	0.853

[a]All = all four crops grown from 1997 DS to 1999 WS; year 1 = 1997 DS and 1998 WS; year 2 = 1998 DS and 1999 WS; WS = 1998 WS and 1999 WS; DS = 1997 DS and 1998 DS. [b]FFP = farmers' fertilizer practice; SSNM = site-specific nutrient management. [c]Δ = SSNM − FFP. $P > \|T\|$ = probability of a significant mean difference between SSNM and FFP. [d]Source of variation of analysis of variance of the difference between SSNM and FFP by farm; $P > \|F\|$ = probability of a significant F-value. [e]Year refers to two consecutive cropping seasons.

unsustainable over a longer period so that the difference for SSNM is likely to be smaller under more normal socioeconomic conditions.

After an initially high dose applied to the 1997 DS crop to adjust for low IPS and IKS, the P and K rates in the SSNM treatment declined and the difference in fertilizer cost decreased from $34 ha^{-1} in year 1 to just $13 ha^{-1} in year 2. This was accompanied by a shift in profitability, although the average differences in gross returns above fertilizer costs (ΔGRF) were not statistically significant because of the large variability among farms (Table 9.6). In the first year, average ΔGRF was −$8 ha^{-1} crop^{-1}, but this number increased to +$34 ha^{-1} crop^{-1} in the second year. This crop-year effect was significant ($P = 0.006$), suggesting a trend toward SSNM becoming profitable after the first year.

Factors affecting the performance of SSNM

Village effects on grain yield, nutrient uptake, N-use efficiency, and profit were mostly not significant (Table 9.5), suggesting that the SSNM approach worked similarly under different agroecological and cropping conditions at the three villages. However, significant yield losses caused by factors other than nutrients occurred during the experimental period and affected the performance of SSNM. From 1997 to 1999, average actual yield in the SSNM was 65% of the model-predicted target yield in the DS, but 76% in WS crops. Actual SSNM yields were within 90% to 100% of the yield target for only 18% of all cases. In the most extreme season, the 1998 DS, yields at Karang Hegar averaged only 42% of the target yield compared with 90% in the year before at the same sites. The yield targets set for the SSNM treatment assumed average climatic conditions and yield potential, good water supply, and no significant yield losses from pests, conditions that were achieved only in a few cases. Key problems observed during the 1997-99 period were (1) climatic yield constraints caused by El Niño–La Niña cycle and (2) poor quality of crop management. The latter mainly included low plant density and pest problems. These effects are difficult to quantify, but also varied depending on water release and planting schedules in the large irrigation scheme of this area. Sites at Karang Hegar belong to irrigation group I, where rice is always planted first. Soils are porous and it is difficult to keep standing water in the field, causing both severe weed and rat control problems. Rat damage is usually high during the dry season because of the migration of rats from sugarcane plantation areas located next to the rice fields. Sites at Bojong Jaya belong to irrigation group II, areas where rice is planted around 2 weeks after group I. Under normal conditions, Bojong Jaya has the fewest production problems, but farmers used wider plant spacing than at Karang Hegar and Sukareja. Sites at Sukareja belong to irrigation group III. Farmers in this area prefer long-duration rice varieties such as Ketan, IR42, or Muncul rather than IR64. During the wet season, standing water in fields is typically very high. Stem borer damage is widespread because of the late planting.

The experiments were conducted during El Niño–La Niña climatic cycle (1997-99), which often accelerated such management problems. In the 1997 DS crop, many of the experimental fields at Karang Hegar had severe weed and rat damage, leading to ratoon crops with uneven ripening. Insufficient water availability, weeds, and rats also affected some experiments at Sukareja. El Niño delayed planting of the 1997-98 WS crop by 4 months. Although the prolonged dry fallow period increased the indigenous nutrient supply, pest problems (rats, insects) occurred on many farms. Unusual rainy, humid, and hot weather and delayed planting (by 2–4 wk) of the 1998 DS crop also caused severe pest infestation and low yields (Fig. 9.4). Weeds, rats, black bugs, leaffolder, stem borer, and BPH caused severe yield losses on practically all farms, but particularly at Karang Hegar.

Table 9.7. Influence of crop management on grain yield, N-use efficiency, and profit increase by site-specific nutrient management (SSNM) at Sukamandi, Indonesia. Farms were grouped into farms with no severe crop management problems (SSNM+) and farms in which one or more severe constraints (mainly water and pests) occurred (SSNM–).[a]

	Grain yield (t ha^{-1})		AEN (kg grain kg^{-1} N)		REN (kg N kg^{-1} N)		ΔProfit (US$ ha^{-1})	
	SSNM+	SSNM–	SSNM+	SSNM–	SSNM+	SSNM–	SSNM+	SSNM–
1997 dry season								
Mean	5.09 a	3.73 b	12.6 a	10.0 a	0.31 a	0.23 b	–16.3 a	–86.9 b
N	13	5	13	4	13	4	13	5
1998 wet season								
Mean	5.83 a	4.75 b	15.4 a	11.6 a	0.67 a	0.44 b	16.9 a	–8.4 a
N	12	8	12	8	11	8	12	8
1999 wet season								
Mean	5.70 a	4.19 b	17.4 a	15.9 a	0.58 a	0.51 a	45.9 a	2.8 b
N	9	11	9	11	9	11	9	11

[a]Within each row (season), means of SSNM+ and SSNM– followed by the same letter are not significantly different using LSD (0.05%).

9.3 Future opportunities for improving SSNM

Implementing a new nutrient management strategy such as SSNM must involve changes in other crop management practices to become successful in West Java. Table 9.7 indicates what can be achieved with improved water management and good pest control. Average grain yields in SSNM plots with less pest incidence and water management problems were 1.1 t ha^{-1} (1998 WS) to 1.5 t ha^{-1} (1999 WS) larger than in SSNM plots with severe problems. This was also associated with much increased REN and profit. Rat and weed control as well as insect and disease management remain key problems to solve if SSNM is to have more widespread success in this area.

However, results from 1997 to 1999 also suggested that transplanting was typically done at too wide plant spacing, mostly resulting in only 13 to 14 hills m^{-2}, which delayed canopy closure and resulted in a low number of panicles m^{-2}. Low planting density is mainly a question of labor cost and availability, but farmers also fear greater rat problems if planting density were to be increased. To test the hypothesis that SSNM can be greatly improved by planting at higher density, an additional SSNM treatment (SSNM2) was tested in the 2000 DS and 2001 WS crops, in which planting was done at recommended levels of 21 hills m^{-2}. Grain yield, plant nutrient uptake, and GRF of this treatment were compared with the standard FFP and SSNM (SSNM1) treatments, which were both planted at an average density of 14 hills m^{-2}. Fertilizer management was the same in both SSNM1 and SSNM2.

Planting at the recommended density greatly increased yields, plant nutrient uptake, and profit in both the DS and WS (Fig. 9.7) although excessive rain during the 2001 WS reduced yields substantially compared with previous years. On average, there was no significant yield difference between SSNM1 and FFP at the low plant density, but a 0.8 t ha^{-1} yield difference between SSNM2 and FFP. Plant N, P, and K uptake were increased by planting at higher density in combination with SSNM. Average N uptake in SSNM2 was 115 kg N ha^{-1} vis-à-vis 95 kg N ha^{-1} in SSNM1 or 101 kg N ha^{-1} in the FFP, indicating much-increased N-use efficiency because of optimal planting. Average N-use efficiencies were significantly higher in SSNM1 than in the FFP (REN 0.54 vs. 0.45 kg kg^{-1}, AEN 32 vs. 22 kg kg^{-1}) because of the efficient use of applied fertilizer N and favorable climatic conditions, particularly in the 2000 DS (Fig. 9.2). Note that N-use efficiencies were substantially higher than in previous years (Table 9.5).

Nitrogen-use efficiencies were not calculated for SSNM2 because a separate –F plot was not included in this treatment. The greater planting density in SSNM2 likely resulted in a greater uptake of indigenous soil nutrients, which may have contributed substantially to the greater plant nutrient uptake in fertilizer treatments of SSNM2 compared with SSNM1 and the FFP. This would affect the calculation of N-use efficiencies following the difference method. Given the already high N-use efficiencies in SSNM1, we suspect that the yield increase in SSNM2 at the same fertilizer rate was caused by an increase in INS rather than by yet another increase in N-use efficiencies. This was supported by the internal nutrient efficiencies (kg grain kg^{-1} plant nutrient), which were similar for all three treatments (52–55 kg kg^{-1} for N, 283–293 kg kg^{-1} for P, and 47–49 kg kg^{-1} for K).

Plant P uptake increased by about 21% over both SSNM1 and the FFP. As in previous years, SSNM greatly reduced the amount of fertilizer N applied by about 45 kg N ha^{-1} crop^{-1} (37%) vis-à-vis the FFP (Table 9.7), but involved higher P and K rates. However, average P and K rates in both SSNM treatments (13 kg P ha^{-1} and 22 kg K ha^{-1}) represented the minimum amount needed to replenish most of the crop removal and should be considered essential for sustaining soil productivity at these sites. Nevertheless, because of the savings in N use under SSNM, the total fertilizer cost was the same in all treatments and averaged $40 ha^{-1} crop^{-1}. Whereas SSNM1 did not increase the profit over FFP, SSNM2 resulted in a significant profit increase of, on average, $129 ha^{-1} crop^{-1}.

9.4 Conclusions

Irrigated rice yields in West Java are stagnating, but population growth continues and some fertile rice land is being converted into nonagricultural uses. Deteriorating soil quality, unbalanced fertilizer use, and the virtual abandonment of organic manuring have been proposed as hypothetical causes for stagnant or even declining productivity in intensive irrigated rice systems. The previously promoted national fertilizer recommendations for large areas may not be suitable for regaining momentum in

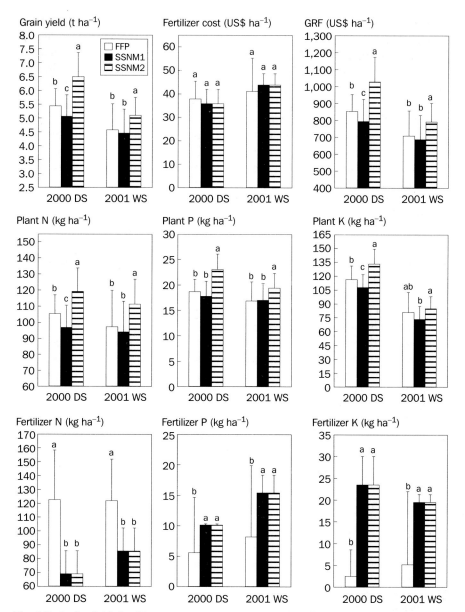

Fig. 9.7. Grain yield, fertilizer cost, gross return above fertilizer cost, plant nutrient uptake, and fertilizer use in the farmers' fertilizer practice (FFP) with regular planting density (14 hills m^{-2}) and two site-specific nutrient management treatments (SSNM1 with 14 hills m^{-2} and SSNM2 with 21 hills m^{-2}) at Sukamandi, Indonesia (2000 DS and 2001 WS; bars: mean; error bars: standard deviation).

yield growth because indigenous nutrient supplies vary from field to field and from season to season.

Site-specific nutrient management increased nutrient uptake and N-use efficiency on 20 farms with transplanted rice. It was associated with much-decreased fertilizer-N use, but a more balanced NPK fertilization. However, significant increases in yield and profit were achieved only when other cropping practices, particularly pest control and plant density, were improved as well. A single measure such as optimizing planting had a large effect on the performance of SSNM. Research and extension activities should therefore focus on developing a site-specific crop management concept that takes into account such key factors and their interactions should be demonstrated at pilot sites to researchers, extension workers, and farmers. If labor cost prohibits planting at higher density, more research on alternative crop establishment techniques such as direct seeding should be conducted to offer choices to farmers in this region. A key issue for more widespread use of the SSNM concept is to simplify crop N status monitoring by replacing the chlorophyll meter with a leaf color chart (LCC). On-farm tools with a strong real-time management component such as the LCC are probably required at sites such as West Java, where yield stability appears to be low. The large variation in yield potential and factors constraining yield such as water availability and pest incidences make it difficult to develop standard fertilizer recommendations based on preselected yield targets. For instance, the attainable yield potential of the DS crops was clearly limited by factors other than climate during the period of measurements in 1997-99. Further research is needed to estimate the season-specific yield potential of varieties used by farmers and to quantify nutrient availability as affected by soil drying before the WS. However, strategies with a strong real-time nutrient management component such as the LCC are probably sufficiently robust to offer practical solutions to farmers for the efficient use of fertilizers. Finally, future research should also study the question whether SSNM with a balanced combination of inorganic and organic fertilizer offers additional benefits in terms of profitability and sustainability of irrigated lowland rice farming.

References

Abdulrachman S, Witt C, Dobermann A, Balasubramanian V, Buresh RJ, Susanti Z, Pahim, Djatiharti A. 2002. Increasing productivity in intensified lowland irrigated rice in Indonesia: use of site-specific nutrient management. Paper presented at National Rice Week, Research Institute for Rice, Sukamandi, Indonesia, 5 March 2002.

Anonymous. 1998. Statistical yearbook of Indonesia. Central Bureau of Statistics, Jakarta, Indonesia. 594 p.

Bouman BAM, Kropff MJ, Tuong TP, Wopereis MCS, ten Berge HFM, van Laar HH. 2001. ORYZA2000: modeling lowland rice. Los Baños (Philippines): International Rice Research Institute, and Wageningen (Netherlands): Wageningen University and Research Centre. 235 p.

Cabrera JMC, Roetter RP, van Laar HH. 1998. Preliminary results of crop model development and evaluation of rice. In: Roetter RP, Hoanh CT, Teng PS, editors. A systems approach to analyzing land use options for sustainable rural development in South and Southeast Asia. IRRI Discussion Paper Series No. 28. Manila (Philippines): International Rice Research Institute (IRRI). p 40-57.

Cassman KG, Dobermann A. 2001. Evolving rice production systems to meet global demand. In: Rice research and production in the 21st century. Proceedings of a symposium honoring Robert F. Chandler, Jr. Los Baños (Philippines): International Rice Research Institute. p 79-100.

Dobermann A , Fairhurst TH. 2000. Rice: nutrient disorders and nutrient management. Singapore, Makati City (Philippines): Potash and Phosphate Institute, International Rice Research Institute. 191 p.

Frère M, Popov GF. 1979. Agrometeorological crop monitoring and forecasting. FAO Plant Production and Protection Paper 17. Rome (Italy): FAO.

Isgiyanto, Suwasik, Karsono, Munip dan Riwanodjo A. 1992. Penggunaan pupuk organik dan pengelolaannya pada padi sawah. In: Penelitian mendukung peningkatan produksi pangan. p 14-21.

Ismunadji and Sudjadi. 1983. Lahan bermasalah dan produksi padi. Dalam. Masalah dan hasil penelitian padi. Puslitbangtan, Bogor, Indonesia.

Olk DC, Cassman KG, Simbahan GC, Sta. Cruz PC, Abdulrachman S, Nagarajan R, Tan PS, Satawathananont S. 1999. Interpreting fertilizer-use efficiency in relation to soil nutrient-supplying capacity, factor productivity, and agronomic efficiency. Nutr. Cycl. Agroecosyst. 53:35-41.

Sri Adiningsih J, Santoso D, Sudjadi M. 1991. The status of N, P, K and S of lowland rice soils in Java. In: Blair G, editor. Sulfur fertilizer policy for lowland and upland rice cropping systems in Indonesia. Melbourne (Australia): ACIAR. p 68-76.

Sta. Cruz PC, Simbahan GC, Hill JE, Dobermann A, Zeigler RS, Du PV, dela Pena FA, Samiayyan K, Suparyono, Tuat NV, Zhong Z. 2001. Pest profiles at varying nutrient input levels. In: Peng S, Hardy B, editors. Rice research for food security and poverty alleviation. Proceedings of the International Rice Research Conference, 31 March-3 April, Los Baños, Philippines: International Rice Research Institute. p 431-440.

Supit I, Hooijer AA, Van Diepen CA. editors. 1994. System description of the WOFOST 6.0 crop simulation model implemented in CGMS. Volume 1. Theory and algorithms. Luxemburg: European Commission, Directorate-General XIII Telecommunications, Information Market and Exploitation of Research. p 1-144.

Witt C, Dobermann A, Abdulrachman S, Gines HC, Wang GH, Nagarajan R, Satawathananont S, Son TT, Tan PS, Tiem LV, Simbahan GC, Olk DC. 1999. Internal nutrient efficiencies of irrigated lowland rice in tropical and subtropical Asia. Field Crops Res. 63:113-138.

Notes

Authors' addresses: S. Abdulrachman, Z. Susanti, Pahim, A. Djatiharti, Indonesian Institute for Rice Research (IIRR), Jalan Raya No 9, Sukamandi 41256, Subang, West Java, Indonesia, e-mail: sarlan@indosat.net.id; A. Dobermann, University of Nebraska, Lincoln, Nebraska, USA; C. Witt, International Rice Research Institute, Los Baños, Philippines.

Acknowledgments: We thank Atim, Ujang Sutaryo, Iwan Juliardi, Putu Wardana, Supena, Suhana, and Dede Subarja for their help collecting data and the farmers at Karang Hegar, Sukareja, and Bojong Jaya who participated in this project. Dr. A.M. Fagi, Dr. A. Hasanuddin, and Dr. Irsal Las are thanked for their support as directors of IIRR during the various stages of the project. We also thank Mr. Gregorio Simbahan, Ms. Arlene Adviento, Mr. Rico Pamplona, Ms. Olivyn Angeles, Ms. Julie Mae Criste Cabrera-Pasuquin, and Mr. Edsel Moscoso (all IRRI) for their guidance and help in data analysis. The authors are grateful to the Swiss Agency for Development and Cooperation (SDC) for its support.

Citation: Dobermann A, Witt C, Dawe D, editors. 2004. Increasing productivity of intensive rice systems through site-specific nutrient management. Enfield, N.H. (USA) and Los Baños (Philippines): Science Publishers, Inc., and International Rice Research Institute (IRRI). 410 p.

0 | Site-specific nutrient management in irrigated rice systems of the Mekong Delta of Vietnam

Pham Sy Tan, Tran Quang Tuyen, Tran Thi Ngoc Huan, Trinh Quang Khuong, Nguyen Thanh Hoai, Le Ngoc Diep, Ho Tri Dung, Cao Van Phung, Nguyen Xuan Lai, and A. Dobermann

10.1 Characteristics of rice production in the Mekong Delta

Trends

The South Vietnamese project sites are located in Cantho Province, the center of the Mekong River Delta (MRD) of Vietnam (Fig. 10.1). The MRD accounts for about 50.2% of the national rice production and almost all of Vietnamese rice exports. Biophysical conditions in the MRD are well suited to rice cultivation. It has a very flat topography with many fertile alluvial soils and abundant freshwater sources. The climate is monsoon with high temperatures throughout the year (average 27 °C, mini-

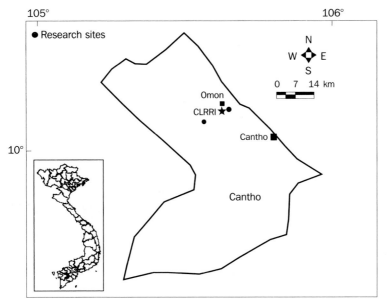

Fig. 10.1. Location of the experimental sites at Omon, Cantho Province, Vietnam.

193

mum 25 °C, maximum 33 °C) and high solar radiation (Fig. 10.2). There are two distinct seasons in the Mekong Delta, a wet season (May to October) and a dry season (November to April). Annual rainfall ranges from 1,500 to 2,000 mm.

Rice production in the Mekong Delta has changed rapidly during the past three decades (Table 10.1). Before 1975, local rice varieties with long growth duration (180–200 days) covered most of the rice area and only one crop per year was grown. During 1976-80, high-yielding semidwarf rice varieties with less than 150 d growth duration were introduced and gradually replaced the local varieties. From 1981 to 1990, improved irrigation systems were developed and double cropping of rice became common. In 1991 to 1995, short-duration varieties (approximately 100 d) were released and rapidly adopted by farmers. Since 1995, several varieties with ultrashort duration (85–90 d) have been released to meet the requirements of triple-crop systems with minimum land preparation. In the past 10 years, many farmers have moved to systems in which four to seven rice crops are grown per two years. The area under triple cropping has increased from about 75,000 ha in the early 1990s to about 300,000 ha now, although the government has tried to discourage farmers from planting three rice crops. Some regional statistics suggest that farmers usually harvest 12 to 15 t of rice ha^{-1} y^{-1} with triple rice cropping, but data are lacking on productivity and profitability comparing triple and double rice-cropping systems. There is also great variation in "triple" rice-cropping systems. Some farmers ratoon one of the rice crops,

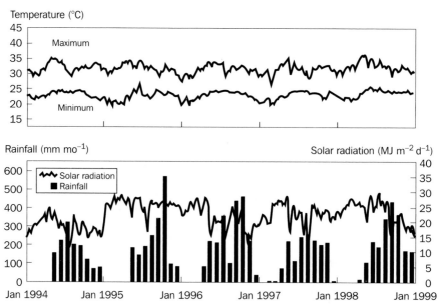

Fig. 10.2. Climatic conditions at Omon, Cantho Province, Vietnam, during 1994 to 1998. Solar radiation and temperature data are 7-day moving averages; rainfall is monthly total. Solar radiation was obtained from sunshine hours using the standard coefficients for dry tropical conditions of 0.25 for a and 0.45 for b (Frère and Popov 1979) for Omon (10°02′N, 105°47′E).

Table 10.1. Changes in rice production in the Mekong Delta and the province of Cantho, South Vietnam, 1975-99.

Item	1976-80	1981-85	1986-90	1991-95	1996-99
Mekong Delta					
Harvested area (1,000 ha)	1,885	2,248	2,367	2,991	3,666
Rice yield (t ha [1])	2.22	2.77	3.32	3.82	4.03
Rice production (million t)	4.21	6.22	7.96	11.46	14.82
Cantho Province[a]					
Harvested area (1,000 ha)	–	–	473	606	762
Rice yield (t ha [1])	–	–	3.09	3.74	4.13
Rice production (million t)	–	–	1.47	2.27	3.15

[a]Including the province of Soc Trang, which was part of Cantho before 1991.
Source: Statistical Data of Vietnam Agriculture, Forestry, and Fishery, 1975-2000.

whereas others grow five crops in two years (alternating triple and double rice cropping) or even seven crops in two years (continuous rice cropping).

The move from single- to double- and triple-crop rice systems was the driving force for increasing rice production in the MRD (Table 10.1). Production growth rates accelerated from 0.29 million t y^{-1} in 1975-87 to 0.73 million t y^{-1} in 1988-99. From 1988 to 1999, rice production rose from 7.6 to 16.3 million t of rice (6.5% y^{-1}), in part because of a steep increase in harvested rice area from about 2.3 million ha in 1988 to 4.0 million ha in 1999, which marks an annual increase of 4.7% or 142,000 ha y^{-1}. Although yields increased from 1.9 t ha^{-1} in 1975 to 4.0 t ha^{-1} in 1999 (3.4% y^{-1}), yield growth has been stagnating since 1994 (0.4%). The decline in yield growth rates is likely at least partly caused by the increasing adoption of ultrashort-duration rice varieties, but their yield potential is not sufficiently understood yet. There is anecdotal evidence that susceptibility to pests and diseases of ultrashort-duration varieties increases if farmers try to maximize yields of these varieties. Rice harvest area continues to increase at more than 4% y^{-1} because of both intensification of existing land and reclamation of land such as in the Plain of Reeds. Cantho Province is a representative site for many of the recent trends in rice production occurring in the whole MRD. The physical land area occupied by rice systems in Cantho is about 170,000 ha, but most farmers grow two to three crops per year so that the harvested area increased by about 70% from 480,200 to 823,000 million ha from 1985 to 1999. In recent years, average rice yields in Cantho ranged from 4.0 to 4.2 t ha^{-1} and the province (including Soc Trang, which was part of Cantho until 1991) produced a record of 3.49×10^6 t of rice in 1999. In the same year, about 60% of the rice area in Cantho was under triple cropping.

World market rice prices have been declining steadily for the past five years and, if they remain at current low levels, farmers will be increasingly encouraged to diversify out of rice and into other crops such as soybean, maize, watermelon, and vegetables within cropping systems that remain based on rice. Other land may be converted to orchards with fruit trees such as citrus, durian, or longan. Thus, triple cropping is likely to become less important in the future.

Rice production in the MRD is an important factor for food security in Vietnam and rice exports. However, little is known about the sustainability of the current production systems, particularly systems with triple cropping and minimum tillage. Intensive rice monoculture may lead to increased weed, disease, and insect pressure. Poor seed quality, low N-use efficiency, deteriorating soil fertility, and stagnating rice productivity are other major concerns (Hoa et al 1998, Lai and Tuan 1997, Phung et al 1998, Tan et al 1995, Tan 1997). During the past 25 years, the fluxes of nutrients within a typical irrigated rice field have increased 5- to 7-fold and cannot be met by natural sources such as sediments provided by the Mekong River alone. Mineral fertilizer inputs have become dominant factors of the overall nutrient balance, but their use is often unbalanced and their efficiency remains below optimum levels. Managing the variability in soil nutrient supply that has resulted from intensive rice cropping is one of the major challenges to sustaining and increasing rice yields in the Mekong Delta (Dobermann et al 1996).

Current biophysical and socioeconomic farm characteristics

Two long-term fertility experiments with the rice-rice system were established at the Cuu Long Delta Rice Research Institute in 1986 and 1994 and on-farm monitoring on rice farms started in 1994 to establish trends in productivity and soil quality. In 1997-99, 24 farms (6 villages with 4 farms each) were monitored in Omon District (10°08′N, 105°32′E) of Cantho Province (Fig. 10.1). Omon is located on alluvial lowland soils about 25 km northwest of Cantho. All sites have a high population density and very intensive double and triple cropping. Fallow periods between the harvest of a crop and planting of the succeeding crop are often only 1–2 days, except in between the late wet season (LWS) and DS, when most of the land remains flooded for about 2 months because of the annual fall flood near the Mekong River.

Biophysical and socioeconomic data collection started on all farms with the 1995 DS crop (see Chapter 2). Farm sizes range from 0.3 to 2.2 ha, with a median of about 0.9 ha (Table 10.2). Typically, about 90% of the land is used for growing rice, mostly following a rice-rice-rice triple-crop system. Farmers rent their land on a long-term basis (50 years) from the government. Farmers in our sample had a median age of 45, 6 years of school education, and a family of six people. Major planting seasons for rice are the dry season (DS) from November-December to February-March, the early wet season (EWS) from March-April to June, and the LWS from June to September. The yield potential of rice decreases in the order DS > EWS ≥ LWS. As an example, the following simulated, average yield potentials (± standard deviation) of IR72 were based on climate data from Omon (1989-99) using the model ORYZA1: 9.8 ± 0.9 t ha^{-1} for a DS crop sown on 1 December, 8.9 ± 0.7 t ha^{-1} for an EWS crop sown on 1 March, and 8.5 ± 0.8 t ha^{-1} for a LWS crop sown on 1 June. Note that solar radiation was obtained from sunshine hours using the standard Ångström coefficients for dry tropical conditions (Frère and Popov 1979, Supit et al

Table 10.2. Demographic and economic characteristics of rice production on 20 farms at Omon, Mekong Delta, Vietnam. Values shown are based on socioeconomic farm surveys conducted for whole farms.

Production characteristics	Min.	25%	Median	75%	Max.
Total cultivated area (ha)	0.26	0.5	0.87	1.42	2.2
Age of household head (y)	30	39	45	58.5	67
Education of household head (y)	3	5	6	12	12
Household size (persons)	2	4	6	6	13
1996 dry-season crop					
Rice area (ha)[a]	0.23	0.33	0.48	0.69	1.95
Yield (t ha^{-1})	4.67	5.72	6.10	6.80	7.56
N fertilizer (kg ha^{-1})	76	83	104	122	176
P fertilizer (kg ha^{-1})	7	13	18	21	36
K fertilizer (kg ha^{-1})	0	4	7	11	17
Insecticide (kg ai ha^{-1})	0.01	0.13	0.47	1.29	5.28
Herbicide (kg ai ha^{-1})	0	0.07	0.42	0.62	1.25
Other pesticides (kg ai ha^{-1})[b]	0	0.04	0.10	0.20	1.56
Total labor (8-h d ha^{-1})	32	53	68	84	128
Net return from rice (US$ ha^{-1} crop^{-1})	386	528	635	712	999
1997 wet-season crop					
Rice area (ha)[a]	0.23	0.30	0.50	0.66	1.30
Yield (t ha^{-1})	2.40	3.38	3.71	4.31	8.08
N fertilizer (kg ha^{-1})	74	89	102	139	230
P fertilizer (kg ha^{-1})	6	13	20	26	60
K fertilizer (kg ha^{-1})	0	4	13	36	57
Insecticide (kg ai ha^{-1})	0	0	0.17	0.50	2.42
Herbicide (kg ai ha^{-1})	0	0.24	0.53	0.93	2.80
Other pesticides (kg ai ha^{-1})[b]	0	0	0	0.17	0.29
Total labor (8-h d ha^{-1})	21	46	91	118	213
Net return from rice (US$ ha^{-1} crop^{-1})	145	225	269	344	928

[a]Rice area in which the treatment plots were embedded in subsequent years. The total rice area may even be larger.
[b]Includes fungicide, molluscicide, rodenticide, and crabicide.

1994), and that some uncertainties are associated with the use of standard coefficients. Simulations showed that potential production would differ by up to 7% depending on the choice of the coefficients (Cabrera et al 1998). Furthermore, growth characteristics of short- and ultrashort-duration cultivars grown in the Mekong Delta (see Section 10.2) are not available, and we can only suspect that the yield potential of these varieties would be 1–1.5 t ha^{-1} lower than that of IR72. With very good management, dry-season yields in farmers' fields can reach 7 to 8 t ha^{-1}, whereas wet-season yields barely exceed 5 t ha^{-1}. Although the DS is very favorable for growing rice, both WS crops face numerous problems. Drought often affects the EWS, whereas floods are common in the LWS. In both the EWS and LWS, weeds, diseases, and insect pests are common and cloudy weather limits yields. Therefore, average WS yields are around only 3.5 t ha^{-1} vis-à-vis DS yields of about 6 t ha^{-1} (Table 10.2).

Wet broadcast seeding with high seed rates is a common practice in the DS and WS crops grown in the MRD. In our sample, no farmer transplants rice anymore. However, compared to mechanized direct seeding in areas such as Central Thailand (see Chapter 7), the total labor input remains relatively high at an average of about 70 8-hour person-days ha^{-1} in the DS or 90 8-hour person-days ha^{-1} in the WS (Table 10.2). Labor inputs varied widely, however, from 21 to 213 8-hour person-days ha^{-1}. About 60% of this is manual family labor, mainly for harvest. Zero tillage is popular in EWS and LWS. After harvesting, rice straw is spread out over the surface of the field and burned. One day after burning, the fields are irrigated to moisten the soil and then pregerminated seeds are broadcast on the field without land preparation. Soil tillage, including puddling, is done in the DS by using buffaloes or small tractors. Surface water originating from the Mekong River is used for irrigation. Nitrogen and phosphorus are applied in similar amounts of about 100 kg N and 20 kg P ha^{-1} crop^{-1} in both DS and WS crops, despite much lower WS yields. Potassium application is not common and averages about 10 kg K ha^{-1} crop^{-1}. Farmers usually apply N in three to four splits, and P and K in two splits, mostly starting about 2 weeks after sowing. Overall, the profitability of rice production is low in the WS crops because of low yields, greater labor requirements, and high fertilizer rates that are not adjusted to the lower yield potential (Table 10.2). Almost all farmers regularly use herbicides and insecticides. In recent years, many farmers have also used foliar fertilizers containing some NPK plus micronutrients, often mixed with fungicides.

Soils at Omon are mostly acid Entisols with silty clay to clay texture (Table 10.3). Some possess acid-sulfate characteristics. The pH is about 4.2 to 4.4 (25th and 75th percentiles) prior to the WS, but increases to 4.8 to 5.0 (25th and 75th percentiles) just before the DS crop after soils have been flooded for several months with inputs from sedimentation. General soil properties depend on the distance to the Mekong River, with heavy, ill-drained clay soils mostly found farther away from it. In general, soils at Omon are high in clay and organic matter, whereas CEC and exchangeable bases are in the moderate range (Table 10.3). Most soils are severely P-deficient, as indicated by very low Bray-2 P of less than 20 mg kg^{-1} and very low Olsen-P of less than 5 mg kg^{-1} on all farms surveyed.

On about 75% of all farms sampled, extractable soil K was above 0.22 cmol$_c$ kg^{-1}. No yield response to K was observed in many fertilizer trials and it was always believed that the regular supply of nutrients and sediments by the Mekong water would replenish K removed by rice. However, the effects of sedimentation on soil fertility and its recent changes because of the regulation of water flows are not well studied. Recent measurements indicate that depletion of soil K occurs and mainly depends on cropping intensity. For example, average resin-extractable K on 32 farms was only 48% of that measured in a +NPK (balanced nutrition) treatment of the new long-term experiment at CLRRI, Omon (Dobermann et al 1996). In ten soils of the

Table 10.3. General soil properties on 24 rice farms at Omon, Vietnam.[a]

Soil properties	Min.	25%	Median	75%	Max.
Clay content (%)	38.0	49.3	54.5	65.3	70.0
Silt content (%)	29.0	34.8	45.0	50.5	61.0
Sand content (%)	0.0	0.0	1.0	1.0	1.0
Soil organic C (g kg^{1})	8.2	15.2	18.3	19.9	32.0
Total soil N (g kg^{1})	1.3	1.9	2.1	2.3	3.7
Soil pH (1:1 H$_2$O)[b]	4.3	4.8	4.9	5.0	5.1
Soil pH (1:1 H$_2$O)	3.9	4.2	4.3	4.4	4.5
Cation exchange capacity (cmol$_c$ kg^{1})	10.9	15.1	17.6	19.6	23.1
Exchangeable K (cmol$_c$ kg^{1})	0.10	0.18	0.19	0.22	0.40
Exchangeable Na (cmol$_c$ kg^{1})	0.16	0.29	0.36	0.43	0.49
Exchangeable Ca (cmol$_c$ kg^{1})	7.42	8.38	9.15	11.33	14.10
Exchangeable Mg (cmol$_c$ kg^{1})	2.44	3.08	3.98	4.86	7.12
Extractable P (Olsen-P, mg kg^{1})	0.85	2.80	3.20	3.73	5.30
Extractable P (Bray-2, mg kg^{1})	5.60	8.35	10.50	12.25	18.00
Extractable Zn (0.05N HCl, mg kg^{1})	0.90	1.48	1.80	2.50	3.80

[a]Measured on soil samples collected in May 2002 before the wet season. [b]Measured on soil samples collected in December 1996 before the dry season.

Mekong Delta, K exhaustion in greenhouse experiments proceeded fast and only a saline soil did not require the addition of fertilizer K to maintain rice growth (Hoa et al 1998).

During 1997 to 1998, N, P, and K omission plots were established to obtain estimates of the indigenous N (INS = plant N accumulation in a zero-N plot), P (IPS = plant P accumulation in a zero-P plot), and K supply (IKS = plant K accumulation in a zero-K plot) in rice. The median INS was 50–55 kg N ha^{-1} in DS crops and 30–35 kg N ha^{-1} in WS (EWS) crops (Fig. 10.3). This large difference in INS is likely caused by the long (flooded) fallow period before the DS vis-à-vis a very short fallow period before the WS. However, poor growth in the WS may also affect this measure. Similarly, the median IPS was about 14 kg P ha^{-1} in the 1998 DS, but only about 8 kg P ha^{-1} in the 1998 WS. The IKS values fluctuated widely among seasons (medians ranging from 60 to 90 kg K ha^{-1}) and farms (3-fold in each season), but reasons for this remain unclear. Assuming nutrient requirements of 14.7 kg N, 2.6 kg P, and 14.5 kg K per 1,000 kg grain yield (Witt et al 1999), the current average INS, IPS, and IKS in the DS are sufficient for achieving rice yields of about 3.7 t ha^{-1} without applying N, 5.4 t ha^{-1} without applying P, and 5 t ha^{-1} without applying K. In the WS, attainable rice yields are only 2.5 t ha^{-1} without applying N, 3.1 t ha^{-1} without applying P, and 5 t ha^{-1} without applying K. These numbers illustrate how important balanced NPK fertilizer application is at this site and that both P and K appear to limit further yield increases on many farms. Moreover, managing the large variation

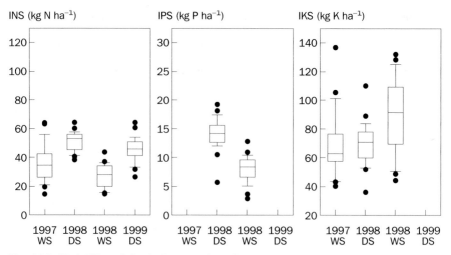

Fig. 10.3. Variability of the indigenous N (INS), P (IPS), and K (IKS) supply among 20 farmers' fields at Omon, Vietnam (1997-99). Median with 10th, 25th, 75th, and 90th percentiles as vertical boxes with error bars; outliers as bullets.

in the indigenous nutrient supply among farms and growing seasons requires many more season- and location-specific approaches than are currently used.

Detailed baseline agronomic data were measured in the farmers' fields (FFP) in the 1995 WS and 1996 DS (Table 10.4). Yields were higher in the DS, mainly because of better climatic conditions. Adequate irrigation and sunlight during ripening resulted in better grain filling (83%) vis-à-vis the WS (71%). However, despite the lower yield potential, farmers applied more N in the WS (123 kg N ha^{-1}) than in the DS (115 kg N ha^{-1}) and the same amounts of P and K. Surveys are lacking to explain this, but two hypotheses to test would be that (1) farmers apply more N in the WS because nutrient inputs from floodwater prior to the WS are lower or (2) farmers try to compensate for the visually poor growth appearance in the WS by applying more N. For comparison, the current recommended N rates for Cantho Province are 80–90 kg N ha^{-1} in the WS and 100–120 kg N ha^{-1} in the DS. Farmers applied P similar to the current DS recommendation, but less than what is recommended in the WS. In general, K use was well below recommended rates in all seasons. Although P input-output balance estimates suggested a positive average balance, the current practice of applying little K results in an average K loss of about 30 kg K ha^{-1} crop^{-1}. However, K input from irrigation and sediments is not contained in this estimate so that the actual balance may be slightly less negative. Nitrogen-use efficiencies varied widely among farms. However, an average AEN of 9 to 10 kg kg^{-1} and REN of 0.2–0.3 kg kg^{-1} in the 1995 WS and 1996 DS indicate a generally low efficiency of fertilizer that is comparable with that of other regions of Asia (Olk et al 1999).

Table 10.4. Baseline agronomic characteristics of rice production on 24 farms at Omon, Mekong Delta. Values shown are means and standard deviations (SD) of the farmers' fertilizer practice for two consecutive rice crops monitored before SSNM plots were established.

Agronomic characteristics	1995 WS crop		1996 DS crop	
	Mean	SD	Mean	SD
Grain yield (t ha^{-1})	3.49	0.58	5.31	0.87
Harvest index	0.46	0.01	0.45	0.02
No. of panicles m^{-2}	620	97	696	113
Total no. of spikelets m^{-2}	22,526	3,507	27,650	4,567
Total no. of spikelets panicle^{-1}	37	8	40	7
Filled spikelets (%)	71	7	83	7
Fertilizer N use (kg ha^{-1})	123.4	27.6	114.9	31.0
Fertilizer P use (kg ha^{-1})	19.3	11.2	19.5	7.8
Fertilizer K use (kg ha^{-1})	8.9	5.3	8.4	4.8
N uptake (kg ha^{-1})	49.6	9.9	82.1	12.1
P uptake (kg ha^{-1})	15.9	2.7	16.1	5.3
K uptake (kg ha^{-1})	73.5	16.8	69.1	20.5
Input-output N balance (kg ha^{-1} crop^{-1})	19.7	9.6	−4.9	11.9
Input-output P balance (kg ha^{-1} crop^{-1})	5.8	11.8	4.5	11.3
Input-output K balance (kg ha^{-1} crop^{-1})	−31.4	9.8	−33.2	10.3
Partial productivity of N (kg kg^{-1})	29.8	9.3	48.9	13.0
Agronomic efficiency of N (kg kg^{-1})	9.9	3.5	8.8	5.2
Recovery efficiency of N (kg kg^{-1})	0.18	0.06	0.28	0.12
Physiological efficiency of N (kg kg^{-1})	55.9	11.7	29.8	11.7

10.2. Effect of SSNM on productivity and nutrient-use efficiency

Management of the SSNM plots

Beginning with the 1997 WS crop, a site-specific nutrient management (SSNM) plot was established in each of the 24 farmers' fields as a comparison with the farmers' fertilizer practice (FFP, see Chapter 2). Results of the four rice crops grown from the 1997 WS to the 1999 DS are presented in this paper. The size of the SSNM plots was 300 m^2 in the 1997 WS and 1998 DS and increased to 1,000 m^2 in the 1998 WS and 1999 DS. Only two crops per year were sampled. In triple-crop systems (about 60% of all farms), WS crops with SSNM refer to the EWS period, whereas LWS crops were managed by farmers as the FFP for the whole field. Rice varieties were chosen by the farmers. During the experimental period, farmers planted 12 different semidwarf modern varieties. The most popular varieties were IR50404 (about 30% of all farmers), IR59656, IR59606, IR62032, OM1327, OM1490, OM1704, OM1706, OM1708, OM2031, OMCS94, and OMCS97. The varieties IR50404, OM1490, OMCS94, and OMCS97 are ultrashort-duration varieties with a growth of about 90–

100 days, and are preferred in all seasons in both double- and triple-cropping systems. All other varieties require about 100–110 days to maturity (short-duration varieties), and farmers more commonly grow them in the DS. All rice was direct-seeded at high seed rates of 200–250 kg ha^{-1}. In the 1998 DS, collaborating farmers received certified seeds from the researchers. Farmers did all water management and pest control following their commonly used methods. Water level was kept at 5–15 cm deep throughout the crop growth period by weekly rotational irrigation; however, the rice field was kept at saturation in the first 3–5 days after sowing for better crop establishment. Permanent flooding starts by the second week after sowing. LWS rice straw was incorporated by tractor, but DS and EWS rice straw was burned after hand harvesting.

Fertilizer applications for SSNM were prescribed on a field- and crop-specific basis following the approach described in Chapter 5 and descriptions given elsewhere (Dobermann and White 1999, Witt et al 1999). In the 1997 WS and 1998 DS, initial model input values were average INS based on N uptake in N omission plots, and average IPS and IKS estimated from P and K uptake in FFP plots in the 1995 WS and 1996 DS. In the 1998 WS and 1999 DS crops, estimates of INS, IPS, and IKS were adjusted for measurements made in 1997-98 and the actual P and K balance (see Chapter 5). The climatic yield potential was assumed to be 9 t ha^{-1} in the DS and 7 t ha^{-1} in the WS. Yield targets were 7 t ha^{-1} in the DS and 5 t ha^{-1} in WS crops. First-crop recovery fractions of 0.40–0.55, 0.30–0.35, and 0.40–0.50 kg kg^{-1} were assumed for fertilizer N, P, and K, respectively.

Fertilizer sources used in SSNM were urea (46% N), single superphosphate (6.5% P), and muriate of potash (50% K). All P was applied topdressed on the soil surface 10 days after sowing (DAS), whereas K was split into 50% at 10 DAS and 50% at panicle initiation (PI). In the 1997 WS and 1998 DS crops, a chlorophyll meter-based N management approach was tested in SSNM. Plots received an initial dose of 20–30 kg N ha^{-1} at 10 DAS. After 20 DAS, N was applied any time SPAD values measured weekly fell below 33. Typically, this resulted in two to four N applications from 10 to about 50 DAS at 30 to 40 kg N ha^{-1} per application. For comparison, farmers applied N in one to four splits, but with widely varying dates and amounts. In the 1998 WS crop, a modified SPAD-based N management was practiced in the SSNM:

FN1	10 DAS	Fixed application	20–30 kg N ha^{-1}
FN2	25–35 DAS (MT)	If SPAD >33	No N
		If SPAD <33	30 kg N ha^{-1}
FN3	40–50 DAS (PI)	If SPAD >35	20–30 kg N ha^{-1}
		If SPAD 33–35	30–40 kg N ha^{-1}
		If SPAD <33	40–50 kg N ha^{-1}

This approach was further refined in the 1999 DS to move to a simplified, growth-stage-oriented N management with less frequent SPAD readings:

FN1	10 DAS	Fixed application	30 kg N ha^{-1}
FN2	20–25 DAS (ET)	If SPAD >37	No N
		If SPAD 35–37	20 kg N ha^{-1}
		If SPAD 33–35	30 kg N ha^{-1}
		If SPAD <33	40 kg N ha^{-1}
FN3	40–45 DAS (PI)	If SPAD >35	40 kg N ha^{-1}
		If SPAD 33–35	50 kg N ha^{-1}
		If SPAD <33	60 kg N ha^{-1}
FN4	60 DAS (H-F)	If SPAD >33	No N
		If SPAD <33	20 kg N ha^{-1}

Effect of SSNM on grain yield and nutrient uptake

Grain yields were higher in most SSNM plots than in the FFP. Averaged over all farms and crops, SSNM resulted in a yield increase of 0.33 t ha^{-1} (8%, Table 10.5). However, average yield increases in the WS (0.19 t ha^{-1}) were not significant, whereas DS yields increased by 0.46 t ha^{-1} ($P = 0.000$). The largest yield increase over the FFP was observed in the 1999 DS (0.54 t ha^{-1}, 10%), probably because of the improvements made in the N management strategy (Fig. 10.4). In the 1999 DS, yields of >6 t ha^{-1} and yield increases by SSNM of >0.6 t ha^{-1} were measured on 10 of the 24 farms. In three cases with very good field management, yield increases exceeded 1 t ha^{-1}, which illustrates the great potential that exists. On average, DS yields under SSNM were about 65% of the climatic yield potential and about 2 t ha^{-1} greater than WS yields. Low WS yields were mainly caused by climatic, water, and pest stresses, and lodging, with perhaps little scope for increase because of the fine-tuning of nutrients alone. This is further illustrated by the large fluctuation in yield differences between SSNM and FFP (DGY) among farms, which ranged from –0.6 to +1.1 t ha^{-1} in the 1997 WS or –0.4 to +1.9 t ha^{-1} in the 1998 WS. For comparison, DGY was positive on all 24 farms in the 1998 DS (0.1 to 0.7 t ha^{-1}) and on 23 out of 24 farms in the 1999 DS (0 to 1.2 t ha^{-1}). Between years, the differences in the yield performance of SSNM were not statistically significant.

Site-specific nutrient management increased plant uptake of N, P, and K in most crops, but the increases were mostly significant in the DS only (Fig. 10.4). Total N uptake in DS crops increased by 8.6 kg ha^{-1} (10%, $P = 0.003$), whereas N uptake in the WS was not significantly increased by SSNM. Average K uptake increased by 18 kg K ha^{-1} (21%, $P = 0.000$), reflecting the much larger amounts of fertilizer K applied in the SSNM plots (Table 10.6). Assuming an internal efficiency of 69 kg grain kg^{-1} K under conditions of optimal growth (Witt et al 1999), this increase in K uptake is equivalent to a potential yield gain of about 1.2 t ha^{-1} or three times the actual yield gain realized. Further analysis showed that DS yields were mainly N-limited, which may only be overcome by a more integrated approach of reduced seeding rate and further improvements in N management. Under conditions of optimal growth, the average actual uptake of N, P, and K in the DS was sufficient to potentially attain yields of 6.6, 8.1, and 7.1 t ha^{-1}, respectively. This compares with average actual yields of 5.7 t ha^{-1} and indicates yield losses of about 0.9 t ha^{-1}

Table 10.5. Effect of site-specific nutrient management on agronomic characteristics at Omon, Vietnam (1997-99).

Parameters	Levels[a]	Treatment[b] SSNM	Treatment[b] FFP	Δ[c]	$P>|T|$[c]	Effects[d]	$P>|F|$[d]
Grain yield (GY)	All	4.77	4.43	0.33	0.079	Village	0.059
(t ha^{-1})	Year 1	4.71	4.44	0.27	0.335	Year[e]	0.109
	Year 2	4.82	4.43	0.40	0.130	Season[e]	0.001
	DS	5.74	5.28	0.46	0.000	Year × season[e]	0.603
	WS	3.63	3.45	0.19	0.395	Village × crop	0.655
Plant N uptake (UN)	All	82.4	76.5	5.9	0.078	Village	0.350
(kg ha^{-1})	Year 1	86.8	82.6	4.3	0.413	Year	0.086
	Year 2	78.1	70.6	7.5	0.062	Season	0.000
	DS	97.4	88.8	8.6	0.003	Year × season	0.247
	WS	64.9	62.2	2.7	0.531	Village × crop	0.143
Plant K uptake (UK)	All	102.4	84.7	17.8	0.000	Village	0.064
(kg ha^{-1})	Year 1	99.1	82.3	16.8	0.002	Year	0.722
	Year 2	105.7	86.9	18.8	0.000	Season	0.821
	DS	102.7	84.4	18.3	0.000	Year × season	0.047
	WS	102.1	85.0	17.2	0.009	Village × crop	0.500
Agronomic efficiency	All	19.9	14.9	5.0	0.000	Village	0.278
of N (AEN)	Year 1	19.3	14.3	5.0	0.001	Year	0.972
(kg grain kg^{-1} N)	Year 2	20.5	15.5	5.0	0.000	Season	0.980
	DS	22.2	16.9	5.3	0.000	Year × season	0.106
	WS	17.2	12.6	4.6	0.003	Village × crop	0.495
Recovery efficiency	All	0.44	0.34	0.10	0.000	Village	0.129
of N (REN)	Year 1	0.45	0.36	0.09	0.023	Year	0.809
(kg plant N kg^{-1} N)	Year 2	0.42	0.32	0.10	0.000	Season	0.409
	DS	0.50	0.39	0.11	0.000	Year × season	0.006
	WS	0.36	0.28	0.08	0.013	Village × crop	0.592
Partial productivity	All	49.6	42.1	7.5	0.001	Village	0.099
of N (PFPN)	Year 1	50.0	41.6	8.5	0.020	Year	0.344
(kg grain kg^{-1} N)	Year 2	49.1	42.6	6.5	0.024	Season	0.151
	DS	58.6	51.9	6.7	0.005	Year × season	0.020
	WS	39.0	30.6	8.4	0.001	Village × crop	0.130

[a]All = all four crops grown from 1997 WS to 1999 DS; year 1 = 1997 WS and 1998 DS; year 2 = 1998 WS and 1999 DS; WS = 1997 WS and 1998 WS; DS = 1998 DS and 1999 DS. [b]FFP = farmers' fertilizer practice; SSNM = site-specific nutrient management. [c]Δ = SSNM – FFP. $P>|T|$ = probability of a significant mean difference between SSNM and FFP. [d]Source of variation of analysis of variance of the difference between SSNM and FFP by farm; $P>|F|$ = probability of a significant F-value. [e]Year refers to two consecutive cropping seasons.

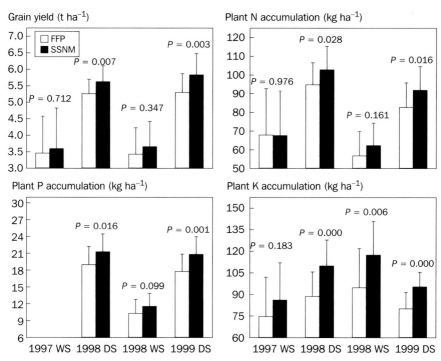

Fig. 10.4. Grain yield and plant N accumulation in the farmers' fertilizer practice (FFP) and site-specific nutrient management (SSNM) plots at Omon, Vietnam (1997-99; bars: mean; error bars: standard deviation).

because of factors other than N, P, or K. Clearly, P supply in the DS was sufficient for yields close to the climatic yield potential at this site. A similar analysis for the two WS crops indicates that the average actual uptake of N, P, and K was sufficient to potentially attain yields of 4.4, 4.4, and 7.0 t ha^{-1}, respectively. Yields of 3.6 t ha^{-1} were achieved in the WS so that the average yield loss because of other factors was about 0.8 t ha^{-1}. Moreover, both N and P supply are major limiting factors in the WS. The latter is probably related to differences in the length of the fallow period between DS and WS crops, which is very short before a WS crop.

Effect of SSNM on nitrogen-use efficiency

Nitrogen-use efficiencies in SSNM were significantly greater than in the FFP. Across all four crops grown and compared to the FFP, AEN increased by 5 kg kg^{-1} (34%, P = 0.000; Table 10.5). Likewise, REN increased by an average of 0.10 kg kg^{-1} (29%, P = 0.000) and PFPN increased by an average of 7.5 kg kg^{-1} (18%, P = 0.001). SSNM had no effect on PEN (Fig. 10.5), probably because PEN is mainly controlled by the general conditions affecting plant growth (Dobermann and Fairhurst 2000), which were the same for the FFP and SSNM. There were no significant crop-year or crop-season effects, that is, similar increases in N efficiency were achieved in DS

Table 10.6. Effect of site-specific nutrient management on fertilizer use, fertilizer cost, and gross returns over fertilizer costs from rice production at Omon, Vietnam (1997-99).

Parameters	Levels[a]	Treatment[b] SSNM	Treatment[b] FFP	Δ[c]	P> |T|[c]	Effects[d]	P> |F|[d]
N fertilizer (FN)	All	98	111	−13.2	0.000	Village	0.084
(kg ha[1])	Year 1	97	113	−16.6	0.001	Year[e]	0.109
	Year 2	99	109	−9.9	0.019	Season[e]	0.000
	DS	101	106	−5.1	0.222	Year × season[e]	0.079
	WS	94	117	−22.7	0.000	Village × crop	0.239
P fertilizer (FP)	All	22.0	19.5	2.5	0.101	Village	0.031
(kg ha[1])	Year 1	27.6	19.1	8.5	0.000	Year	0.000
	Year 2	16.5	19.8	−3.4	0.095	Season	0.006
	DS	23.0	18.3	4.7	0.014	Year × season	0.839
	WS	20.9	20.9	0.0	0.999	Village × crop	0.023
K fertilizer (FK)	All	62.4	20.6	41.7	0.000	Village	0.022
(kg ha[1])	Year 1	75.3	19.3	55.9	0.000	Year	0.000
	Year 2	49.8	21.9	27.8	0.000	Season	0.037
	DS	65.9	20.7	45.2	0.000	Year × season	0.000
	WS	58.2	20.6	37.6	0.000	Village × crop	0.049
Fertilizer cost	All	90.3	79.9	10.5	0.000	Village	0.040
(US$ ha[1])	Year 1	100.2	80.2	20.0	0.000	Year	0.000
	Year 2	80.7	79.6	1.1	0.750	Season	0.000
	DS	94.0	76.3	17.7	0.000	Year × season	0.003
	WS	86.0	84.1	1.9	0.673	Village × crop	0.039
Gross returns over	All	551	516	34.2	0.171	Village	0.016
fertilizer costs	Year 1	533	517	15.7	0.671	Year	0.025
(US$ ha[1])	Year 2	568	516	52.3	0.125	Season	0.097
	DS	677	634	43.9	0.003	Year × season	0.657
	WS	402	379	23.0	0.416	Village × crop	0.501

[a]All = all four crops grown from 1997 WS to 1999 DS; year 1 = 1997 WS and 1998 DS; year 2 = 1998 WS and 1999 DS; WS = 1997 WS and 1998 WS; DS = 1998 DS and 1999 DS. [b]FFP = farmers' fertilizer practice; SSNM = site-specific nutrient management. [c]Δ = SSNM − FFP. P> \|T\| = probability of a significant mean difference between SSNM and FFP. [d]Source of variation of analysis of variance of the difference between SSNM and FFP by farm; P> \|F\| = probability of a significant F-value. [e]Year refers to two consecutive cropping seasons.

and WS crops. The only exception was the first SSNM crop grown (1997 WS, Fig. 10.5), probably because the initial N management strategy chosen did not result in an improved congruence of N supply and plant N demand. However, the improvements in N management rules made over time resulted in consistent increases in N efficiencies in three consecutive crops grown thereafter (Fig. 10.5). At an absolute level, the average DS AEN of 22 kg kg^{-1} and REN of 0.50 kg kg^{-1} achieved under SSNM (Table 10.5) were close to what is normally achieved only in field experiments with good management.

Compared with the initial AEN and REN measured in farmers' fields before intervention (Table 10.4), the average AEN and REN in the SSNM crops (Table

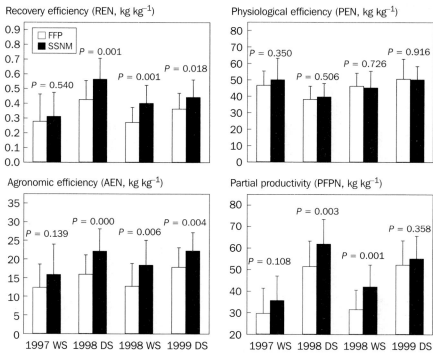

Recovery efficiency (REN, kg kg^{-1})

Physiological efficiency (PEN, kg kg^{-1})

Agronomic efficiency (AEN, kg kg^{-1})

Partial productivity (PFPN, kg kg^{-1})

Fig. 10.5. Fertilizer nitrogen-use efficiencies in the farmers' fertilizer practice (FFP) and site-specific nutrient management (SSNM) plots at Omon, Vietnam (1997-99; bars: mean; error bars: standard deviation).

10.5) represent an increase of about 200%. For example, the average REN increased from 0.23 kg kg^{-1} (FFP) in 1995-96 to 0.44 kg kg^{-1} in 1997-99 (SSNM). Note that REN in the FFP also increased to 0.34 kg kg^{-1} in 1997-99, which was probably related to some copying of the N management practiced in SSNM by farmers. However, the grain yields in the baseline data set (Table 10.4) were identical to those achieved in the FFP in 1997 to 1999, indicating little difference because of climate between those two periods. Therefore, the 200% increase in N-use efficiency probably represents a good measure of what can be achieved rather quickly by moving to N management schemes that are more plant-based and real-time. The crop stage-specific N management strategy employed in the 1999 DS resulted in an AEN of ≥25 kg kg^{-1} and REN of ≥0.50 kg kg^{-1} on seven farms, illustrating the potential that exists to increase N efficiency at the farm level. There is probably little scope for achieving similar increases through blanket recommendations for N use.

Effect of SSNM on fertilizer use and profit

SSNM significantly reduced N fertilizer use but increased K use (Fig. 10.6 and Table 10.6). Nitrogen rates in SSNM plots were about 13 kg ha^{-1} lower than in the FFP (13%), but most of this difference was due to much lower N rates in WS crops (25%

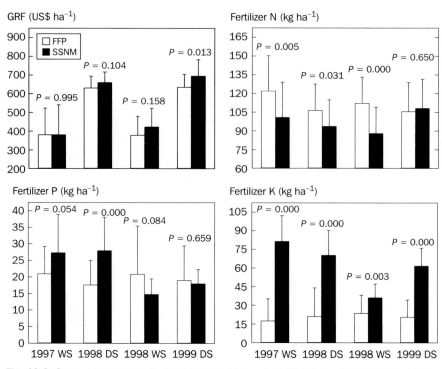

Fig. 10.6. Gross returns over fertilizer costs (GRF) and fertilizer use in the farmers' fertilizer practice (FFP) and site-specific nutrient management (SSNM) plots at Omon, Vietnam (1997-1999; bars: mean; error bars: standard deviation).

less than in the FFP). This did not lower the WS yields in SSNM treatments (Fig. 10.3), indicating that farmers could save substantial amounts of fertilizer N in the WS. Over the whole 2-year period, the total amount of P applied to SSNM plots was not significantly larger than that applied by the farmers. P doses were larger in the initial two crops to compensate for low IPS at many sites, but decreased thereafter to levels closer to actual crop removal (Fig. 10.6). However, about 42 kg K ha^{-1} crop^{-1} more were applied under SSNM than in the FFP plots (300% increase). This increase in K use was largest in the first year (56 kg ha^{-1}), but declined to only 28 kg K ha^{-1} in year 2 when IKS increases were taken into account (see Chapter 5). Interestingly, farmers also more than doubled their K use from less than 9 kg ha^{-1} in 1995-96 (Table 10.4) to 21 kg ha^{-1} in 1997-99 (Table 10.6). To some degree, this was related to first-time observations of visual symptoms of K deficiency at several sites during the 1998 DS crop, which prompted farmers to apply more K.

Averaged over four seasons, fertilizer cost in SSNM plots was US$10.50 ha^{-1} crop^{-1} higher than in the FFP, mostly because of very high K rates in 1997 and 1998. In year 1, the fertilizer cost difference between SSNM and the FFP was $20 ha^{-1} ($P$ = 0.000), but it decreased to only $1 ha^{-1} in year 2 and became insignificant (P = 0.750). The average profit increase because of SSNM was $34 ha^{-1} crop^{-1} (Table

10.6), but the increase in the profit difference from $16 ha^{-1} crop^{-1} in year 1 to $52 ha^{-1} crop^{-1} in year 2 was significant (crop-year P = 0.025), indicating an improvement in the performance of SSNM over time. This is probably related to the gradual change in N management as well as the buildup in soil P and K fertility. The profit increase was not significant in WS crops ($23 ha^{-1}), but highly significant in DS crops ($44 ha^{-1}, P = 0.003).

Examples of the performance of SSNM over time

To illustrate some of the principles of SSNM applied in our study, Table 10.7 shows the performance of SSNM and the FFP in the field of Mr. P.V. Nam at Thitran. Mr. Nam may be considered a good farmer in that area. Like many others, before intervention he tended to apply much N and P in the WS, he applied very little K, and his N-use efficiencies were well below achievable levels. We applied relatively large doses of P and K in the first two SSNM crops to at least partly replenish depleted soil reserves and because the initial yield goals were set too high. However, this P and K was not a lost investment and was properly taken into account in succeeding crops, resulting in a decrease in rates over time. In the first SSNM crop, no yield increase was achieved, but the three consecutive rice crops grown thereafter showed clear yield advantages. Those are probably related to the buildup in soil fertility as well as improvements in N management and the use of certified seeds since 1998.

Compared to the period before SSNM, Mr. Nam also increased his K use by almost 400% and his P use by about 30%, perhaps in an attempt to copy what was done in the SSNM plot. However, his N use did not change much and, because he had no means of adjusting N better to the actual plant N status, AEN and REN achieved in his FFP plot remained similar to those in 1995-96 and well below those in the SSNM plot.

The results of this case demonstrate that development and field testing of a new approach such as SSNM is a learning exercise that must be conducted over a succession of crops and years at many representative locations. It is important to be able to simultaneously manipulate medium-term changes in soil P and K as well as short-term fluctuations in N supply and demand. Neglecting either one increases the risk of making wrong or premature conclusions.

Factors affecting the performance of SSNM

Average actual yields in SSNM were only 68% of the model-predicted target yield in the WS and 82% in DS crops, indicating that factors other than N, P, and K caused significant yield reductions. The yield targets set for the SSNM treatment assumed average climatic conditions and yield potential, good water supply, and no pest losses, conditions that were achieved on only a few farms. Some of the yield losses were due to uncontrollable factors such as climate, others were due to variation in the quality of crop management. The experiments were conducted during El Niño–La Niña climatic cycle (1997-99), which particularly resulted in lower than normal climatic yield potential in the 1998 DS because of very high temperatures around flowering. An

Table 10.7. Performance of site-specific nutrient management (SSNM) and farmers' fertilizer practice (FFP) on the farm of P.V. Nam at Thitran, Omon District.[a]

Season	Grain yield (t ha^{-1})		FN (kg ha^{-1})		FP (kg ha^{-1})		FK (kg ha^{-1})		AEN (kg kg^{-1})		REN (kg kg^{-1})	
	FFP	SSNM	FFP	SSNM	FFP	SSNM	FFP	SSNM	FFP	SSNM	FFP	SSNM
Baseline data, 1995-96												
1995 DS	6.50	–	114	–	22	–	15	–	9	–	0.38	–
1995 WS	3.56	–	113	–	28	–	12	–	13	–	0.20	–
1996 DS	4.82	–	90	–	16	–	5	–	11	–	0.35	–
Mean	4.96	–	106	–	22	–	11	–	11	–	0.31	–
SSNM testing period, 1997-99												
1997 WS	5.08	5.07	122	87	32	30	58	116	16	22	0.42	0.47
1998 DS	4.87	5.44	99	94	35	29	50	76	8	14	0.30	0.51
1998 WS	3.97	4.43	108	86	24	19	27	30	17	27	0.34	0.54
1999 DS	5.46	6.52	105	114	25	15	36	30	11	19	0.23	0.35
Mean	4.85	5.37	109	95	29	23	43	63	13	21	0.32	0.47

[a]FN = fertilizer N, FP = fertilizer P, FK = fertilizer K, AEN = agronomic N-use efficiency, REN = recovery efficiency of applied N.

Table 10.8. Influence of crop management on grain yield, N-use efficiency, and profit increase by site-specific nutrient management (SSNM) at Omon, Vietnam. Farms were grouped into farms with no severe crop management problems (SSNM+) and farms in which one or more severe constraints (water, pests, crop establishment) occurred (SSNM−).[a]

	Grain yield (t ha⁻¹)		AEN (kg grain kg⁻¹ N)		REN (kg N kg⁻¹ N)		ΔProfit (US$ ha⁻¹)	
	SSNM+	SSNM−	SSNM+	SSNM−	SSNM+	SSNM−	SSNM+	SSNM−
1997 Wet season								
Mean	4.54 a	2.45 b	19.4 a	11.5 b	0.40 a	0.20 b	17.0 a	−20.1 a
N	11	9	11	9	11	9	11	9
1998 Dry season								
Mean	5.80 a	5.00 b	22.6 a	20.4 a	0.57 a	0.56 a	32.8 a	12.7 a
N	19	5	19	5	19	5	19	5
1998 Wet season								
Mean	4.06 a	3.01 b	20.3 a	15.4 a	0.43 a	0.36 a	44.1 a	45.2 a
N	13	8	13	8	13	8	13	8
1999 Dry season								
Mean	6.07 a	4.97 b	23.3 a	18.3 b	0.46 a	0.37 a	73.5 a	4.6 b
N	19	5	19	5	19	5	19	5

[a]Within each row (season), means of SSNM+ and SSNM− followed by the same letter are not significantly different using LSD (0.05%).

indicator for this was low grain filling (77%) and low grain yields even in 0-N plots (3.6 t ha⁻¹ vis-à-vis about 4.5 t ha⁻¹ in normal years).

Major crop management problems reduced yields on most farms in WS crops, but also occasionally in DS crops. They mainly included poor seed quality, heterogeneous crop establishment, snail damage, weeds, poor water management, and diseases. Among the latter, yellow leaf or red stripe disease was a common problem, but the actual causes of it are not yet known. In all four cropping seasons, a comparison of farms with better management and those with poor field management shows yield differences of 1–2 t ha⁻¹ (Table 10.8). Differences among villages in crop management practices also affected the gains from SSNM. For example, the village effect on grain yield increase was almost significant ($P = 0.059$, Table 10.5) and further analysis showed differences among villages in the DS. Grain yield increases by SSNM tended to be largest at Thitran and Dinhmon, whereas poorer performance was always observed at Thoilong and Thoithan (Fig. 10.7). This was apparently related to less crop care and interest in increasing productivity in the latter two villages.

The yields obtained on farms with better management, particularly in the WS, reflect the potential that exists. In most seasons, better crop management was also associated with greater N-use efficiency and increase in profit, although the large variation among farms made those differences not statistically significant (Table 10.8). Our data indicate that adopting an approach such as SSNM requires a minimum of

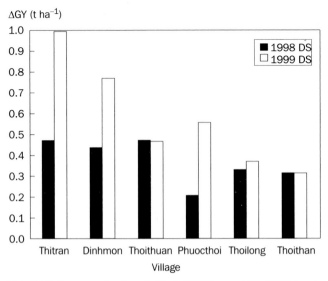

ΔGY (t ha^{-1})

■ 1998 DS
□ 1999 DS

Thitran Dinhmon Thoithuan Phuocthoi Thoilong Thoithan

Village

Fig. 10.7. Increase in rice yield of SSNM over FFP (ΔGY) in six villages near Omon, Vietnam. Values shown are means of four farms in each village.

overall improvements in crop management to be successful. Providing farmers with certified seeds, improving weed control in zero-tillage systems, improving disease control, and further refinement of N management appear to be the key measures for increasing nutrient efficiency in the double- and triple-rice systems of the MRD.

10.3. Future opportunities for adopting SSNM

Recognizing the good opportunities for increasing productivity and yield through improved nutrient management, the SSNM approach needs to be simplified for wider-scale extension in the Mekong River Delta. Nitrogen management strategies will have to be locally adapted, including the use of simpler on-farm tools for real-time N management such as the leaf color chart. Omission plots can be used to evaluate current farmers' fertilizer P and K rates, and improved nutrient management strategies may have to be integrated with other guidelines and technologies aiming at improved crop management to fully exploit the synergy that occurs if more than one factor is improved. As a starting point, the adoption potential of such a technology package was already evaluated in farmer cooperatives of 30–50 ha at pilot sites in five different provinces (An-giang, Cantho, Dong-thap, Tien-giang, and Soc Trang) in 1997-2000. Extension staff worked closely with cooperative leaders and farmers, and each cooperative nominated about 30–50 farmers with three levels of farming experience or degree of crop care (50% good, 25% medium, and 25% poor). Test fields were selected by farmers depending on the uniformity of plant growth in their fields. Fields were divided into two plots with standard practices and an improved integrated crop management approach, including variety selection, row seeding of 100 kg seeds ha^{-1} using a newly developed plastic drum seeder based on an IRRI prototype, improved N splitting using a leaf color chart, and reduced pesticide use in the first 40 days. Farmers received an initial training and continuing guidance by researchers, but implemented treatments themselves, and evaluated yield and profit together with extension workers. Yield and profit increases were comparable with those of the SSNM technique, and initial feedback showed that there was much interest among farmers and extension workers. The general impression was that the technology was simple and feasible enough. The initial records suggested that the integrated crop management technology increased grain yields by 10–12% with 15–20 kg N ha^{-1} lower N rates and reduced variation in fertilizer-N rates among farms, 50% lower seed rates, 35% lower costs for chemicals and labor for application of pesticides, but 20% higher labor cost for crop establishment because of row seeding. With this technology, farmers were able to reduce input costs on average by about 1.0 to 1.5 million Vietnamese dong per ha (about $70–100 ha^{-1}). In response to these encouraging results, extension workers in many provinces in the MRD have requested CLRRI to assist in training and the development and implementation of SSNM approaches. The delivery of such packaged technologies through the existing extension system seems feasible, and a first introduction of the technology in farmers' cooperatives for larger-scale demonstration appears to generate sufficient interest and support to, we hope, also reach farmers with smaller landholdings in the Mekong Delta.

Key problems to study in further extrapolation and extension work include the development of feasible methods for estimating INS, IPS, and IKS, and the simplification of the SSNM concept. This could include site- and season-specific ranges of fertilizer-N rates (i.e., apply x-y kg N ha^{-1} in a particular season), guidelines for fine tuning of N management using a leaf color chart (Balasubramanian et al 1999), and recommended site- and season-specific minimum rates for fertilizer P and K. Guidelines on splitting and timing of fertilizer P and K applications may include general recommendations such as applying fertilizer P in two equal doses at 10 and 25 DAS, and to apply more P in the WS because of the drying period between the DS and WS that reduces the indigenous soil P supply. Fertilizer K should be applied in equal splits at 10 and 45 DAS. Furthermore, any attempts to extend or extrapolate SSNM in the MRD must be part of a broader approach for improving rice systems considering the strong interactions with other crop management factors. Of particular importance will be improvements in seed quality and disease control.

10.4 Conclusions

Rice yield growth rates in the Mekong Delta have slowed down in recent years, and intensification of cropping systems continues despite a diversification out of rice into other crops. Nitrogen-use efficiencies in farmers' fields with rice are well below attainable levels and wider-scale reevaluation of plant-based P and K requirements in different parts of the Mekong Delta is badly needed. Much uncertainty exists about the sustainability of triple-crop rice systems.

Our data indicate potential for yield increases of at least 10% by following an SSNM approach in DS crops. We have demonstrated that the AEN of >20 kg kg^{-1} and REN of >0.50 kg kg^{-1} can be achieved under average farm conditions and that even greater gains are feasible with very good crop management. Profit increases of about \$50 ha^{-1} per DS crop and more balanced NPK fertilizer rates that sustain soil fertility are additional attractive features of SSNM. Although SSNM also increased N-use efficiency in the WS, yield and profit gains were small and highly variable because of a multitude of other factors.

A combination of a knowledge-intensive SSNM approach in the DS with a simplified fertilizer recommendation in the WS may currently work best in the MRD. However, even the latter should include tools for adjusting N rates to crop demand to avoid excessive N use in the low-yielding season. Although the SSNM approach was tested on only 24 farms, it has already contributed much to our knowledge about developing new packages of technologies to improve rice productivity and profitability in the Mekong River Delta of Vietnam.

References

Balasubramanian V, Morales AC, Cruz RT, Abdulrachman S. 1999. On-farm adaptation of knowledge-intensive nitrogen management technologies for rice systems. Nutr. Cycl. Agroecosyst. 53:59-69.

Cabrera JMC, Roetter RP, van Laar HH. 1998. Preliminary results of crop model development and evaluation of rice. In: Roetter RP, Hoanh CT, Teng PS, editors. A systems approach to analyzing land use options for sustainable rural development in South and Southeast Asia. IRRI Discussion Paper Series No. 28. Manila (Philippines): International Rice Research Institute. p 40-57.

Dobermann A, Cassman KG, Peng S, Tan PS, Phung CV, Sta. Cruz PC, Bajita JB, Adviento MAA, Olk DC. 1996. Precision nutrient management in intensive irrigated rice systems. In: Proceedings of the International Symposium on Maximizing Sustainable Rice Yields Through Improved Soil and Environmental Management, 11-17 November 1996, Khon Kaen, Thailand. Bangkok: Department of Agriculture, Soil and Fertilizer Society of Thailand, Department of Land Development, ISSS. p 133-154.

Dobermann A, Fairhurst TH. 2000. Rice: nutritional disorders and nutrient management. Singapore, Makati City (Philippines): Potash and Phosphate Institute, International Rice Research Institute. 191 p.

Dobermann A, White PF. 1999. Strategies for nutrient management in irrigated and rainfed lowland rice systems. Nutr. Cycl. Agroecosyst. 53:1-18.

Frère M, Popov GF. 1979. Agrometeorological crop monitoring and forecasting. FAO Plant Production and Protection Paper 17. Rome (Italy): FAO.

Hoa NM, Singh U, Samonte HP. 1998. Potassium supplying capacity of some lowland rice soils in the Mekong Delta (Vietnam). Better Crops Int. 12:11-15.

Lai NX, Tuan TQ. 1997. Reversing trends of declining productivity in intensive irrigated rice systems: the socioeconomic component. Omonrice 5:82-87.

Olk DC, Cassman KG, Simbahan GC, Sta. Cruz PC, Abdulrachman S, Nagarajan R, Tan PS, Satawathananont S. 1999. Interpreting fertilizer-use efficiency in relation to soil nutrient-supplying capacity, factor productivity, and agronomic efficiency. Nutr. Cycl. Agroecosyst. 53:35-41.

Phung CV, Tan PS, Olk DC, Cassman KG. 1998. Indigenous nitrogen supply and N use efficiency in rice farms of the Mekong delta in Vietnam. In: Proceedings of the 3rd Asian Crop Science Conference, Taichung, Taiwan. p 279-295.

Supit I, Hooijer AA, Van Diepen CA, editors. 1994. System description of the WOFOST 6.0 crop simulation model implemented in CGMS. Volume 1. Theory and algorithms. Luxemburg: European Commission, Directorate-General XIII Telecommunications, Information Market and Exploitation of Research. p 1-144.

Tan PS. 1997. Nitrogen use efficiency in relation to indigenous soil N supply in Mekong Delta intensive lowland rice system. Omonrice 5:14-22.

Tan PS, Anh TN, Luat NV, Puckridge DW. 1995. Yield trends of a long-term NPK experiment for intensive rice monoculture in the Mekong river delta of Vietnam. Field Crops Res. 42:101-109.

Witt C, Dobermann A, Abdulrachman S, Gines HC, Wang GH, Nagarajan R, Satawathananont S, Son TT, Tan PS, Tiem LV, Simbahan GC, Olk DC. 1999. Internal nutrient efficiencies of irrigated lowland rice in tropical and subtropical Asia. Field Crops Res. 63:113-138.

Notes

Acknowledgments: We are grateful to the farmers cooperating with us in this project. The authors are much obliged to Dr. Nguyen Van Luat, formerly director of CLRRI and Irrigated Rice Research Consortium steering committee member, for his support of our research. Sincere thanks are due to Dr. Bui Ba Bong, vice minister of the Ministry of Agriculture and Rural Development (MARD) and former director of CLRRI for his constant support and encouragement during this study. We wish to thank Piedad Moya, Don Pabale, and Marites Tiongco for help with the socioeconomic farm monitoring, Ma. Arlene Adviento for help with the soil and plant analysis work, and Gregorio C. Simbahan, Rico Pamplona, Olivyn Angeles, Julie Mae Criste Cabrera-Pasuquin, and Edsel Moscoso (all IRRI) for help with data management and statistical analysis. Appreciation and thanks go to Truong Thi Canh, Nguyen Thi Hinh, and Nguyen Thi Reo for their assistance in laboratory work. The SysNet project coordinated by Dr. Reimund Roetter (IRRI) provided data for modeling long-term trends of yield potential. We acknowledge contributions made by other researchers involved in this work, including Christian Witt, David Dawe, and Pompe Sta. Cruz of IRRI.

Authors' addresses: Pham Sy Tan, Tran Quang Tuyen, Tran Thi Ngoc Huan, Trinh Quang Khuong, Nguyen Thanh Hoai, Le Ngoc Diep, Ho Tri Dung, Cao Van Phung, Nguyen Xuan Lai, Cuu Long Delta Rice Research Institute, Omon, Cantho, Vietnam, e-mail: pstan@hcm.vnn.vn; A. Dobermann, International Rice Research Institute, Los Baños, Philippines, and University of Nebraska, Lincoln, Nebraska, USA.

Citation: Dobermann A, Witt C, Dawe D, editors. 2004. Increasing productivity of intensive rice systems through site-specific nutrient management. Enfield, N.H. (USA) and Los Baños (Philippines): Science Publishers, Inc., and International Rice Research Institute (IRRI). 410 p.

11 | Site-specific nutrient management in irrigated rice systems of the Red River Delta of Vietnam

Tran Thuc Son, Nguyen Van Chien, Vu Thi Kim Thoa, A. Dobermann, and C. Witt

11.1 Characteristics of rice production in the Red River Delta

Trends

The North Vietnamese project sites are located in the Red River Delta (Fig. 11.1), the most intensively cropped agricultural land in Vietnam. The Red River Delta (RRD) accounts for about 20% of the national rice production and rice produced there is mainly used for local consumption. The total harvested rice area in the RRD is about 1 million hectares (Table 11.1), with a physical land area of about 575,000 to 600,000 ha, mostly found in the lowlands along the Red River and in coastal regions. About 75% of the harvested rice was grown with irrigation in the late 1990s. The sub-

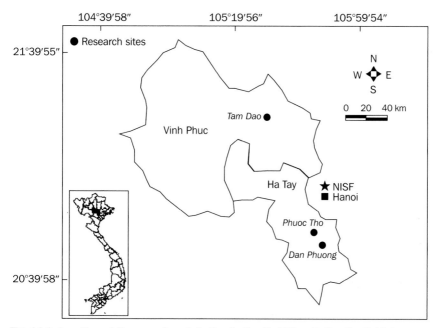

Fig. 11.1. Location of the experimental sites in the Red River Delta, North Vietnam.

Table 11.1. Rice area, yield, and production in the Red River Delta, North Vietnam.

Item	1975-79	1980-84	1985-89	1990-94	1995-99
Red River Delta					
Rice area (1,000 ha)	1,028	1,030	1,047	1,031	1,041
Irrigated rice area (%)	–	60[a]	65[a]	70[a]	73[a]
Rice yield (t ha^{-1})	2.42	2.74	3.07	3.81	4.91
Rice production (1,000 t)	2,474	2,832	3,235	3,937	5,114
Ha Tay Province (alluvial soil)					
Rice area (1,000 ha)	–	–	157	164	167
Rice yield (t ha^{-1})	–	–	2.64	3.40	4.42
Rice production (1,000 t)	–	–	415	559	737
Vinh Phuc Province (degraded soil)					
Rice area (1,000 ha)	–	–	127	140	140
Rice yield (t ha^{-1})	–	–	2.45	2.53	3.20
Rice production (1,000 t)	–	–	312	354	448

[a]Data refer to 1980, 1985, 1989, 1990, and 1995, respectively.
Sources: Statistical data of the Socialist Republic of Vietnam, 1975 to 2001, Statistical Publishing House, Hanoi, 2001.

tropical climate in this region is characterized by a cool winter season and a hot and humid summer (Fig. 11.2). Rainfall averages about 1,700 mm per year, most of this occurring from May to September. Farmers typically grow three crops per year, spring or early rice (ER) from February to June, summer or late rice (LR) from June to September, and a winter crop from October to January. The potential land area for growing winter crops is about 60% of the rice area, but only 40% of the rice area is currently cultivated with winter crops. The yield potential of rice is greater in the ER crop than in LR because the latter is often affected by heavy rainfall or storms and farmers grow varieties with shorter growth duration in summer to be able to plant the winter crop early enough. Recently, some farmers started growing hybrid rice with greater yield potential in both rice crops.

The RRD is probably one of the most intensively cultivated agricultural areas in the world, in terms of both cropping intensity and the cumulative amount of grain produced per year. Good lowland farmers routinely produce 15 to 18 t of grain ha^{-1} each year in a system that is traditionally community-based, very labor-intensive, and involving much recycling of nutrients (Bray 1998). Wet cultivation of transplanted rice started about 4,000 years ago and has intensified since then because of the dramatic increases in population density in the fertile RRD, which now exceeds 2,000 people km^{-2} in several provinces.

The recent changes in rice production in the RRD (Table 11.1) reflect the socioeconomic changes in agriculture. Until 1988, in a cooperative and state farm system, crop land was redistributed to farmers' families every year and the cooperatives controlled most farming operations. The new economic policies introduced in 1986

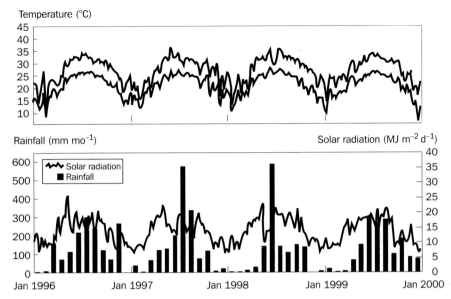

Temperature (°C)

Rainfall (mm mo⁻¹) Solar radiation (MJ m⁻² d⁻¹)

Solar radiation
Rainfall

Jan 1996   Jan 1997   Jan 1998   Jan 1999   Jan 2000

Fig. 11.2. Climatic conditions at Hanoi, Red River Delta, Vietnam, 1996 to 1999. Solar radiation and temperature data are 7-day moving averages; rainfall is monthly total. Solar radiation was obtained from sunshine hours using the standard coefficients for dry tropical conditions of 0.25 for a and 0.45 for b (Frère and Popov 1979) for Hanoi (21°1′N, 105°50′E).

now allow farmers to rent the same land for at least 10 years and give them more flexibility in deciding what and how to grow, but cooperatives still have major functions in irrigation management and the dissemination of new varieties and technical information (Bray 1998). During the past 20 years, short-duration high-yielding rice and maize varieties were rapidly adopted by farmers. Rice-rice-maize cropping became widespread, also because of the large labor surplus in the rural areas of the RRD and because of good irrigation systems.

The rice area has remained constant in the last 25 years (Table 11.1). Large increases in rice yields, from 2.8 t ha⁻¹ in 1987 to 5.4 t ha⁻¹ in 2000, were the driving force for sustaining growth rates of total rice production on the order of 5.6% y⁻¹ from 1988 to 2000. The exceptionally high yield growth rates were associated with significant increases in the irrigated area (from 65% in the late 1980s to about 75% in the late 1990s) and the use of agrochemicals in addition to the traditional prevailing use of farmyard manure, green manure, and human excrement. Total N fertilizer consumption in the Red River Delta nearly doubled from 1980 to 1997 (data not shown). Strong efforts were made to promote a more balanced use of fertilizers as farmers intensified their systems. Consumption of P fertilizer and K fertilizer increased significantly after the introduction of the new agricultural policies. For example, K use has grown at annual rates of about 20% since 1990 (data not shown), a factor that is probably essential for sustaining the productivity of land used at the cropping intensity found in the RRD.

Current biophysical and socioeconomic farm characteristics

The experimental domain is located in the Red River Delta around Hanoi ($21°1'N$, $105°53'E$) and includes 24 farms in three districts: Dan Phuong (4 farms) and Phuc Tho (8) in Ha Tay Province and Tam Dao in Vinh Phuc Province (Fig. 11.1). Tam Dao is located on degraded soils about 50 km northwest of Hanoi, whereas Phuc Tho and Dan Phuong are about 25 km southwest of Hanoi on alluvial lowland soils. All sites have a high population density, very small farm and field sizes, and intensive triple cropping in which fallow periods between harvest of a crop and planting of the succeeding crop barely exceed a few days.

Biophysical and socioeconomic data collection started on all farms with the 1997 ER crop (see Chapter 2). Farm sizes range from 0.1 to 0.6 ha, with a median of 0.3 ha (Table 11.2). On average, individual fields are only about 500 m^2 (range 200–1,000 m^2). Farmers rent their land on a long-term basis (10 years) from the government, but each farmer typically manages about 6–7 small plots that are not adjacent to each other. Recent policies encourage farmers to exchange their parcels to create somewhat larger land areas managed by the same family. Farmers in our sample had a median age of 49, 7 years of school education, and a family of six people.

All rice is transplanted and the total labor input is probably the highest in irrigated rice areas of Asia. It typically ranges from 220 to 280 8-hour person-days ha^{-1} $crop^{-1}$ (Table 11.2), almost all of it manual family labor. Interestingly, the median labor input in our study (230 labor days ha^{-1} $crop^{-1}$) was the same as measured in a socioeconomic study of Nguyen Xa village in Thai Binh Province (Le et al 1993), suggesting that these numbers are fairly representative for large parts of the RRD. Soil tillage is done by using buffalos or small tractors. Surface water originating from the Red River is used for irrigation. Many farmers follow the existing fertilizer recommendations so that relatively balanced amounts of N, P, and K are applied. Compared with other farmers in Asia, farmers in the RRD apply only moderate amounts of N (an average of about 100 kg N ha^{-1} $crop^{-1}$) and P and K at rates that are probably close to what is removed with the harvest product and not recycled to the field. Average K rates of 40 to 50 kg K ha^{-1} $crop^{-1}$ are common, about double those of farmers in South Vietnam (see Chapter 10). Farmers usually apply N in three to four splits and fungicide sprays (mostly Validamycin) are common, particularly in the humid summer-autumn (LR) crop. At harvest, rice or maize plants are cut at the soil surface by hand and all straw or maize stalks are completely removed to facilitate quick planting of the next crop.

In general, soil fertility declines in the order Dan Phuong (alluvial soil, 4 farms) > Phuc Tho (alluvial soil, 8 farms) > Tam Dao (degraded soil, 12 farms) and pest infestations are severe. Land at Tam Dao is very favorable for growing maize but less favorable for rice, whereas the reverse is true for the lowlands at Phuc Tho and Dan Phuong. Winter maize at Tam Dao is planted in early September so that tasseling and grain filling occur during warmer periods. In contrast, on the heavier soils at Dan Phuong, anaerobic conditions prevail longer and reduce maize growth. The degraded soils at Tam Dao are low in clay, organic matter, CEC, and exchangeable bases, whereas Olsen-P varied from 5 to 21 mg kg^{-1} (Table 11.3). The wide Ca:Mg ratio

Table 11.2. Demographic and economic characteristics of rice production on 24 farms in the Red River Delta, Vietnam. Values shown are based on socioeconomic farm surveys conducted for whole farms.

Production characteristics	Min.	25%	Median	75%	Max.
Total cultivated area (ha)	0.14	0.18	0.32	0.39	0.55
Age of household head (y)	27	39	49	57	68
Education of household head (y)	1	6	7	10	10
Household size (persons)	2	5	6	7	10
1998 spring rice					
Yield (t ha [1])	4.67	5.41	6.08	6.41	7.50
N fertilizer (kg ha [1])	63	79	98	126	167
P fertilizer (kg ha [1])	0	9	18	20	40
K fertilizer (kg ha [1])	0	42	50	73	110
Insecticide (kg ai ha [1])	0	0	0.02	0.70	2.27
Herbicide (kg ai ha [1])	0	0	0.77	1.76	2.61
Other pesticides (kg ai ha [1])[a]	0	0.25	0.37	1.00	2.55
Total labor (8-h d ha [1])	201	219	233	286	352
Net return from rice (US$ ha [1] crop [1])	746	898	1,052	1,095	1,273
1998 summer rice					
Yield (t ha [1])	4.10	4.57	5.50	6.30	7.45
N fertilizer (kg ha [1])	64	90	102	115	146
P fertilizer (kg ha [1])	0	6	15	19	29
K fertilizer (kg ha [1])	20	41	42	73	132
Insecticide (kg ai ha [1])	0	0	0.64	1.02	3.94
Herbicide (kg ai ha [1])	0	0	0.29	1.73	3.91
Other pesticides (kg ai ha [1])[a]	0	0.14	0.23	0.36	4.58
Total labor (8-h d ha [1])	188	216	228	270	310
Net return from rice (US$ ha [1] crop [1])	670	860	997	1,172	1,411

[a]Includes fungicide, molluscicide, rodenticide, and crabicide.

(usually 10 to 15:1, in one case even 40:1) presents a specific problem. Considering that for very high rice yields this ratio should be about 4:1, Mg deficiency is likely to limit yields. The alluvial soils at Phuc Tho and Dan Phuong have a higher CEC and clay and organic matter content and a narrower Ca:Mg ratio (4 to 6:1). However, Olsen-P was less than 10 mg kg^{-1} on 8 of the 12 farms at these sites. Available Zn was mostly >1 mg kg^{-1} so that no deficiency is likely to occur. Soil fertility varied widely among farms (Table 11.3).

In 1997 to 1998, N, P, and K omission plots were established to obtain estimates of the indigenous N (INS = plant N accumulation in a 0-N plot), P (IPS = plant P accumulation in a 0-P plot), and K supply (IKS = plant K accumulation in a 0-K plot) in rice. In general, threefold ranges in indigenous N, P, and K supply occurred among the 12 farmers' fields at each site with alluvial or degraded soil (Fig. 11.3). The INS, IPS, and IKS followed the expected gradient in soil fertility (Dan Phuong ≥

Table 11.3. General soil properties on 24 rice farms of the Red River Delta, Vietnam.

Soil properties[a]	Alluvial soil[b]		Degraded soil[c]	
	Mean	SD	Mean	SD
Clay content (%)	32.3	5.7	11.0	5.4
Silt content (%)	63.8	5.0	64.8	4.3
Sand content (%)	3.9	2.2	24.2	7.6
Soil organic C (g kg^{-1})	18.7	4.0	10.8	2.0
Total soil N (g kg^{-1})	2.2	0.4	1.2	0.2
Soil pH (1:1 H$_2$O)	5.0	0.2	4.9	0.3
Cation exchange capacity (cmol$_c$ kg^{-1})	11.4	1.2	5.0	0.9
Exchangeable K (cmol$_c$ kg^{-1})	0.19	0.04	0.13	0.05
Exchangeable Na (cmol$_c$ kg^{-1})	0.15	0.05	0.07	0.03
Exchangeable Ca (cmol$_c$ kg^{-1})	6.89	1.27	2.80	0.64
Exchangeable Mg (cmol$_c$ kg^{-1})	1.22	0.23	0.16	0.08
Extractable P (Olsen-P, mg kg^{-1})	8.92	3.18	10.28	4.62
Extractable Zn (0.05N HCl, mg kg^{-1})	1.92	0.46	1.53	0.36

[a]Measured on initial soil samples collected before the 1997 ER. [b]Mean and standard deviation (SD) of 12 farms near Dan Phuong (4) and Phuc Tho (8). [c]Mean and standard deviation (SD) of 12 farms near Tam Dao.

Phuc Tho > Tam Dao). However, within each site, CVs of INS, IPS, and IKS were typically 15–30%. In most crops sampled on alluvial soil, median nutrient supplies were about 60 kg N ha^{-1} for INS, about 16 kg P ha^{-1} for IPS, and about 75 kg K ha^{-1} for IKS (Fig. 11.3). On degraded soil, the corresponding median values were much lower, with only about 50 kg N ha^{-1} for INS, about 13 kg P ha^{-1} for IPS, and about 60 kg K ha^{-1} for IKS. Note that IKS was very low (<40 kg K ha^{-1}) on about 10% of all farms. The current average indigenous nutrient supplies on alluvial soil were sufficient for achieving rice yields of about 5 t ha^{-1} without applying N and 6 t ha^{-1} without applying P or K. The corresponding nutrient-limited yields on degraded soil were 3.5 t ha^{-1} without applying N, 5 t ha^{-1} without applying P, and 4.5 t ha^{-1} without applying K. Potassium appears to be the second most yield-limiting nutrient on degraded soils. Grain yield measured in omission plots is probably a more practical and sufficiently robust indicator of soil nutrient supplies than plant nutrient uptake (Dobermann et al 2003). The limited data set from the RRD showed a strong correlation between indigenous nutrient supplies measured as plant nutrient uptake (y) and grain yield (x) for N (y = 12.7x, r^2 = 0.57), but the relationship was weaker for P (y = 2.6x, r^2 = 0.20) and K (y = 13.1x, r^2 = 0.27). However, the slopes of these regressions correspond well with the plant nutrient requirements of 14.7 kg N, 2.6 kg P, and 14.7 kg K per t grain yield estimated using QUEFTS (Witt et al 1999). There were no consistent differences in measurements of indigenous nutrient supplies among ER and LR seasons, but more research is needed to clarify this. Note that the omission plots did not receive farmyard manure (FYM), which needs to be taken into

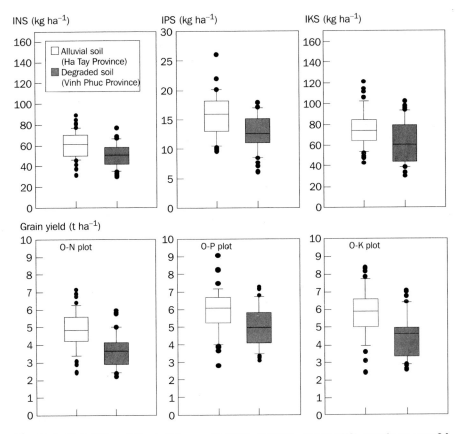

Fig. 11.3. Variability of the indigenous N (INS), P (IPS), and K (IKS) supply among 24 farmers' fields on alluvial ($n = 12$) and degraded soil ($n = 12$) of the Red River Delta, North Vietnam (1997-98). Median with 10th, 25th, 75th, and 90th percentiles as vertical boxes with error bars; outliers as bullets.

account when developing fertilizer recommendations. For practical reasons, we recommend applying FYM in nutrient omission plots (Witt et al 2002).

Baseline agronomic data were measured in the farmers' fertilizer practice (FFP) in 1997 (Table 11.4). Yields were highest in the longer-duration spring rice crop, mainly because of better grain filling (88%) than in the summer rice crop (75%). Grain yield and plant N and P uptake in the FFP typically increased in the order Dan Phuong (alluvial soil) > Phuc Tho (alluvial soil) > Tam Dao (degraded soil). Pest incidence was highest on the degraded soil. Farmers applied relatively high amounts of K fertilizer (50 kg K ha^{-1}) to each rice crop, but less N and P in LR than in ER because of the lower yield potential. A specific feature of rice-rice-maize cropping in the RRD is that practically all farmers apply farmyard manure and often also human excrement ("night soil"). Rates are difficult to measure exactly because, particularly in maize, manure application is often done frequently. Our data suggest that farmers

Table 11.4. Baseline agronomic characteristics of rice production on 24 farms at Hanoi, North Vietnam. Values shown are means and standard deviations (SD) of the farmers' fertilizer practice for two consecutive rice crops monitored before SSNM plots were established.

Agronomic characteristics[a]	1997 spring rice		1997 summer rice	
	Mean	SD	Mean	SD
Grain yield (t ha $^{-1}$)	6.07	1.45	4.74	1.00
Harvest index	0.51	0.05	0.45	0.05
No. of panicles m $^{-2}$	358	71	356	58
Total no. of spikelets m $^{-2}$	25,906	7,696	27,489	4,331
Total no. of spikelets panicle $^{-1}$	73	17	79	14
Filled spikelets (%)	88	4	75	7
Fertilizer N use (kg ha $^{-1}$)	97.2	21.5	83.3	13.5
Fertilizer P use (kg ha $^{-1}$)	18.7	8.4	13.3	9.3
Fertilizer K use (kg ha $^{-1}$)	50.9	20.0	50.2	18.7
N uptake (kg ha $^{-1}$)	96.1	17.2	85.1	17.6
P uptake (kg ha $^{-1}$)	13.5	4.4	18.8	3.7
K uptake (kg ha $^{-1}$)	90.8	15.5	102.6	22.8
Input-output N balance (kg ha $^{-1}$ crop $^{-1}$)[a]	11.5	15.1	31.9	10.0
Input-output P balance (kg ha $^{-1}$ crop $^{-1}$)[a]	22.1	9.5	11.5	9.9
Input-output K balance (kg ha $^{-1}$ crop $^{-1}$)[a]	13.3	23.7	1.0	30.1
Partial productivity of N (kg kg $^{-1}$)	60.0	16.3	57.8	13.0
Agronomic efficiency of N (kg kg $^{-1}$)	11.5	13.5	18.5	8.9
Recovery efficiency of N (kg kg $^{-1}$)	0.37	0.18	0.50	0.14
Physiological efficiency of N (kg kg $^{-1}$)	21.9	19.0	36.2	14.4

[a]Includes estimated nutrient input from farmyard manure.

typically apply about 10–12 t of fresh manure ha^{-1} crop^{-1}, which is equivalent to nutrient inputs of about 30–35 kg N, 20–25 kg P, and 50–60 kg K ha^{-1} crop^{-1} (Table 11.5). Strategies for site-specific nutrient management (SSNM) must take into account nutrient input and residual effects of farmyard manure.

Unlike in most other parts of Southeast Asia, the combined use of mineral fertilizer and FYM in the RRD appears to sustain positive nutrient balances, despite the complete removal of all crop residues. In some years, even roots are pulled out and used to feed cattle or as fuel. Another impressive feature of intensive rice farming in the RRD is the relatively high N-use efficiencies achieved by many farmers, which probably result from the intensive crop care. Although recovery efficiency of N (REN) and agronomic efficiency of N (AEN) varied widely among farms, the average REN of 0.39 kg kg^{-1} and AEN of 17.9 kg kg^{-1} were among the highest measured in farmers' fields at our project sites in Asia (Dobermann et al, Chapter 15, this volume). Moreover, the average partial factor productivity of N (PFPN) was 58–60 kg kg^{-1} in both seasons. Note, however, that we probably overestimated AEN and REN because no FYM was applied to the 0-N plots, whereas farmers applied FYM in the FFP. Again, we strongly suggest installing omission plots in farmers' fields with FYM when estimating indigenous nutrient supplies for the development of fertilizer recommendations.

Table 11.5. Rates and estimated nutrient input of farmyard manure (FYM) applied to site-specific nutrient management (SSNM), the farmers' fertilizer practice (FFP) in the same field, and a neighboring field operated by the same or a different farmer (FFP2).

	1998		1999		
	SSNM	FFP	SSNM	FFP	FFP2
Spring rice					
FYM (t fresh weight ha [1])					
Mean	8.5	11.2	8.5	12.5	8.5
SD	0.5	2.3	0.5	3.9	1.4
Min.	8.0	7.7	8.0	8.0	6.0
Max.	9.0	16.5	9.0	19.0	11.0
Average nutrient input (kg ha [1])[a]					
N	26	34	26	38	26
P	17	22	17	25	17
K	43	56	43	63	43
Summer rice					
FYM (t fresh weight ha [1])					
Mean	8.5	12.7	8.4	11.8	10.2
SD	0.5	2.6	0.5	4.1	2.2
Min.	8.0	8.3	7.0	8.0	8.0
Max.	9.0	18.6	9.0	25.0	15.0
Average nutrient input (kg ha [1])[a]					
N	26	38	25	35	31
P	17	25	17	24	20
K	43	64	42	59	51

[a]Assuming an average nutrient content of 3 kg N, 2 kg P, and 5 kg K per ton of FYM.

In summary, farmers in the RRD always had to cope with the challenge of maximizing food production without compromising fertility because of the scarcity of land. In the past, they have done so by implementing a community-based farming system with intensive crop care, the careful use of natural resources, and recycling of nutrients and organic matter. With increasing population growth and industrialization, further fine-tuning of crop management is required to account for variation in soil fertility, but productivity increases are likely to become smaller than those achieved during the past 20 years because many farmers routinely achieve high yields already.

11.2 Effect of SSNM on productivity and nutrient-use efficiency

Management of the SSNM plots

Beginning with the 1998 ER crop, a site-specific nutrient management plot (SSNM) was established on each of the 24 farm fields as a comparison with the farmers' fertilizer practice (FFP, see Chapter 2). Only results of the four rice crops grown in

1998 and 1999 will be presented in this paper. The size of the SSNM plots ranged from 70 to 160 m^2 (average 110 m^2) in 1998 and from 150 to 370 m^2 (average 220 m^2) in 1999, depending on the size of the farmers' fields in which the SSNM and FFP plots were located. Rice varieties were chosen by the farmers. During the experimental period, farmers planted 13 different semidwarf modern Chinese and Vietnamese varieties. The most popular ER varieties were DT10 (about 60% of all farmers), C70, C71, DT13, Q5, and Khang Dan. Popular LR varieties included Quangte, Vi Di, CR203, CN2, and Khang Dan. All rice was transplanted at a density of about 45–50 hills m^{-2} on alluvial soil and 50–55 hills m^{-2} on degraded soil using 50–60-d-old seedlings in ER and 15–20-d-old seedlings in LR. Some farmers also use the dapog method to produce seedlings in early (spring) rice, planted at an age of 15–20 d. Farmers did all water management and pest control following their commonly used methods in both the FFP and SSNM. Rice straw was completely removed from the fields after harvest by hand.

Fertilizer applications for SSNM were prescribed on a field- and crop-specific basis following the approach described in Chapter 5 and descriptions given elsewhere (Dobermann and White 1999, Witt et al 1999). In 1998 ER, values of the INS, IPS, and IKS measured in the 1997 ER cropping season were used as model inputs and the target yield was set to 6.5 t ha^{-1} at Tam Dao, 7 t ha^{-1} at Phuc Tho, and 9 t ha^{-1} at Dan Phuong. In 1998 LR, average INS, IPS, and IKS measured during the 1997 ER and LR crop were used as model inputs and the target yield ranged from about 5 t ha^{-1} at Tam Dao to 6–6.5 t ha^{-1} at Phuc Tho and Dan Phuong. The lower yield target at Tam Dao was chosen because farmers there aim at growing a short-duration LR crop to be able to plant maize as early as possible. In 1999 ER, average INS, IPS, and IKS measured in 1997 and 1998 ER were used as model inputs, with IPS and IKS also adjusted for the nutrient balance measured in the 1998 ER crop. Target yields were 6.5 t ha^{-1} at Tam Dao, 7.5 t ha^{-1} at Phuc Tho, and 8 t ha^{-1} at Dan Phuong. In 1999 LR, average INS, IPS, and IKS measured in 1997 and 1998 ER and 1998 LR were used as model inputs, with IPS and IKS also adjusted for the nutrient balance measured in the 1998 ER and LR crops. Target yields for 1999 LR were 5.5 t ha^{-1} at Tam Dao and 6.5 t ha^{-1} at Phuc Tho and Dan Phuong. First-crop recovery fractions of 0.5, 0.2–0.25, and 0.4–0.6 kg kg^{-1} were assumed for fertilizer N, P, and K, respectively. The latter was assumed to be the highest (0.5–0.6 kg kg^{-1} on the degraded soils at Tam Dao). Nutrient addition from FYM was treated as an increase in indigenous nutrient supply to reduce mineral fertilizer rates accordingly. We used average contents of 3 kg N, 2 kg P, and 5 kg K per ton fresh manure and assumed recovery efficiencies of 0.5, 0.2, and 0.5 kg kg^{-1}, respectively. No crop simulation data were available to estimate the climatic yield potential. The yield potential at Dan Phuong and Phuc Tho was assumed to be 9.5–10 t ha^{-1} in ER and 7.5 t ha^{-1} in LR. At Tam Dao, yield potentials of 8.5 t ha^{-1} in ER and 6.5 t ha^{-1} in LR were used, taking into account that the genetic yield potential at this site is limited not only by climate but also by soil fertility. Fertilizer sources used in SSNM were urea (46% N), single superphosphate (8.3% P), and muriate of potash (50% K). All P fertilizer was incorporated into the soil before transplanting (100% basal). K fertilizer was split into 25% basal, 25% 20–25 days after transplanting (DAT), and 50% at panicle initiation (PI) in 1998, or 50% basal and 50% at PI in 1999.

In 1998, fertilizer-N applications followed a modification of available local recommendations and adjustments based on visual judgment of growth. In addition, chlorophyll meter readings were collected to determine the time of N application at critical growth stages. The regime typically followed in the 1998 ER crop was

FN1	Basal	20–30 kg N ha^{-1}	
FN2	20–25 DAT	30–40 kg N ha^{-1}	if SPAD <35
FN3	40–50 DAT	40–50 kg N ha^{-1}	if SPAD <35
FN4	60–80 DAT	10–20 kg N ha^{-1}	if SPAD <35

A similar approach was used in 1998 LR, with N applied in four splits (basal, 15–20 DAT, 30–35 DAT, and 40–50 DAT). For comparison, farmers also applied N in 3–4 splits, but with widely varying dates and amounts. Beginning with the 1999 ER crop, SPAD-based N management was practiced in SSNM. In 1999 ER, N was applied as two fixed applications and up to three more topdressings depending on SPAD readings at critical growth stages:

FN1	Basal		20 kg N ha^{-1}
FN2	20–25 DAT		30 kg N ha^{-1}
FN3	28–35 DAT (MT)	If SPAD >37	20 kg N ha^{-1}
		If SPAD 35–37	30 kg N ha^{-1}
		If SPAD 33–35	40 kg N ha^{-1}
		If SPAD <33	50 kg N ha^{-1}
FN4	40–50 DAT (PI)	If SPAD >35	40 kg N ha^{-1}
		If SPAD 33–35	50 kg N ha^{-1}
		If SPAD <33	60 kg N ha^{-1}
FN5	55–65 DAT (H-F)	If SPAD >36	No N
		If SPAD 33–36	20 kg N ha^{-1}
		If SPAD <33	30 kg N ha^{-1}

In 1999 LR, N was applied only topdressed at critical growth stages:

FN1	15–20 DAT		20 kg N ha^{-1}
FN2	30 DAT (LT)	If SPAD >37	20 kg N ha^{-1}
		If SPAD 35–37	30 kg N ha^{-1}
		If SPAD <35	40 kg N ha^{-1}
FN3	40–45 DAT (PI)	If SPAD >37	20 kg N ha^{-1}
		If SPAD 35–37	30 kg N ha^{-1}
		If SPAD <35	40 kg N ha^{-1}

In interpreting the results, note that the first SSNM crop did not receive FYM, whereas the FFP did (Table 11.5) because our initial goal was to substitute nutrients with mineral fertilizer alone. Because FYM application is a standard practice in the RRD, beginning with the 1998 LR crop, FYM was applied to both SSNM and the FFP. Manure rates in SSNM were typical average rates used by farmers at each site, but there were clear signs that the farmers tried to compete with SSNM by adding much more FYM to the FFP. For example, in the 1999 ER crop, average FYM use was 8.5 t ha^{-1} in SSNM but 12.5 t ha^{-1} in the FFP, and the latter rate was also about 50% higher that that used in neighboring farmers' fields.

Effect of SSNM on grain yield and nutrient uptake

Averaged over all sites and crops, SSNM resulted in a small yield increase of 0.19 t ha^{-1} (3%) over the FFP, which was not statistically significant (Table 11.6). However, the performance of SSNM improved with time and the comparison with FFP was confounded by the different rates of FYM used in both treatments. Researchers also observed that farmers put special emphasis on crop care in the FFP plot sampled by the researchers. On average, nutrient input from manure was 10–12 kg N ha^{-1}, 7–8 kg P ha^{-1}, and 15–20 kg K ha^{-1} greater in the FFP than in SSNM (Table 11.5), plus extra benefits such as micronutrients, organic matter, etc. This must be taken into account when comparing the two treatments, even though the exact direct and indirect benefits of this extra manure cannot be quantified because of varying composition and lack of detailed measurements. Therefore, we will discuss the performance of SSNM in absolute terms, not purely based on statistical comparison with the FFP.

In the first crop, 1998 ER, average grain yields were similar in SSNM and the FFP (about 5.8 t ha^{-1}), even though no FYM was applied in SSNM, which led to a reduction in N and P uptake (Fig. 11.4). Small yield increases from SSNM were observed on 10 farms, but yields at Dan Phuong decreased by about 2.5 t ha^{-1} because of a storm causing lodging and sheath blight during grain filling. Subsequently, major improvements made in SSNM included application of FYM and adjustment of the mineral fertilizer rates for it (beginning with 1998 LR) as well as SPAD-based N management (1999). The yield increase over the FFP in 1999 (0.31 t ha^{-1}) was significantly larger ($P = 0.026$) than that achieved in 1998 (0.05 t ha^{-1}), indicating that these improvements improved the yield performance over time despite the strong "competition" by the farmers.

Since the 1998 LR crop, average SSNM yields were consistently high (Fig. 11.4), close to the yield target, and significantly larger than the yields measured in 1997 before SSNM was introduced (Table 11.4). For example, average yield was 5.73 t ha^{-1} in 1998 LR (96% of the average target yield) and yields on 18 farms were higher than in the FFP. In 1999 ER, climate was favorable and the average yield reached 6.70 t ha^{-1} (94% of the average target yield), with little variation among farms (CV = 13%). In 1999 LR, the average yield in SSNM was 6.62 t ha^{-1} or 109% of the target yield. The correlation between predicted and actual SSNM yield was r = 0.68 for pooled data from 1998 LR, 1999 ER, and 1999 LR. These results confirm the suitability of the QUEFTS model for predicting fertilizer needs when constraints other than N, P, and K are minimized and the great yield stability achieved with SSNM.

Table 11.6. Effect of site-specific nutrient management on agronomic characteristics on 24 rice farms of the Red River Delta, North Vietnam (1998-99).

Parameters	Levels[a]	Treatment[b] SSNM	Treatment[b] FFP	Δ[c]	$P>|T|$[c]	Effects[d]	$P>|F|$[d]
Grain yield (GY)	All	6.24	6.06	0.19	0.202	Village	0.342
(t ha[1])	Year 1	5.82	5.76	0.06	0.780	Year[e]	0.026
	Year 2	6.66	6.34	0.31	0.105	Season[e]	0.640
	ER	6.25	6.07	0.18	0.350	Year × season[e]	0.100
	LR	6.23	6.05	0.18	0.388	Village × crop	0.601
Plant N uptake (UN)	All	95.6	94.2	1.4	0.517	Village	0.391
(kg ha[1])	Year 1	97.1	98.9	−1.8	0.569	Year	0.003
	Year 2	94.2	89.5	4.7	0.125	Season	0.000
	ER	97.4	99.8	−2.4	0.453	Year × season	0.102
	LR	93.9	88.5	5.4	0.072	Village × crop	0.203
Plant P uptake (UP)	All	18.5	19.1	−0.6	0.397	Village	0.028
(kg ha[1])	Year 1	16.2	18.6	−2.4	0.025	Year	0.000
	Year 2	20.8	19.6	1.2	0.122	Season	0.001
	ER	18.7	22.4	−3.6	0.000	Year × season	0.000
	LR	18.3	15.7	2.4	0.000	Village × crop	0.001
Plant K uptake (UK)	All	109.0	114.9	−6.0	0.021	Village	0.001
(kg ha[1])	Year 1	102.5	112.2	−9.7	0.006	Year	0.109
	Year 2	115.4	117.6	−2.3	0.528	Season	0.653
	ER	109.2	114.7	−5.5	0.125	Year × season	0.487
	LR	108.8	115.2	−6.4	0.089	Village × crop	0.525
Agronomic efficiency	All	17.9	13.9	4.0	0.000	Village	0.057
of N (AEN)	Year 1	17.7	15.2	2.5	0.143	Year	0.045
(kg grain kg[1] N)	Year 2	18.1	12.7	5.5	0.000	Season	0.031
	ER	17.4	14.6	2.8	0.090	Year × season	0.007
	LR	18.4	13.2	5.2	0.000	Village × crop	0.077
Recovery efficiency	All	0.39	0.33	0.06	0.029	Village	0.141
of N (REN)	Year 1	0.42	0.41	0.01	0.755	Year	0.002
(kg N kg[1] N)	Year 2	0.35	0.25	0.10	0.004	Season	0.000
	ER	0.41	0.42	−0.01	0.879	Year × season	0.004
	LR	0.36	0.24	0.12	0.000	Village × crop	0.012
Partial productivity	All	69.9	60.6	9.4	0.000	Village	0.025
of N (PFPN)	Year 1	65.4	59.6	5.8	0.109	Year	0.016
(kg grain kg[1] N)	Year 2	74.4	61.5	12.9	0.000	Season	0.000
	ER	65.3	61.2	4.1	0.224	Year × season	0.010
	LR	74.6	59.9	14.8	0.000	Village × crop	0.000

[a]All = all four crops grown from 1998 ER to 1999 LR; year 1 = 1998 spring or early rice (ER) and 1998 summer or late rice (LR); year 2 = 1999 ER and 1999 LR; ER = 1998 ER and 1999 ER; LR = 1998 LR and 1999 LR. [b]FFP = farmers' fertilizer practice; SSNM = site-specific nutrient management. [c]Δ = SSNM − FFP. $P>|T|$ = probability of a significant mean difference between SSNM and FFP. [d]Source of variation of analysis of variance of the difference between SSNM and FFP by farm; $P>|F|$ = probability of a significant F-value. [e]Year refers to two consecutive cropping seasons.

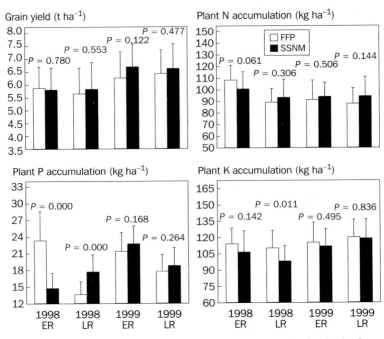

Fig. 11.4. Grain yield and plant N, P, and K accumulation by rice in the farmers' fertilizer practice (FFP) and site-specific nutrient management (SSNM) plots at Hanoi, Vietnam (1998-99; bars: mean; error bars: standard deviation).

Average N uptake differed little between SSNM and the FFP, but the significant crop-year effect ($P = 0.003$) suggests an increase in the difference between N uptake in SSNM and the FFP over time (Table 11.6). Effects on P uptake were year-, season-, and village-specific. SSNM increased plant P accumulation by 17% in LR ($P = 0.000$), whereas it was significantly lower than the FFP in the 1998 ER crop because no manure was applied (Fig. 11.4). Interestingly, the increase in plant P uptake in LR occurred even though total P input was more than 10 kg P ha^{-1} less in SSNM than in the FFP. Significant village ($P = 0.028$) and village × crop ($P = 0.000$) effects on P uptake may be related to differences in FYM rates among the sites and in different seasons. Total K input from mineral fertilizer and FYM was about 40 kg K ha^{-1} less in SSNM than in the FFP, but, with the exception of the 1998 LR crop, plant K accumulation was not statistically different between the two treatments. The significant village effect ($P = 0.001$) on K uptake is probably related to the differences in soil types.

Effect of SSNM on nitrogen-use efficiency

Nitrogen-use efficiencies in SSNM were significantly larger than in the FFP, except in the first SSNM crop grown (Fig. 11.5). In 1998 ER, N management in the SSNM plot was not yet based on SPAD and the AEN, REN, and PFPN tended to be lower than in the FFP because more mineral N and no FYM were applied in SSNM. How-

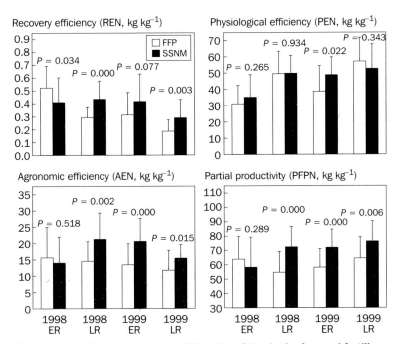

Fig. 11.5. Fertilizer nitrogen-use efficiencies of rice in the farmers' fertilizer practice (FFP) and site-specific nutrient management (SSNM) plots at Hanoi, Vietnam (1998-99; bars: mean; error bars: standard deviation).

ever, the use of FYM in SSNM (since 1998 LR) and improved rules for N management using SPAD (since 1999 ER) led to significant increases in AEN, REN, and PFPP vis-à-vis the farmers' N management.

Across all four crops grown and compared with the FFP, AEN increased by 4.0 kg kg^{-1} (29%, P = 0.000; Table 11.6) and reached high average levels of about 20 kg kg^{-1} in the 1998 LR and 1999 ER crops (Fig. 11.5). Note, however, that the average AEN increase was 5.5 kg kg^{-1} in 1999 (43%) compared with only 2.5 kg kg^{-1} in 1998 (16%), confirming that the use of the chlorophyll meter led to a significant improvement in the congruence of N supply and crop N demand. Likewise, REN increased by an average of 0.06 kg kg^{-1} (18%, P = 0.029), but 0.10 kg kg^{-1} in 1999 (40%, P = 0.004). Physiological N-use efficiency was among the highest measured at the different sites in the RTDP project and averaged about 50 kg kg^{-1} in 1998 LR and the 1999 ER and LR crops, a level that indicates efficient plant internal use of fertilizer N because of excellent crop management in combination with favorable climatic conditions. On average, SSNM increased PFPN by 9 kg kg^{-1} (15%, P = 0.000), but the increase was particularly high in the second year (13 kg kg^{-1}, 21%, P = 0.000).

The different parameters characterizing N-use efficiency indicate the potential for improving it at this site, even though farmers already achieve relatively high AEN and REN in many years and many farmers in our study tried to compete with the N management practiced in the SSNM plots. In many crops sampled in 1998 and 1999,

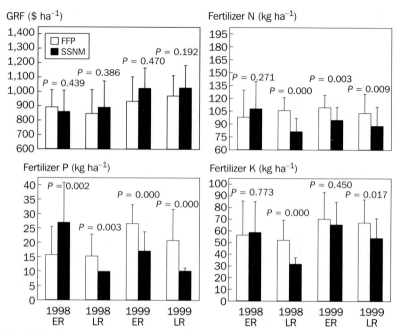

Fig. 11.6. Gross returns above fertilizer costs (GRF) and fertilizer use in the farmers' fertilizer practice (FFP) and site-specific nutrient management (SSNM) plots at Hanoi, Vietnam (1998-99; bars: mean; error bars: standard deviation).

REN exceeded 0.5 kg kg^{-1} and AEN exceeded 25 kg kg^{-1} in the SSNM plots of about 25% farms. With fertilizer rates adjusted properly for the N input from manure, we achieved a very high PFPN of fertilizer N, which averaged about 70 kg kg^{-1}, but exceeded 80 kg kg^{-1} in about 25% of all cases. Not that this kind of fertilizer adjustment is currently not done properly by the farmers. Table 11.6 also indicates significant village × crop effects on N-use efficiency that need further study.

Effect of SSNM on fertilizer use and profit

Site-specific nutrient management decreased the total fertilizer cost by about US$2 ha^{-1} crop^{-1} in 1998 (nonsignificant) and by $22 ha^{-1} crop^{-1} in 1999 (28% less than FFP, $P = 0.000$; Table 11.7) because fertilizer rates were adjusted according to the indigenous N, P, and K supply as well as the nutrient input from FYM (Table 11.5). Because of the variable nature of manure use and its varying composition, this kind of adjustment is currently not practiced by most farmers or is not part of the existing fertilizer recommendations. With the exception of the first crop (1998 ER), rates of mineral N, P, and K applied in the SSNM plots were generally significantly lower than in the FFP (Fig. 11.6), but yields were sustained at the same high levels or even increased (Fig. 11.4). For example, in 1999, the average amount of fertilizer N in SSNM was only 91 kg N ha^{-1} crop^{-1} vis-à-vis 106 kg N ha^{-1} crop^{-1} in the FFP (16%

Table 11.7. Effect of site-specific nutrient management on fertilizer use, fertilizer cost, and gross returns above fertilizer costs from rice production on 24 rice farms of the Red River Delta, North Vietnam (1998-99).

Parameters	Levels[a]	Treatment[b] SSNM	Treatment[b] FFP	Δ[c]	$P > \|T\|$[c]	Effects[d]	$P > \|F\|$[d]
N fertilizer (FN)	All	93	104	−10.8	0.001	Village	0.001
(kg ha^{-1})	Year 1	95	102	−6.9	0.198	Year[e]	0.046
	Year 2	91	106	−14.7	0.000	Season[e]	0.000
	ER	101	104	−2.3	0.651	Year × season[e]	0.006
	LR	85	104	−19.6	0.000	Village × crop	0.000
P fertilizer (FP)	All	16.1	19.6	−3.5	0.018	Village	0.000
(kg ha^{-1})	Year 1	18.7	15.5	3.2	0.167	Year	0.000
	Year 2	13.6	23.7	−10.1	0.000	Season	0.000
	ER	22.0	21.2	0.9	0.704	Year × season	0.011
	LR	10.1	18.1	−8.0	0.000	Village × crop	0.000
K fertilizer (FK)	All	52.8	61.6	−8.9	0.009	Village	0.000
(kg ha^{-1})	Year 1	45.6	54.4	−8.7	0.077	Year	0.767
	Year 2	59.8	68.8	−9.0	0.032	Season	0.000
	ER	62.3	63.4	−1.2	0.820	Year × season	0.164
	LR	42.8	59.3	−16.5	0.000	Village × crop	0.085
Fertilizer cost	All	81.0	93.2	−12.2	0.000	Village	0.037
(US$ ha^{-1})	Year 1	82.2	84.4	−2.2	0.627	Year	0.000
	Year 2	79.8	102.0	−22.2	0.000	Season	0.000
	ER	95.6	96.0	−0.4	0.918	Year × season	0.002
	LR	66.5	90.4	−23.9	0.000	Village × crop	0.000
Gross returns above	All	951	910	46	0.089	Village	0.167
fertilizer costs	Year 1	876	869	7	0.828	Year	0.002
(US$ ha^{-1})	Year 2	1,026	952	74	0.018	Season	0.111
	ER	942	912	31	0.339	Year × season	0.029
	LR	959	909	51	0.158	Village × crop	0.214

[a]All = all four crops grown from 1998 ER to 1999 LR; year 1 = 1998 spring or early rice (ER) and 1998 summer or late rice (LR); year 2 = 1999 ER and 1999 LR; ER = 1998 ER and 1999 ER; LR = 1998 LR and 1999 LR. [b]FFP = farmers' fertilizer practice; SSNM = site-specific nutrient management. [c]Δ = SSNM − FFP. $P > \|T\|$ = probability of a significant mean difference between SSNM and FFP. [d]Source of variation of analysis of variance of the difference between SSNM and FFP by farm; $P > \|F\|$ = probability of a significant F-value. [e]Year refers to two consecutive cropping seasons.

less, $P = 0.000$). Similarly, 74% less P and 15% less K were applied in 1999 in SSNM than in the FFP, but nutrient balance estimates suggested that a close-to-neutral P and K balance was maintained in SSNM (data not shown).

The average profit increase over FFP was $41 ha^{-1} crop^{-1} ($P = 0.089$, Table 11.7), mainly because of the poorer performance of the first SSNM crop (1998 ER, see Fig. 11.6), when nutrient management was not yet optimized. However, profit increases over FFP averaged $74 ha^{-1} crop^{-1} in year 2 ($P = 0.018$), when nutrient management became more fine-tuned. Moreover, GRF estimates as used here do not account for differences in manure use between SSNM and FFP (Table 11.5) and the associated extra cost in the FFP so that the true profit increase is probably even larger.

Farm examples

To illustrate some of the principles of SSNM applied in our study, Table 11.8 shows the performance of SSNM and the FFP in two farmers' fields at Phuc Tho and Tam Dao for four rice crops. Note that both fields are located about 75 km apart from each other on very different soil types.

Mr. Cap's farm is on a very fertile alluvial soil with high INS (75 kg N ha^{-1} crop^{-1}) and he consistently achieved very high rice yields of, on average, 6.9 t ha^{-1}. Nevertheless, average yield in SSNM was 7.2 t ha^{-1} even though the average fertilizer N rate in SSNM was 41 kg N ha^{-1} crop^{-1} less than in the FFP. This yield gain was associated with an almost 100% increase in AEN and REN. Whereas Mr. Cap's N management resulted in an REN of less than 20%, the same was close to 40% in the SSNM plot, where it even improved in 1999 when the chlorophyll meter was introduced. On a cumulative basis for all four crops, SSNM produced 1.32 t ha^{-1} more rice during 1998 to 1999. The average profit increase over FFP was $77 ha^{-1} crop^{-1} or a total of $308 ha^{-1} for all four crops grown.

The farm of Mr. Cay (no. 310) is farther upstream, on a much poorer, degraded soil close to upland areas. Following official recommendations and campaigns for balanced nutrition in this area, he has been using relatively high P and K rates as well as FYM so that IPS and IKS levels in his field were similar to those of Mr. Cap (Table 11.8). However, INS was only about 50 kg N ha^{-1}, which would be sufficient to support yields of 3–3.5 t ha^{-1} without fertilizer application. Yield goals of 80% of the genetic climate-adjusted yield potential at this site could probably not be reached because of constraints other than climate (e.g., low organic matter, low CEC, Mg and Si deficiency, greater incidence of diseases). Therefore, SSNM increased yields only slightly (by 0.2 t ha^{-1} crop^{-1}), but with very large savings in N, P, and K fertilizer because we were able to properly account for the indigenous supplies, manure input, and crop requirements for the achievable yield target. Moreover, high AEN (>20 kg kg^{-1}) and REN (about 0.55 kg kg^{-1}) were achieved in practically all four crops. Because of large savings in fertilizer cost coupled with a small yield increase, the average profit increase over FFP was $72 ha^{-1} crop^{-1} or a total of $288 ha^{-1} for all four crops grown in the field of Mr. Cay (data not shown).

These two examples show how a strategy such as SSNM can work on two very different soils provided that it is fined-tuned to the local conditions.

Effect of farmyard manure on the performance of SSNM

Beginning in 1999 ER, a second SSNM plot was added to each field to compare a standard SSNM treatment that received FYM with one without FYM. In both treatments, the QUEFTS model was used to estimate fertilizer requirements taking into account the presence or absence of nutrient input from FYM. Our hypothesis was that mineral fertilizers could fully substitute for the additional direct and indirect contributions of manure to rice yield.

Table 11.8. Performance of site-specific nutrient management (SSNM) and farmers' fertilizer practice (FFP) in two farmers' fields of the Red River Delta.[a]

Season	Grain yield (t ha⁻¹)		FN (kg ha⁻¹)		FP (kg ha⁻¹)		FK (kg ha⁻¹)		AEN (kg kg⁻¹)		REN (kg kg⁻¹)	
	FFP	SSNM	FFP	SSNM	FFP	SSNM	FFP	SSNM	FFP	SSNM	FFP	SSNM
N.V. Cap (farm no. 205), Phuc Tho, alluvial soil												
INS = 75 kg N ha⁻¹, IPS = 15 kg P ha⁻¹, IKS = 70 kg K ha⁻¹												
1998 ER	6.75	7.40	128	92	10	15	41	69	9	20	0.15	0.30
1998 LR	5.19	5.67	102	74	19	10	28	30	9	19	0.29	0.29
1999 ER	7.84	7.96	131	80	29	30	58	75	14	24	0.15	0.49
1999 LR	7.60	7.69	150	100	30	10	80	90	10	17	0.13	0.37
Mean	6.85	7.18	128	87	22	16	52	66	11	20	0.18	0.36
N.V. Cay (farm no. 310), Tam Dao, degraded soil												
INS = 52 kg N ha⁻¹, IPS = 14 kg P ha⁻¹, IKS = 70 kg K ha⁻¹												
1998 ER	5.45	5.93	91	80	39	40	69	30	16	24	0.58	0.71
1998 LR	4.24	4.48	117	60	6	10	77	30	9	22	0.28	0.56
1999 ER	5.59	5.92	128	80	20	10	97	50	11	21	0.46	0.52
1999 LR	6.43	6.23	97	80	19	10	122	50	20	21	0.30	0.47
Mean	5.43	5.64	108	75	21	18	91	40	14	22	0.41	0.57

[a]FN = fertilizer N, FP = fertilizer P, FK = fertilizer K, AEN = agronomic N-use efficiency, REN = recovery efficiency of applied N.

Table 11.9. Effect of farmyard manure application on fertilizer use, nutrient uptake, grain yield, and the economics of site-specific nutrient management in rice (means of 24 farms).

Item	1999 spring rice		1999 summer rice	
	+FYM	−FYM	+FYM	−FYM
Fertilizer applied (kg ha [1])				
N	95	112	88	95
P	17	25	10	13
K	66	77	54	74
NPK input from FYM (kg ha [1])				
N	21.3	0	21.3	0
P	17.0	0	17.0	0
K	34.0	0	34.0	0
Total nutrient input (kg ha [1])				
N	116.1	111.7	109.3	95
P	34.1	24.6	27.0	13
K	99.6	77.1	88.0	74
Total plant nutrient uptake (kg ha [1])				
N	104.0	97.6	104.2	98.2
P	25.1	24.0	20.8	20.5
K	116.2	109.9	126.3	111.3
Grain yield (t ha [1])	6.90	6.46	6.83	6.38
Increase in profit because of FYM ($ ha [1])[a]				
Without labor cost for FYM	94	90		
With labor cost for FYM[b]	62	58		

[a]$\Delta = GRF_{+FYM} - GRF_{-FYM}$. [b]Labor cost for FYM application estimated as $1.54 (1 labor day) per 400 kg FYM.

Results from two crops indicate that yields in SSNM crops without FYM were lower than with the application of FYM (Table 11.9). Part of this difference was due to smaller total nutrient input in the treatment without manure, resulting in a lower uptake of N and K. In other words, no full nutrient compensation was achieved. Moreover, FYM application is likely to provide other benefits to soil fertility that were not taken into account. Specifically, on the 12 farms located on degraded soils at Tam Dao, FYM application is important to supply elements such as Si and Mg, but also to maintain soil organic matter content and CEC.

The increase in profit because of FYM application averaged $90 to $94 ha[-1] crop[-1] in both crops sampled in 1999 (Table 11.9). However, manure application in the RRD is very labor-intensive because most of the FYM is carried or carted to the rice fields by hand. Taking into account this extra labor, the profit increase as compared to SSNM without FYM averaged about $60 ha^{-1} crop^{-1}. Considering the high population density, small farm sizes, availability of family labor, intensity of triple cropping, complete removal of straw, and the soil types prevailing in the RRD, applying FYM appears to remain an essential component for sustaining rice-based cropping systems in this region.

Opportunities for SSNM in the RRD

The opportunities for site-specific nutrient management in the Red River Delta are probably greater than a comparison of SSNM and FFP treatments suggests. During four cropping seasons in 1999-2000, basic agronomic and economic data were collected not only from fields of collaborating farmers (FFP1) and embedded researchers' plots (SSNM) but also from surrounding fields (FFP2) managed by other farmers not involved in our project (Table 11.10, Fig. 11.7). On average and across sites with alluvial and degraded soil, grain yield with SSNM was only 0.31 t ha^{-1} greater than in FFP1, but 0.84 t ha^{-1} greater than in FFP2. Collaborating farmers gave special attention to crop care and invested additional resources in their FFP1 fields, possibly in an attempt to compete with researchers for higher yields. Besides attempts to copy certain activities in the researchers' SSNM practice (e.g., timing of fertilizer N applications), farmers also often followed recommendations by researchers to use better seeds and seedlings. At Tam Dao, farmers applied FYM twice versus only once in the SSNM treatment. Compared to noncollaborators (FFP2) and across the two sites, collaborating farmers used 5%, 10%, 27%, and 25% more fertilizer N, P, K, and FYM, respectively, in their fields (FFP1). These differences in management practices probably explain the significantly greater nutrient uptake and yield in FFP1 vis-à-vis FFP2, which was consistent among sites with alluvial and degraded soil (Table 11.10). Nutrient efficiencies are not presented because we did not install omission plots in FFP2, which would be needed as a separate reference.

It is obvious that the FFP1 plot is not very useful as a baseline to assess the potential impact of SSNM because that site has a high labor input and frequent field visits by farmers. In 1999-2000, yields with SSNM were on average and across sites 15% higher than yields in FFP2 despite a decrease in the use of fertilizer N (–9%), P (–37%), and farmyard manure (–10%). There were significant differences in fertilizer K use at sites with alluvial and degraded soil but yield increases largely contributed to an improved match of fertilizer N application with crop N demand. A more realistic estimate of profit increases with SSNM across sites is probably the observed +$153 ha^{-1} crop^{-1} over FFP2 than the +$70 ha^{-1} crop^{-1} over FFP1. Savings in fertilizer cost contributed only 8% to the profit increase over FFP2, but 32% over FFP1 because of the greater fertilizer use in the latter.

The yield and profitability advantage with SSNM was slightly higher on degraded soil than on alluvial soil and this was consistent throughout four cropping seasons in 1999-2000 (Table 11.10, Fig. 11.7). In SSNM, the yield difference was only 0.70 t ha^{-1} comparing alluvial and degraded soil and therefore smaller than the

Table 11.10. Agronomic and economic performance of site-specific nutrient management (SSNM) and farmers' fertilizer practice of collaborating farmers (FFP1) versus noncollaborating farmers (FFP2). (Means of each 12 farms on alluvial and degraded soil during four cropping seasons in 1999-2000.)

Parameter	Unit	SSNM	FFP1	FFP2
Alluvial soil (Ha Tay Province)				
Grain yield	t ha^{-1}	6.95 a[a]	6.70 b	6.26 c
Nitrogen fertilizer	kg ha^{-1}	103.2 b	109.7 a	101.2 b
Phosphorus fertilizer	kg ha^{-1}	16.0 c	25.9 a	22.3 b
Potassium fertilizer	kg ha^{-1}	69.8 a	61.7 b	46.0 c
Farmyard manure	t ha^{-1}	8.93 b	9.66 a	8.46 c
Total N uptake	kg ha^{-1}	96.5 a	91.9 b	85.9 c
Total P uptake	kg ha^{-1}	22.9 a	22.1 b	20.4 c
Total K uptake	kg ha^{-1}	121.1 a	117.8 a	110.3 b
Harvest index	kg kg^{-1}	0.53 a	0.53 a	0.52 b
Fertilizer cost	$ ha^{-1}	91.7 b	104.0 a	90.5 b
Gross return over fertilizer use	$ ha^{-1}	1,062.7 a	1,008.7 b	950.0 c
Degraded soil (Vinh Phuc Province)				
Grain yield	t ha^{-1}	6.25 a	5.93 b	5.25 c
Nitrogen fertilizer	kg ha^{-1}	80.8 b	102.3 a	101.7 a
Phosphorus fertilizer	kg ha^{-1}	11.3 b	21.6 a	20.6 a
Potassium fertilizer	kg ha^{-1}	49.4 c	78.4 a	64.5 b
Farmyard manure	t ha^{-1}	8.0 c	14.0 a	10.3 b
Total N uptake	kg ha^{-1}	91.7 a	89.8 b	76.8 c
Total P uptake	kg ha^{-1}	20.5 a	19.5 b	16.9 c
Total K uptake	kg ha^{-1}	125.5 b	131.7 a	111.8 c
Harvest index	kg kg^{-1}	0.50 a	0.48 b	0.48 b
Fertilizer cost	$ ha^{-1}	68.5 c	100.8 a	94.7 b
Gross return over fertilizer use	$ ha^{-1}	969.5 a	883.9 b	776.7 c

[a]Means with the same letter in each row are not statistically significant at $P = 0.05$ (LSD).

yield difference in the farmers' fertilizer practice (0.77 t ha^{-1} in FFP1 and 1.01 t ha^{-1} in FFP2) or in the N unfertilized plots (1.21 t ha^{-1}). This is consistent with the provincial statistics showing a yield difference of 1.22 t ha^{-1} between the provinces of Ha Tay (alluvial soil) and Vinh Phuc (degraded soil) in the late 1990s (Table 11.1). Some of the differences in farmers' management and indigenous nutrient supplies between the two provinces affected yield, and were apparently compensated through site-specific fertilizer strategies. However, the SSNM strategy did not attempt to compensate for differences in soil fertility between the two sites with alluvial and degraded soil (Fig. 11.3). Instead, fertilizer rates with SSNM in 1999-2000 were slightly higher on alluvial soil than on degraded soil for N (103 vs 81 kg N ha^{-1}), P (16 vs 11 kg P ha^{-1}), and K (70 vs 49 kg K ha^{-1}) as shown in Table 11.10. Higher fertilizer rates on alluvial soil were chosen assuming that the attainable yield increase over the unfertilized omission plot was greater on alluvial soil than on degraded soil for N. This was apparently not the case. The yield increase with SSNM

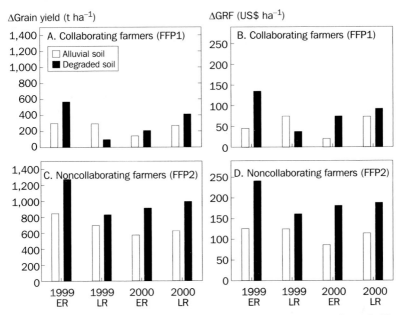

ΔGrain yield (t ha⁻¹)

A. Collaborating farmers (FFP1)
☐ Alluvial soil
■ Degraded soil

C. Noncollaborating farmers (FFP2)

ΔGRF (US$ ha⁻¹)

B. Collaborating farmers (FFP1)

D. Noncollaborating farmers (FFP2)

1999 ER   1999 LR   2000 ER   2000 LR

Fig. 11.7. Average difference in grain yield (ΔGY) and gross return above fertilizer costs (ΔGRF) between the site-specific nutrient management treatment and the fertilizer practice of collaborating (FFP1) and noncollaborating farmers (FFP2) on alluvial (_n_ = 12) and degraded soil (_n_ = 12) near Hanoi, Vietnam, 1999-2000 (bars: mean).

over the 0-N plot was greater on degraded soil than on alluvial soil (1.94 vs 1.43 Δt ha⁻¹), with corresponding differences in AEN (24 vs 14 kg kg⁻¹) and REN (0.42 vs 0.24 kg kg⁻¹). We can only speculate that in general the application of FYM and improved N management strategies had a greater effect on yield on degraded soil than on alluvial soil. Note that the 0-N plots did not receive any FYM and that plant growth and yield may be more affected by FYM application on degraded soil than on alluvial soil. However, noncollaborating farmers seem to have tried to overcome the limited nutrient supply in Vinh Phuc by applying more K (64 vs 46 kg K ha⁻¹) and FYM (10.3 vs 8.5 t ha⁻¹) on degraded soil than on alluvial soil, while the consumption of fertilizer N (about 100 kg N ha⁻¹) and P (about 22 kg P ha⁻¹) was the same at both sites. Farmers' efforts in Vinh Phuc, however, were not much rewarded, and the yield with SSNM on degraded soil was 1 t ha⁻¹ greater than in FFP2 despite a reduction in the use of fertilizer N, P, K, and FYM (Table 11.10).

In summary, the results presented suggest substantial opportunities for farmers to increase yield and profit on their already intensively managed farm land through a more efficient use of fertilizer and improved crop management practices (variety selection, seed quality). The SSNM strategies developed for the research sites have probably evolved to optimal levels, although based on more recent data, fertilizer K rates on alluvial soil could be lowered to levels similar to those used on degraded soil. The SSNM strategy will have to be further simplified for wider-scale dissemination in the RRD and we consider the following issues to be essential in this:

- On both alluvial and degraded soil, fertilizer application should be based on plant need, that is, fertilizer rates are determined by the expected yield gain (difference between plant nutrient demand of a specified yield target and soil nutrient supply).
- Fertilizer P and K recommendations should be developed for recommendation domains, that is, larger areas that are mainly based on a broad classification of indigenous soil P and K supply. Existing information on soil indigenous nutrient supplies (e.g., soil chemical and physical properties) can be verified using the omission plot technique.
- Fertilizer N management strategies need to include a strong real-time N management component to efficiently match the seasonal variation in plant N demand and avoid lodging and pest infestation. This is particularly valid for the RRD, considering the substantial small-scale variation in indigenous nutrient supplies as affected by amount and quality of FYM used by farmers, crop rotation, cropping history, and variation in more stable soil properties (e.g., soil texture, soil organic matter).
- Fertilizer recommendations for dissemination can include guidelines valid for larger areas (variety-adjusted guidelines for LCC-based N management, application of fertilizer K in two equal splits—basal and at PI), regional recommendations (domain-specific P and K recommendations), and local adaptations (adjustment of fertilizer P and K rates to the use of FYM and crop rotations).
- Fertilizer recommendations should be disseminated in combination with a campaign promoting the use of appropriate varieties and good-quality seeds.

11.3 Conclusions

Rice yields in the Red River Delta have increased at high growth rates since the late 1980s. However, smaller increments are likely to occur in future attempts to close the already narrow yield gap. Fine-tuning of nutrient and pest management will play a pivotal role in this. Previous research and extension work have led to a relatively balanced use of mineral and organic fertilizers by most farmers in the RRD. The high labor availability per unit land area has favored the widespread adoption of cropping systems that are characterized by high cropping intensity and good-quality crop care. In addition to known gradients in soil fertility and suitability to grow different lowland and upland crops along the Red River, indigenous N, P, and K supply varied widely among rice fields in the Red River Delta, indicating considerable potential for farm- or field-specific nutrient management.

The SSNM concept used in our study evolved and improved over a period of six rice-cropping seasons. True grain yield increases were difficult to quantify because of the lack of a true reference basis with the same basic soil management. However, using significantly smaller amounts of FYM and N, P, and K fertilizer than currently applied by the farmers, SSNM promises substantial yield (+15%) and profit increases (+$150 ha⁻¹ crop⁻¹) based on a comparison with data from fields of noncollaborating farmers. Because of the improved timing of N applications, N-use efficiency increased to high levels. Realistic goals for intensive rice cultivation in the RRD are an REN of about 0.5 kg kg⁻¹, PEN of 50 kg kg⁻¹, and AEN of about 20–25 kg kg⁻¹. When predictable climatic factors or pests constrained yields, model-predicted yields were closely correlated with the actual yields, suggesting that the modified QUEFTS model used in the SSNM approach provides accurate fertilizer recommendations. However, given the prevailing quality of crop management and the scarcity of land, the use of diagnostic tools such as the chlorophyll meter (Peng et al 1996) or leaf color charts (Balasubramanian et al 1999) is likely to show particular promise in environments such as the RRD, where variability in soil indigenous nutrient supplies is further enhanced by farmers applying FYM in varying amounts and quality. We consider the estimation of indigenous P and K supplies a key component in the development of meaningful fertilizer P and K recommendations for reasonably large areas. The omission plot technique was sufficiently robust to detect differences in nutrient supply among areas with degraded and alluvial soil and it can also be used to assess the combined nutrient-supplying power of FYM and soil indigenous nutrient sources.

References

Balasubramanian V, Morales AC, Cruz RT, Abdulrachman S. 1999. On-farm adaptation of knowledge-intensive nitrogen management technologies for rice systems. Nutr. Cycl. Agroecosyst. 53:59-69.

Bray F. 1998. A stable landscape? Social and cultural sustainability in Asian rice systems. In: Dowling NG, Greenfield SM, Fischer KS, editors. Sustainability of rice in the global food system. Davis, Calif. (USA), Manila (Philippines): Pacific Basin Study Center, International Rice Research Institute. p 45-66.

Dobermann A, White PF. 1999. Strategies for nutrient management in irrigated and rainfed lowland rice systems. Nutr. Cycl. Agroecosyst. 53:1-18.

Dobermann A, Witt C, Abdulrachman S, Gines HC, Nagarajan R, Son TT, Tan PS, Wang GH, Chien NV, Thoa VTK, Phung CV, Stalin P, Muthukrishnan P, Ravi V, Babu M, Simbahan GC, Adviento MA, Bartolome V. 2003. Estimating indigenous nutrient supplies for site-specific nutrient management in irrigated rice. Agron. J. 95:924-935.

Frère M, Popov GF. 1979. Agrometeorological crop monitoring and forecasting. FAO Plant Production and Protection Paper 17. Rome (Italy): FAO.

Le TC, Rambo AT, Gillogly K, editors. 1993. Too many people, too little land: the human ecology of a wet rice-growing village in the Red River Delta of Vietnam. East-West Center Occasional Papers of the Program on Environment No. 15. Honolulu, Haw. (USA): East-West Center, University of Hawaii.

Peng S, Garcia FV, Laza RC, Sanico AL, Visperas RM, Cassman KG. 1996. Increased N-use efficiency using a chlorophyll meter on high-yielding irrigated rice. Field Crops Res. 47:243-252.

Witt C, Dobermann A, Abdulrachman S, Gines HC, Wang GH, Nagarajan R, Satawathananont S, Son TT, Tan PS, Tiem LV, Simbahan GC, Olk DC. 1999. Internal nutrient efficiencies of irrigated lowland rice in tropical and subtropical Asia. Field Crops Res. 63:113-138.

Witt C, Balasubramanian V, Dobermann A, Buresh RJ. 2002. Nutrient management. In: Fairhurst T, Witt C, editors. Rice: a practical guide for nutrient management. Singapore and Los Baños (Philippines): Potash and Phosphate Institute & Potash and Phosphate Institute of Canada and International Rice Research Institute. p 1-45.

Notes

Authors' addresses: Tran Thuc Son, Nguyen Van Chien, Vu Thi Kim Thoa, National Institute for Soils and Fertilizers (NISF), Chem, Tu Liem, Hanoi, Vietnam, email: tsonnisfacvn@hn.vnn.vn; A. Dobermann, C. Witt, International Rice Research Institute, Los Baños, Philippines; A. Dobermann, University of Nebraska, Lincoln, Nebraska, USA.

Acknowledgments: We are grateful to the farmers cooperating with us in this project. The authors are much obliged to Dr. Nguyen Van Bo, formerly director of NISF and IRRC steering committee member, for his support of our research. We wish to thank Piedad Moya and Marites Tiongco (IRRI) for help with the socioeconomic farm monitoring, Ma. Arlene Adviento (IRRI) for help with the soil and plant analysis work, and Gregorio Simbahan, Rico Pamplona, Olivyn Angeles, Julie Mae Criste Cabrera-Pasuquin, and Edsel Moscoso (IRRI) for help with data management and statistical analysis. We acknowledge contributions made by other researchers involved in this work, including David Dawe and Pompe Sta. Cruz of IRRI.

Citation: Dobermann A, Witt C, Dawe D, editors. 2004. Increasing productivity of intensive rice systems through site-specific nutrient management. Enfield, N.H. (USA) and Los Baños (Philippines): Science Publishers, Inc., and International Rice Research Institute (IRRI). 410 p.

12 Site-specific nutrient management in irrigated rice systems of Zhejiang Province, China

Wang Guanghuo, Q. Sun, R. Fu, X. Huang, X. Ding, J. Wu, Y. He, A. Dobermann, and C. Witt

12.1 Characteristics of rice production in Zhejiang Province, China

Trends

Zhejiang Province is located in southeastern China, in the southern sector of the Yangtze River delta (Fig. 12.1). It belongs to the subtropical climate zone, with warm temperature (mean temperature 16–18 °C) and adequate rainfall (annual precipitation 1,100–1,900 mm). In 1995, Zhejiang Province had a total of 1.62 million ha of cultivated land, of which 1.27 million ha were used for growing irrigated rice (Zhejiang Statistical Bureau 1998). In the central part of this province is the Jinhua-Quzhou

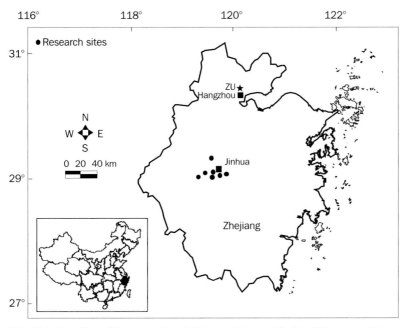

Fig. 12.1. Location of the experimental sites at Jinhua, Zhejiang Province, China.

(Jin-Qu) basin, with about 395,000 ha of irrigated rice land. The Jin-Qu basin is important for commercial food production at both the provincial and national level. There are three main types of rice soils in Zhejiang (Soil Survey Office of Zhejiang 1994):

1. Soils derived from alluvial deposits (40% of rice area). These soils occur in the valley plains along the upper and middle reaches of various rivers in Zhejiang. These soils vary greatly in fertility and productivity.
2. Soils derived from lacustrine deposits, marine deposits, or alluvial-marine deposits (40%). These soils are found in coastal areas and river deltas and usually have high organic matter content and fertility.
3. Various red soils (20%). Because of the relatively low fertility of their parent materials, these soils belong to the medium- or low-yielding rice soils.

Double-rice cropping was adopted in the early 1960s when semidwarf early-maturing rice varieties were adopted in Zhejiang. During the 1970s and '80s, rice-rice-wheat (or oil rape, green manure) triple cropping was the main cropping system because the government promoted measures for increasing the food and oil supply. In the 1990s, the winter crop growth area declined because yield and profit of the winter crop were low and the food supply increased. The two main rice-based cropping systems in Zhejiang Province in the 1990s were (1) early, typically inbred rice (ER) and late, typically hybrid rice (LR) in the central and southern parts and (2) rice–winter crop in the north.

Since the mid-1980s, Zhejiang has gradually become one of the more developed areas in China and off-farm work of farmers has increased. A decline in the ER area at about 2% per year from 1980 to 1997 has been directly or indirectly associated with the rapid industrialization and urbanization processes (Table 12.1). Much of this decline was associated with a shift from rice cropping to growing other crops, particularly in the ER growing season. From 1980 to 1997, the rice harvest area declined by 337,000 ha in ER and by 87,000 ha in LR. In the late 1990s, the overall importance of ER decreased significantly after agricultural production was deregulated through policy change. Farmers now increasingly grow a single LR crop and leave the field as fallow for the rest of the year or grow more profitable cash crops (e.g., vegetables) instead of early rice. Consequently, the rice area cropped to early inbred rice with limited quality and profit declined by 530,000 ha or 64% from 1997 to 2001. At the same time, the area cropped to the more profitable hybrid rice (higher yield and quality) also declined by about 218,000 ha or 17%. Rice imports, particularly of japonica varieties from other provinces, have increased in Zhejiang in recent years.

Rice yields increased at 3% annually until the mid-1980s mainly because of the widespread adoption of hybrid rice grown in the LR cropping season (5% yield growth year^{-1} in LR vs 1.3% yield growth in ER, 1970-83), increasing fertilizer use, and possibly also because of other improvements in crop management. However, average yield growth has slowed since the mid-1980s. From 1984 to 2001, yield growth was still increasing annually at 66 kg ha^{-1} or 1.1% in late (hybrid) rice, but yield growth was negative in early (inbred) rice (–85 kg ha^{-1} or –0.73% y^{-1}) despite

Table 12.1. Rice area, production, and yield in early and late rice-cropping seasons and fertilizer consumption in Zhejiang Province, China, 1962-2000.

Item	1962-65	1966-70	1971-75	1976-80	1981-85	1986-90	1991-95	1996-2000
Early and late rice								
Rice area (10^6 ha)	2.18	2.37	2.53	2.53	2.48	2.37	2.21	1.95
Rice production (10^6 t)	6.65	8.51	9.79	11.45	13.56	13.26	12.68	11.69
Rice yield (t ha^{-1})	3.04	3.59	3.87	4.52	5.47	5.60	5.74	5.99
Early rice								
Rice area (10^6 ha)	–	–	1.25	1.22	1.12	1.04	0.92	0.72
Rice production (10^6 t)	–	–	5.85	6.17	6.44	5.84	4.99	3.85
Rice yield (t ha^{-1})	–	–	4.67	5.08	5.77	5.63	5.40	5.33
Late rice								
Rice area (10^6 ha)	–	–	1.28	1.31	1.36	1.33	1.29	1.23
Rice production (10^6 t)	–	–	3.94	5.27	7.11	7.42	7.69	7.84
Rice yield (t ha^{-1})	–	–	3.08	4.01	5.22	5.57	5.97	6.36
Fertilizer consumption in Zhejiang (1,000 t)								
Total (N-P_2O_5-K_2O)	–	–	–	–	678	884	929	941
N fertilizer (N)	–	–	–	–	543	664	653	639
P fertilizer (P_2O_5)	–	–	–	–	106	128	127	127
K fertilizer (K_2O)	–	–	–	–	15	39	53	57
Complex fertilizers	–	–	–	–	13	54	97	119

Source: Zhejiang Bureau of Statistics: Rural statistical yearbook of Zhejiang. Chinese Statistic Press, Beijing, 1962-2000.

further increases in fertilizer consumption. In the mid-1990s, farmers started shifting from transplanting to direct seeding in ER, which is further evidence that farmers spend less time and resources in growing the lower-yielding ER crop and increasingly focus on maximizing yield in LR.

Total fertilizer consumption in Zhejiang increased from 618,000 t in 1980 to 976,000 t in 1995, but has been declining since then (897,000 t in 2001). The steady loss of rice land in the past 18 years was initially compensated for by an increase in yield ha^{-1}, but total rice production declined by 37% from 1991 to 2001 (–3.1% year^{-1}). The current average rice yield is only about 50% to 60% of the estimated genetic and climatic yield potential of 10 to 12 t ha^{-1} for the modern rice varieties grown in this area (Zheng et al 1997a). In recent years, complex fertilizers have become more attractive to farmers.

The liquidation of the original collective farm system resulted in small family farms, which also created new challenges for the agricultural extension system. Routine soil testing for fertilizer recommendations for rice is now rarely used in Zhejiang. Fertilizer prescriptions are given only for large areas and are not based on a more site-specific knowledge of soil nutrient status. Different crop management practices by individual farmers have already resulted in great diversity in soil fertility and

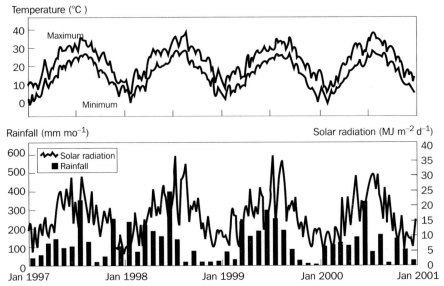

Temperature (°C)

Rainfall (mm mo⁻¹)   Solar radiation (MJ m⁻² d⁻¹)

Jan 1997   Jan 1998   Jan 1999   Jan 2000   Jan 2001

Fig. 12.2. Climatic conditions at Jinhua, Zhejiang Province, China, during 1997 to 2000. Solar radiation and temperature data are 7-day moving averages; rainfall is monthly total. Solar radiation was calculated from sunshine hours using the location-specific Ångström coefficients of a = 0.207 and b = 0.725 for Jinhua (29°12′N, 119°38′E).

productivity, but detailed studies were lacking. Various studies, however, have revealed that the recovery efficiency of applied fertilizer N is often only 30% or less (National Soil and Fertilizer Station 1993, Wang et al 1994). Farmers' income from rice production varies greatly among farms and villages in this province. Environmental pollution by nutrient leaching or runoff from rice fields has become a serious concern.

Current biophysical and socioeconomic farm characteristics

The experimental domain of the RTDP project is located around Jinhua City (29°5′N, 119°47′E) in the center of Zhejiang (Fig. 12.1). This area has a subtropical climate (annual rainfall 1,300–1,500 mm and annual mean temperature of 16.6–17.7 °C, Fig. 12.2). There are 168,000 ha of cultivated land in Jinhua District, of which 146,900 ha are paddy fields. Double rice cropping is the main cropping system and it started about 30 years ago. Early rice is grown from early April to mid- or late July using both hybrids and modern conventional rice varieties. Late rice is 90% hybrid rice grown from mid-July to late October. Land preparation is done on wet soils with tractors. The planting density is usually about 25 hills m⁻² for early rice and 18 hills m⁻² for late rice. To reduce labor cost and time needed for transplanting, seedling throwing has become a popular crop establishment method in Jinhua (about 17,000 ha in 1999), resulting in more variable hill density. Severe pest problems are usually not encountered. Rivers and reservoirs irrigate all the paddy fields but seasonal water stress cannot be totally eliminated. Mid-season drainage is a popular water

Table 12.2. Demographic and economic characteristics of rice production on 21 farms of Jinhua, Zhejiang, China. Values shown are based on socioeconomic farm surveys conducted for whole farms.

Production characteristics	Min.	25%	Median	75%	Max.
Total cultivated area (ha)	0.15	0.27	0.31	2.13	5.87
Age of household head (y)	29	35	43	48	54
Education of household head (y)	1	5	8	8	12
Household size (persons)	3	3	4	4	5
1998 early rice					
Rice area (ha)[a]	0.10	0.19	0.29	1.33	5.00
Yield (t ha^{-1})	4.58	5.26	5.83	6.04	6.51
Insecticide (kg ai ha^{-1})[b]	0.48	0.58	0.75	0.88	1.34
Herbicide (kg ai ha^{-1})	0.09	0.09	0.09	0.10	0.12
Other pesticides (kg ai ha^{-1})[b]	0.01	0.04	0.04	0.04	0.05
Total labor (8-h d ha^{-1})	53	89	109	119	205
Net return from rice (US$ ha^{-1} crop^{-1})	284	401	461	543	669
1998 late rice					
Rice area (ha)[a]	0.13	0.20	0.28	2.13	5.00
Yield (t ha^{-1})	5.71	6.54	6.83	7.09	8.82
Insecticide (kg ai ha^{-1})	5.55	5.76	6.03	6.38	10.40
Herbicide (kg ai ha^{-1})	0.08	0.09	0.09	0.10	0.11
Other pesticides (kg ai ha^{-1})[b]	0.06	0.09	0.09	0.10	0.12
Total labor (8-h d ha^{-1})	48	72	85	98	115
Net return from rice (US$ ha^{-1} crop^{-1})	741	865	916	979	1,323

[a]Rice area in which the treatment plots were embedded in subsequent years. The total rice area may even be larger.
[b]Includes fungicide, molluscicide, rodenticide, and crabicide.

management practice for rice in this area. Harvesting and threshing are done manually or using combines. Straw is normally completely removed from the field after harvesting the early rice, but it often remains in the field and is burned after the late rice season.

To conduct a more detailed analysis of current farm-level productivity and its variation among farms, we established on-farm monitoring experiments in 1997. The 21 farmers belong to seven villages located within 10–20 km of Jinhua City (Fig.12.1) and represent a range of socioeconomic conditions and soil types. All farmers grow rice in an early rice–late rice system. Thirteen farmer families rent 0.2 to 0.5 ha of rice land each, whereas the other eight are so-called "big food production families," which rent 2 to 5 ha of rice land from the local authorities. Soils were alluvial (19) and red soil (2). The experimental approaches were the same as described in Chapter 2. Agronomic and socioeconomic data were collected in farmers' fields for eight successive rice crops grown from 1997 to 2000.

Table 12.2 shows the variation in demographic and economic characteristics of rice production among the 21 farms. Median farm size was small (0.3 ha). Because of longer growth duration, the use of hybrids, and more favorable climate,

Table 12.3. General soil properties on 21 rice farms at Jinhua, Zhejiang, China.[a]

Soil properties	Min.	25%	Median	75%	Max.
Clay content (%)	24.2	30.1	31.2	36.6	45.0
Silt content (%)	39.9	43.3	44.4	49.6	58.1
Sand content (%)	6.2	14.9	24.3	26.2	30.3
Soil organic C (g kg^{-1})	15.5	18.1	18.6	19.8	23.0
Total soil N (g kg^{-1})	1.63	1.92	1.98	2.12	2.57
Soil pH (1:1 H$_2$O)	4.8	5.1	5.3	5.6	6.3
Cation exchange capacity (cmol$_c$ kg^{-1})	6.9	9.6	10.8	12.4	16.3
Exchangeable K (cmol$_c$ kg^{-1})	0.12	0.19	0.23	0.28	0.85
Exchangeable Na (cmol$_c$ kg^{-1})	0.11	0.15	0.22	0.30	0.44
Exchangeable Ca (cmol$_c$ kg^{-1})	3.27	5.62	6.58	7.85	16.88
Exchangeable Mg (cmol$_c$ kg^{-1})	0.25	0.42	0.68	0.88	1.19
Extractable P (Olsen-P, mg kg^{-1})	7.62	11.61	17.44	23.70	60.52
Extractable Zn (0.05N HCl, mg kg^{-1})	1.42	1.65	1.85	2.18	5.43

[a]Measured on initial soil samples collected before the 1997 early rice crop.

average late rice yields in 1998 were about 1 t ha^{-1} higher than early rice yields and 20% less labor was required for LR, resulting in twice as high net returns from rice. Insecticide use was high in LR, but farmers did not spray many herbicides in either season. Of the 21 farmers in our sample, only one to three occasionally apply farmyard manure and one occasionally grows a green manure crop (milk vetch) during winter.

Soils in the study domain are mostly fertile, but variation in soil nutrient levels among the 21 farms was large (Table 12.3). Most farms had a clay loam to silty clay soil texture and relatively high soil organic matter content, but an acid pH and only low to moderate CEC. Contents of available K, P, and Zn were highly variable among the farms (CVs 40–70%). Olsen-P content ranged from 8 to 61 mg P kg^{-1}. About 80% of all farms had Olsen-P contents above the commonly used critical level of >10 mg P kg^{-1}, suggesting that the regular P application has led to soil P accumulation. Potassium levels were in the moderate to low range. The long-term fertility experiment conducted at Jinhua since 1997 showed that soil K depletion might occur very fast at this site. In this experiment, K-deficiency symptoms appeared on rice leaves after only two consecutive crops grown without K fertilizer. Soil organic matter and total N contents were relatively high on most farms and there was no indication of micronutrient deficiencies.

Measurements of the indigenous nutrient supply to rice confirmed the relatively high soil fertility status at this site and the variation among farms (Fig. 12.3). The indigenous N supply (INS = plant N accumulation in 0-N plot) was similar in most cropping seasons sampled (average 70 kg N ha^{-1}, range 50 to 100 kg N ha^{-1}). However, the INS was generally about 10 kg N ha^{-1} lower in the 1999 and 2000 ER crops, probably because of unfavorable climatic conditions for rice growth (see below). The indigenous P supply (IPS = plant P accumulation in 0-P plot) was about 21 kg P ha^{-1} in ER and LR, with a range of 13 to 30 kg P ha^{-1}. The average indigenous

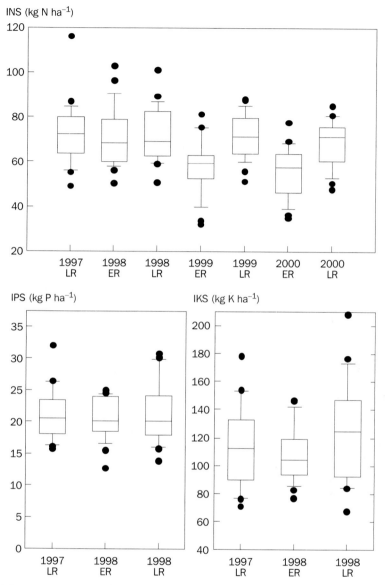

Fig. 12.3. Variability of the indigenous N (INS), P (IPS), and K (IKS) supply among 21 farmers' fields at Jinhua, China (1997-98). Median with 10th, 25th, 75th, and 90th percentiles as vertical boxes with error bars; outliers as bullets.

Table 12.4. Baseline agronomic characteristics of rice production on 21 farms at Jinhua, Zhejiang. Values shown are means and standard deviations (SD) of the farmers' fertilizer practice for two consecutive rice crops.

Agronomic characteristics[a]	1998 early rice crop		1998 late rice crop	
	Mean	SD	Mean	SD
Grain yield (t ha^{-1})	5.92	0.68	6.91	0.61
Harvest index	0.55	0.07	0.53	0.04
No. of panicles m^{-2}	400	65	265	40
Total no. of spikelets m^{-2}	21,830	3,134	29,164	2,980
Total no. of spikelets panicle^{-1}	77	14	111	12
Filled spikelets (%)	73	7	85	3
Fertilizer N use (kg ha^{-1})	158.7	23.1	156.3	28.4
Fertilizer P use (kg ha^{-1})	15.8	6.0	15.6	8.8
Fertilizer K use (kg ha^{-1})	42.6	26.0	63.2	27.3
N uptake (kg ha^{-1})	104.4	13.0	105.1	17.8
P uptake (kg ha^{-1})	20.0	2.6	22.4	3.3
K uptake (kg ha^{-1})	111.5	21.8	144.9	20.6
Input-output N balance (kg ha^{-1} crop^{-1})	−10.1	11.7	−14.3	12.3
Input-output P balance (kg ha^{-1} crop^{-1})	−3.8	7.4	−6.0	10.3
Input-output K balance (kg ha^{-1} crop^{-1})	−35.1	27.5	−36.8	31.4
Partial productivity of N (kg kg^{-1})	38.0	7.0	45.9	10.8
Agronomic efficiency of N (kg kg^{-1})	6.1	3.1	8.2	3.3
Recovery efficiency of N (kg kg^{-1})	0.22	0.11	0.21	0.10
Physiological efficiency of N (kg kg^{-1})	29.1	15.5	43.1	19.8

K supply (IKS = plant K accumulation in 0-K plot) ranged from 107 kg K ha^{-1} in 1998 ER to 125 kg K ha^{-1} in LR. The higher IKS in LR may be due to the more vigorous root system and greater soil K extraction power of hybrid rice, which was mostly grown in LR. Note, however, the very large range of IKS in the 1997 and 1998 LR crops.

Most farmers applied about 16 kg P ha^{-1} and 50 kg K ha^{-1} and nutrient balance estimates suggested a negative input balance for both nutrients (Table 12.4). However, at least over the short term, the current average levels of indigenous N, P, and K supply are sufficient for achieving rice yields of about 5 t ha^{-1} without applying N and 8 t ha^{-1} without applying P and K. These estimates assume optimal balanced nutrient requirements of 14.7 kg N, 2.6 kg P, and 14.5 kg K per 1,000 kg of grain yield (Witt et al 1999).

In 1998, farmers applied 125 to 192 kg N ha^{-1} in ER or 137 to 193 kg N ha^{-1} in LR (Table 12.4). However, the average agronomic N efficiency (AEN) was only 6.1 kg kg^{-1} in ER and 8.2 kg kg^{-1} in LR. Average recovery efficiencies of N (REN) were 22% in ER and 21% in LR, suggesting large gaseous N losses. Low N-use efficiency is probably one of the key constraints to increasing rice yields at Jinhua. Reasons for the low N efficiency include (1) the use of blanket fertilizer recommendations that do not account for the large variation in indigenous nutrient supply, (2) the traditional fertilization practice of applying nearly all the fertilizers within the first 10 d after transplanting, (3) low hill density because of labor shortage at transplanting time, and (4) mid-season drainage and intermittent irrigation during later growth stages. Farmers usually apply all N very early (50% basal, 50% at 5–10 days after transplanting) as NH_4HCO_3 and urea and conduct a mid-season drainage from about 25 to 35 DAT (maximum tillering to panicle initiation, PI, stage) to halt ineffective tillering. This practice also causes losses of remaining fertilizer N from the soil. Moreover, mid-season drainage has to be completed before the PI stage, which is difficult to control because the farmers are often not able to distinguish growth stages of rice crops properly. Very few farmers apply N after PI.

12.2 Effect of SSNM on productivity and nutrient-use efficiency

Management of the SSNM plots

Beginning with the 1998 ER crop, a site-specific nutrient management (SSNM) plot was established in each of the 21 farm fields as a comparison with the farmers' fertilizer practice (FFP, see Chapter 2). Results of the six consecutive rice crops grown from 1998 to 2000 are presented in this paper.

The size of the SSNM plots ranged from 300 to 1,000 m^2 (most were 500 to 1,000 m^2), depending on the size of the farmers' fields in which the SSNM and FFP plots were located. Rice varieties were chosen by the farmers. In general, farmers planted conventional modern varieties in early rice (about 10 different ones such as Jinzhao22, Zhefu802, Zhong903, and Zhe733), but hybrid rice in late rice (e.g., Xieyou64, Ilyou88, Ilyou92, and Xieyou963). Average plant density in the FFP and SSNM plots was 23 hills m^{-2} for ER in both 1998 and 1999. In 1998, late rice was planted at an average density of 19 hills m^{-2} in the FFP and SSNM, but the density differed by about 15% in 1999 (FFP: 19 hills m^{-2}; SSNM: 22 hills m^{-2}). In 2000, the planting density was similar in SSNM and FFP with about 25 hills m^{-2} in ER and 18 hills m^{-2} in LR.

Since the early 1990s, farmers in Zhejiang Province have been shifting from transplanting to direct seeding in ER (inbred varieties). Out of the 21 farms monitored in this project, only one farmer used direct seeding in 1998, followed by 10 farmers in 1999 and 17 farmers in 2000. Farmers did not adjust their fertilizer management strategies when switching to direct seeding, and neither were adjustments made in SSNM because optimal seed rates of 65–70 kg seeds ha^{-1} were followed, resulting in planting densities that were comparable with those of transplanted rice.

Farmers did all water management and pest control in both FFP and SSNM plots following the commonly adopted methods. This included some guidance by a local technician. In general, no severe pest incidence was observed during the experimental period. Rice straw was usually removed or burned in the fields.

Fertilizer applications for SSNM were prescribed on a field- and crop-specific basis following the approach described in Chapter 5 and descriptions given elsewhere (Dobermann and White 1999, Witt et al 1999). In 1998, values of the INS, IPS, and IKS measured during the 1997 LR crop were used as model inputs and the target yield was set to 7.5–8.0 t ha^{-1} in ER and 7.5 to 8.5 t ha^{-1} in LR. In 1999, average values of the INS, IPS, and IKS measured in the 1998 ER and LR crops were used as model inputs and the target yield was set to 7.2–8.0 t ha^{-1} in ER and 8 t ha^{-1} in LR. The IPS and IKS values of 1999 were also used in 2000, but INS values in both ER and LR were based on average values from 1997 to 1999. Yield targets ranged from 6.9 to 7.9 t ha^{-1} in 2000 ER and from 7.3 to 8.6 t ha^{-1} in 2000 LR. First-crop recovery fractions of 0.4, 0.2, and 0.5 kg kg^{-1} were assumed for fertilizer N, P, and K, respectively. The climatic yield potential was set to 9 t ha^{-1} for ER and 10 t ha^{-1} for LR (Zheng et al 1997a).

Fertilizer sources used were urea (46% N), single superphosphate (6.1% P), and muriate of potash (50% K). All P fertilizer was incorporated into the soil before transplanting (100% basal). K fertilizer was split into 50% basal plus 50% at PI stage. In 1998, N (urea) was applied in three splits at fixed growth stages (40% at 1 or 2 days before transplanting incorporated into soil, 20% topdressed at 14 DAT, and 40% topdressed at PI stage). Compared with the practices of most farmers (two early applications only), this splitting represented a more evenly distributed N application scheme. However, results obtained in 1998 suggested that further fine-tuning could be achieved by a more dynamic, plant-based N management. Therefore, in 1999, N was applied as two fixed applications and one or two more topdressings depending on SPAD readings at critical growth stages. For example, the application scheme in the 1999 ER crop was

FN 1	Basal	40% of model-predicted fertilizer N rate	
FN2	10–14 DAT	20% of model-predicted fertilizer N rate	
FN3	35–45 DAT (PI)	If SPAD >36	30 kg N ha^{-1}
		If SPAD 33–36	40 kg N ha^{-1}
		If SPAD <33	50 kg N ha^{-1}

Only small changes occurred in N management in 2000. Fertilizer N was applied in three splits at fixed growth stages as in the previous years and the amount applied with individual splits was based on a predetermined percentage of the total N suggested by the model. Urea was applied at 1 or 2 days before transplanting incorporated into soil (35% in ER and 30% in LR) and as topdressings at 10–14 DAT (20% in ER and 25% in LR) and at PI stage (35% in ER and 40% in LR). The standard rates for the last topdressing were adjusted using the same SPAD ranges as in 1999, when SPAD readings were outside the range of 34 to 36 (+20% at SPAD < 34 and –20% at SPAD > 36).

Effect of SSNM on grain yield and nutrient uptake
Compared with the FFP, SSNM consistently increased grain yields and plant N, P, and K accumulation, although the increases where not always statistically significant (Fig. 12.4). The average yield difference between SSNM and FFP for the six crops grown was 0.40 t ha^{-1} (6.5%, $P = 0.003$) and was similar in the early and late rice crops (Table 12.5). Although the average grain yield increase was 0.43 t ha^{-1} in LR vs 0.36 t ha^{-1} in ER, seasonal differences in performance were not statistically sig-

Table 12.5. Effect of site-specific nutrient management on agronomic characteristics at Jinhua, Zhejiang Province, China (1998-2000).

Parameters	Levels[a]	Treatment[b] SSNM	Treatment[b] FFP	Δ^c	$P>|T|^c$	Effects[d]	$P>|F|^d$
Grain yield (GY)	All	6.41	6.02	0.40	0.003	Village	0.007
(t ha^{-1})	Year 1	6.88	6.41	0.47	0.014	Year[e]	0.016
	Year 2	5.82	5.40	0.43	0.046	Season[e]	0.287
	Year 3	6.54	6.25	0.29	0.198	Year × season[e]	0.197
	ER	5.85	5.48	0.36	0.019	Village × crop	0.006
	LR	6.98	6.55	0.43	0.010		
Plant N uptake (UN)	All	104.0	98.1	5.95	0.012	Village	0.068
(kg ha^{-1})	Year 1	111.1	104.7	6.46	0.092	Year	0.130
	Year 2	100.6	91.7	8.91	0.004	Season	0.738
	Year 3	100.2	97.9	2.41	0.613	Year × season	0.948
	ER	98.0	92.5	5.44	0.075	Village × crop	0.704
	LR	110.0	103.6	6.41	0.052		
Plant P uptake (UP)	All	21.1	19.3	1.8	0.003	Village	0.050
(kg ha^{-1})	Year 1	24.6	21.2	3.3	0.000	Year	0.005
	Year 2	17.2	15.7	1.5	0.052	Season	0.000
	Year 3	21.4	20.9	0.6	0.540	Year × season	0.001
	ER	20.6	18.9	1.7	0.045	Village × crop	0.009
	LR	21.6	19.7	1.9	0.030		
Plant K uptake (UK)	All	129.6	120.7	8.9	0.015	Village	0.286
(kg ha^{-1})	Year 1	139.2	128.1	11.1	0.104	Year	0.075
	Year 2	121.3	108.9	12.4	0.043	Season	0.960
	Year 3	128.2	125.1	3.2	0.557	Year × season	0.767
	ER	118.9	110.7	8.1	0.084	Village × crop	0.305
	LR	140.4	130.7	9.7	0.052		
Agronomic efficiency	All	12.5	6.8	5.7	0.000	Village	0.000
of N (AEN)	Year 1	10.8	7.2	3.7	0.000	Year	0.000
(kg grain kg^{-1} N)	Year 2	11.9	5.5	6.4	0.000	Season	0.014
	Year 3	14.8	7.8	7.0	0.000	Year × season	0.196
	ER	11.3	6.4	4.9	0.000	Village × crop	0.030
	LR	13.7	7.2	6.5	0.000		
Recovery efficiency	All	0.31	0.19	0.12	0.000	Village	0.002
of N (REN)	Year 1	0.27	0.22	0.06	0.031	Year	0.002
(kg plant N kg^{-1} N)	Year 2	0.31	0.15	0.16	0.000	Season	0.379
	Year 3	0.34	0.21	0.14	0.000	Year × season	0.318
	ER	0.29	0.19	0.10	0.000	Village × crop	0.377
	LR	0.33	0.19	0.14	0.000		
Partial productivity	All	52.3	36.8	15.6	0.000	Village	0.000
of N (PFPN)	Year 1	47.3	42.0	5.3	0.022	Year	0.000
(kg grain kg^{-1} N)	Year 2	50.4	31.3	19.2	0.000	Season	0.000
	Year 3	59.2	37.1	22.2	0.000	Year × season	0.003
	ER	47.5	34.7	12.7	0.000	Village × crop	0.014
	LR	57.2	38.8	18.4	0.000		

[a]All = all six crops grown from 1998 ER to 2000 LR; year 1 = 1998 ER and 1998 LR; year 2 = 1999 ER and 1999 LR; year 3 = 2000 ER and 2000 LR; ER = 1998 ER, 1999 ER, and 2000 ER; LR = 1998 LR, 1999 LR, and 2000 LR. [b]FFP = farmers' fertilizer practice; SSNM = site-specific nutrient management. $^c\Delta$ = SSNM – FFP. $P>|T|$ = probability of a significant mean difference between SSNM and FFP. [d]Source of variation of analysis of variance of the difference between SSNM and FFP by farm; $P>|F|$ = probability of a significant F-value. [e]Year refers to two consecutive cropping seasons.

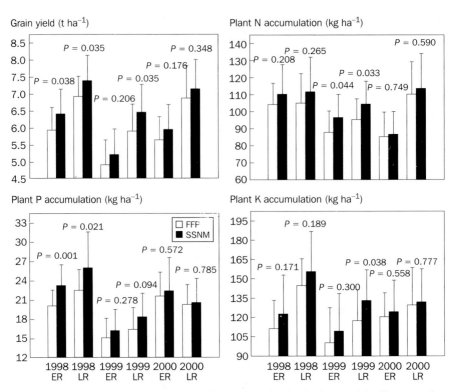

Fig. 12.4. Grain yield and plant N, P, and K accumulation in the farmers' fertilizer practice (FFP) and site-specific nutrient management (SSNM) plots at Jinhua, China (1998-2000; bars: mean; error bars: standard deviation).

nificant. The average yield increase was similar in 1998 (0.47 t ha^{-1}) and 1999 (0.43 t ha^{-1}), but decreased to 0.29 t ha^{-1} in 2000. This decrease was probably caused by increased fertilizer-N use (see below) in attempts by farmers to compete with the SSNM treatment. Note that SSNM and FFP yields in 1999 (year 2) were about 1 t ha^{-1} lower than in 1998 because of less favorable climatic conditions. Yields were highest in the 1998 LR crop, when the average yield in SSNM was 7.4 t ha^{-1}. On five farms, yields in SSNM exceeded 8 t ha^{-1}, with a maximum yield of 8.7 t ha^{-1}. On three farms, yield increases compared with the FFP were greater than 1 t ha^{-1}, showing the potential of SSNM when climatic conditions are favorable.

On average, plant N accumulation increased by 6.0 kg ha^{-1} (6%, $P = 0.012$), P accumulation by 1.8 kg ha^{-1} (9%, $P = 0.003$), and K accumulation by 8.9 kg ha^{-1} (7%, $P = 0.015$). Similar increases in nutrient uptake were achieved in ER and LR crops (Table 12.5). However, increases in N uptake were statistically significant only in 1999 when the N management in SSNM was further improved by using the chlorophyll meter (Figs. 12.4 and 12.5). This increase in N uptake and plant biomass is probably the major reason for sustaining higher P and K uptake in SSNM, even though in 1999 less P and K fertilizer was applied in SSNM than in FFP plots (Figs.

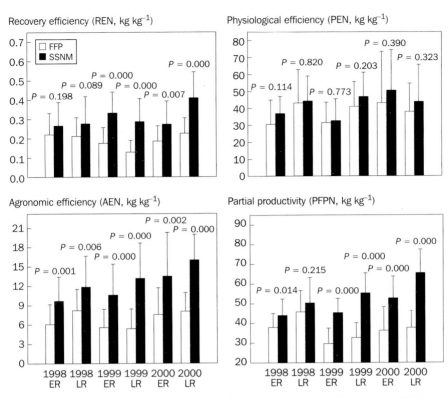

Fig. 12.5. Fertilizer nitrogen-use efficiencies in the farmers' fertilizer practice (FFP) and site-specific nutrient management (SSNM) plots at Jinhua, China (1998-2000; bars: mean; error bars: standard deviation).

12.4 and 12.5). In 2000, differences in N, P, and K uptake between SSNM and FFP were not statistically significant in both ER and LR crops, but rates of fertilizer N, P, and K applied were much smaller in SSNM than in FFP plots (Fig. 12.6).

Effect of SSNM on nitrogen-use efficiency

Large increases in N-use efficiency were achieved through the field- and season-specific N management practiced in SSNM, whereas N-use efficiency remained low in the FFP. In general, compared with the FFP, less N fertilizer was applied in the SSNM plots (Fig. 12.6) and AEN, REN, and the partial factor productivity of N (PFPN) increased significantly (Fig. 12.5). Across all six crops grown and vis-à-vis the FFP, AEN increased by 5.7 kg kg^{-1} (84%, $P = 0.000$), REN by 0.12 kg kg^{-1} (63%, $P = 0.002$), and PFPN by 15.6 kg kg^{-1} (42%, $P = 0.000$; Table 12.5). Increases in N-use efficiency over the FFP were larger in LR than in ER crops (significant crop-season effects for AEN and PFPN, Table 12.5). SSNM had no significant effect on the physiological N efficiency (PEN), which suggests that there was little difference overall in crop management and factors other than N between SSNM and the FFP.

However, significant village effects occurred for all N-use efficiency parameters (Table 12.5). At some sites, recovery efficiencies of applied fertilizer N were particularly low in SSNM treatments (e.g., 0.25 kg kg^{-1} at the Shimen farm), indicating further opportunities to improve N management. There was also evidence that general growing conditions or crop management had positive effects on the performance of SSNM in some villages. Physiological efficiencies were high with SSNM (49–54 kg kg^{-1}) in the villages Qiubin, Bailongqiao, and Jiangtang, indicating excellent growing conditions, whereas PEN in other villages ranged from 34 to 40 kg kg^{-1}. In the FFP, physiological efficiencies were generally below 40 kg kg^{-1} except for the village Qiubin (52 kg kg^{-1}). These differences resulted in a greater variation in agronomic efficiencies among villages with SSNM (9–18 kg kg^{-1}) than with the FFP (6–9 kg kg^{-1}).

Gradual fine-tuning of N management to the local conditions increased N-use efficiency in the SSNM treatment over time, whereas no such changes were observed in the FFP (Fig. 12.5). From 1998 to 2000, average AEN in the SSNM treatment increased from 10.8 to 14.8 kg kg^{-1}, REN from 0.27 to 0.34 kg kg^{-1}, and PFPN from 47.3 to 59.2 kg kg^{-1}. Moreover, the differences in N-use efficiency between SSNM and the FFP also increased with time (significant crop-year effects in Table 12.5). A typical example for the success of in-season adjustment of N rates was the ER crop in 1999, in which unfavorable weather caused a low climatic yield potential. Nevertheless, the farmers applied their usual high amounts of N (average 170 kg ha^{-1}), all at early stages when they could not know the weather conditions and crop yield potential in advance. In contrast, only 115 kg N ha^{-1} was applied in SSNM, resulting in an AEN and REN almost twice those of the FFP. Similarly large differences were obtained in the 1999 and 2000 LR crops (Fig. 12.5).

The current farmers' N management practice appears to be inconsistent with the physiological nutrient requirement of the rice crops and leads to large N losses. Nitrogen supply appears to be excessive during early vegetative growth, whereas N deficiency during later growth stages may limit yields. Moreover, it appears that the timing of N applications is not optimized with regard to the prevailing water management practices so that large N losses occur. Other studies in China have shown that applying more N fertilizer during middle growth stages improved N-use efficiency and increased N uptake and grain yields (Zheng and Xiao 1992, Zheng et al 1997b). At the same site in Jinhua, in an experiment conducted in 1994 LR, skipping the basal N application and applying N in four splits from 8 to 38 DAT increased the grain yield from 6.0 to 6.4 t ha^{-1} and AEN from 7.4 to 11.3 kg kg^{-1} vis-à-vis the current recommended practice (ten Berge et al 1997). Our results provide the first detailed on-farm evidence for the large potential for increasing rice yields with decreasing N fertilizer use in Zhejiang. However, the N-use efficiencies achieved in the third year of SSNM (AEN = 14.8 kg kg^{-1}, REN = 0.34 kg kg^{-1}, Table 12.5) were still below optimal values of about AEN = 20 kg kg^{-1} or REN = 0.50 kg kg^{-1}, which can be achieved with good crop management. There is further scope for improving the N management strategy applied in SSNM toward an even more real-time N management.

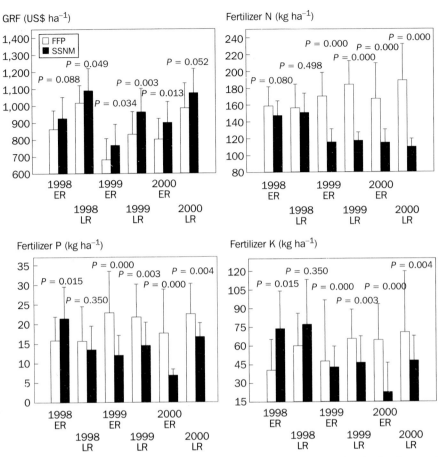

Fig. 12.6. Gross returns above fertilizer costs (GRF) and fertilizer use in the farmers' fertilizer practice (FFP) and site-specific nutrient management (SSNM) plots at Jinhua, China (1998-2000; bars: mean; error bars: standard deviation).

Effect of SSNM on fertilizer use and profit

SSNM significantly reduced fertilizer use and increased profit (Fig. 12.6 and Table 12.6). On average, in each rice crop grown, about 45 kg ha^{-1} less fertilizer N was used in SSNM than in the FFP (-36%, $P = 0.000$), particularly in years 2 (-61 kg N ha^{-1}, -51%) and 3 (-66 kg N ha^{-1}, -58%). Crop-year effects were all significant for NPK fertilizer applications and the general trend was that NPK rates in the SSNM treatment decreased in the order year 1 > year 2 > year 3. Although rates of N, P, and K were similar in SSNM and the FFP in 1998, SSNM fertilizer use was below that of the FFP in 1999 and 2000 (Fig. 12.6). In the final year, 2000, 66 kg ha^{-1} less N, 8 kg ha^{-1} less P, and 32 kg ha^{-1} less K were applied per crop in SSNM compared with the FFP. The gradually decreasing fertilizer rates in SSNM resulted from the high native soil fertility status measured as plant nutrient uptake in omission plots in 1997 and

Table 12.6. Effect of site-specific nutrient management on fertilizer use, fertilizer cost, and gross returns above fertilizer costs from rice production at Jinhua, Zhejiang, China (1998-2000).

Parameters	Levels[a]	Treatment[b] SSNM	Treatment[b] FFP	Δ^c	$P > |T|^c$	Effects[d]	$P > |F|^d$
N fertilizer (FN)	All	126.1	170.9	−44.8	0.000	Village	0.002
(kg ha⁻¹)	Year 1	149.2	157.5	−8.4	0.099	Year[e]	0.000
	Year 2	116.6	177.2	−60.6	0.000	Season[e]	0.022
	Year 3	112.5	178.0	−65.5	0.000	Year × season[e]	0.056
	ER	126.1	165.4	−39.3	0.000	Village × crop	0.042
	LR	126.1	176.5	−50.4	0.000		
P fertilizer (FP)	All	14.2	19.3	−5.2	0.000	Village	0.593
(kg ha⁻¹)	Year 1	17.4	15.7	1.7	0.321	Year	0.000
	Year 2	13.3	22.3	−9.0	0.000	Season	0.647
	Year 3	11.8	20.0	−8.3	0.000	Year × season	0.001
	ER	13.4	18.7	−5.3	0.001	Village × crop	0.009
	LR	14.9	19.9	−5.1	0.000		
K fertilizer (FK)	All	51.9	58.2	−6.3	0.139	Village	0.351
(kg ha⁻¹)	Year 1	75.5	50.3	25.2	0.000	Year	0.000
	Year 2	44.7	56.6	−11.9	0.080	Season	0.530
	Year 3	35.4	67.6	−32.3	0.000	Year × season	0.109
	ER	46.5	50.9	−4.4	0.471	Village × crop	0.965
	LR	57.2	65.5	−8.2	0.155		
Fertilizer cost	All	83.9	109.4	−25.5	0.000	Village	0.102
(US$ ha⁻¹)	Year 1	106.3	98.6	7.7	0.114	Year	0.000
	Year 2	77.5	116.0	−38.5	0.000	Season	0.192
	Year 3	70.8	117.8	−47.0	0.000	Year × season	0.425
	ER	82.1	105.3	−23.2	0.000	Village × crop	0.080
	LR	87.6	116.3	−28.7	0.000		
Gross returns above	All	940.8	851.8	89.0	0.000	Village	0.019
fertilizer costs	Year 1	1,008.0	939.3	68.7	0.029	Year	0.039
(US$ ha⁻¹)	Year 2	965.0	757.2	107.8	0.002	Season	0.157
	Year 3	987.0	893.0	94.0	0.009	Year × season	0.360
	ER	864.2	782.3	81.9	0.001	Village × crop	0.018
	LR	1,042.5	944.1	98.4	0.000		

[a]All = all six crops grown from 1998 ER to 2000 LR; year 1 = 1998 ER and 1998 LR; year 2 = 1999 ER and 1999 LR; year 3 = 2000 ER and 2000 LR; ER = 1998 ER, 1999 ER, and 2000 ER; LR = 1998 LR, 1999 LR, and 2000 LR. [b]FFP = farmers' fertilizer practice; SSNM = site-specific nutrient management. [c]Δ = SSNM − FFP. $P > |T|$ = probability of a significant mean difference between SSNM and FFP. [d]Source of variation of analysis of variance of the difference between SSNM and FFP by farm; $P > |F|$ = probability of a significant F-value. [e]Year refers to two consecutive cropping seasons.

1998. The fertilizer P and K rates used in 2000 are probably the required minimum rates as they reflect a replenishment of crop removal to maintain IPS and IKS.

In contrast, the farmers had no means to estimate the actual indigenous nutrient supplies and adjust their fertilizer rates accordingly. For many years, they have followed existing blanket recommendations for "balanced" fertilizer use, perhaps resulting in an economically unfavorable overuse of fertilizer. Trends in NPK use in the FFP also suggest that at least some farmers attempted to compete with the higher yields observed in SSNM by applying even more fertilizer. Fertilizer-N use in the FFP increased from 157 kg N ha^{-1} per crop in 1998 to 178 kg N ha^{-1} in 2000. Similar increases over time were observed for fertilizer P and K use in the FFP; however, FFP yields remained below those in SSNM plots, providing further evidence that (1) rates of P and K applied in SSNM accurately reflected the high levels of IPS and IKS and (2) increasing the amount of N is less likely to increase yields than optimizing the timing of N supply to achieve better congruence with crop N demand.

Compared with the FFP, the total fertilizer cost in the SSNM decreased by about US$26 ha^{-1} crop^{-1} ($P = 0.000$), but this savings increased from only $8 ha^{-1} in 1998 to $47 ha^{-1} in 2000. The average profit increase in all six crops grown was $89 ha^{-1} crop^{-1} (10.4%, $P = 0.000$), but it was significantly larger in 1999 ($108 ha^{-1}) and 2000 ($94 ha^{-1}) than in 1998 ($69 ha^{-1}). The profit increase in SSNM over the FFP tended to be larger in LR ($98 ha^{-1} crop^{-1}) than in ER ($82 ha^{-1}), but the crop-season effect was not statistically significant. Although SSNM performed better than the FFP in yield and profitability in every village, the significant village effect for profit increase indicated differences in the performance of SSNM among villages (Table 12.6). This was attributed to differences in both fertilizer cost and yield between SSNM and the FFP. Grain yield increases with SSNM over the FFP differed significantly among villages, ranging from 0.3 to 1.0 t ha^{-1} (Table 12.5). Causes were village-specific, including single or multiple effects of factors such as suboptimal N management in SSNM (e.g., low REN on Shimen farm), optimal crop management and growing conditions (e.g., in the villages Qiubin, Bailongqiao, and Jiangtang), and excessive fertilizer use in the FFP (e.g., >200 kg N ha^{-1} in Jiangtang). Fertilizer use in SSNM was adjusted to match the deficit between plant demand and indigenous nutrient supply, whereas this was apparently not done in the FFP. Consequently, fertilizer savings with SSNM compared with the FFP varied from $8 to $63 ha^{-1} crop^{-1} among villages. A better understanding of farmers' decision making in fertilizer use is required, and extension approaches for the dissemination of SSNM may have to be adjusted to differences in farmers' beliefs among villages.

Some differences in crop management between SSNM and FFP plots must be considered when assessing profitability calculations. In the 1999 LR crop, rice in the SSNM plot was planted at about 15% greater density than in the FFP, mainly to increase N-use efficiency during early growth. However, considering the high tillering ability of hybrid rice, the effect on yield was probably not significant. If adoption of SSNM would require planting LR at a density of 20–24 hills m^{-2} rather than the commonly used 17–20 hills m^{-2}, the extra seed cost would amount to about $3 ha^{-1}. Planting at the greater density would, at most, require about 2.5 8-hour labor days ha^{-1} more than in the FFP, which is equivalent to an extra cost of about $11 ha^{-1}. Thus, the total additional crop establishment cost in the 1999 LR season was $14 ha^{-1} vis-à-vis an average gross return over fertilizer cost (GRF) of $132 ha^{-1} measured in the same crop. The yield advantage with SSNM over the FFP decreased slightly in 2000 to an average of +0.29 t ha^{-1} across both seasons, which was mainly attributed to the small difference in plant nutrient uptake between the two treatments compared to previous years (Fig. 12.4). The planting density of 18 hills m^{-2} may have been too low in LR to fully exploit the crop's potential, but farmers also steadily increased fertilizer inputs from year to year, which most likely resulted in increased nutrient uptake and higher yield (Fig. 12.6). Profit increases in 2000 were similar with SSNM to those of previous years because of the higher fertilizer cost in the FFP ($118 vs $71 ha^{-1}).

The real-time N management approach used in SSNM is associated with an extra cost. Using a chlorophyll meter or a simple leaf color chart to gather information about crop N status requires about ½ hour per field. However, if this can be done in a more community-oriented management mode with one person doing it for about 20 fields per day, the cost per hectare becomes small, probably well below $5 ha^{-1} per crop cycle. Another issue is labor for applying N fertilizer because SSNM was often associated with extra topdressing of N. In 1999, the average number of N applications was 2.4 in the FFP vs 3 in SSNM. Assuming that it takes one person about 3 hours to apply N on 1 hectare, the additional cost vis-à-vis the FFP is <$2 ha^{-1}. In summary, although SSNM was associated with an additional cost, those expenses were far below the large increases in GRF measured in all six crops.

Factors affecting the performance of SSNM

Site-specific nutrient management appears to play an important role in increasing productivity in irrigated rice through integrated crop management. Its performance, however, is affected by various other crop management practices. Average actual yields in SSNM were 82% of the model-predicted target yields in 1998 ER, 92% in 1998 LR, 69% in 1999 ER, 80% in 1999 LR, 80% in 2000 ER, and 89% in 2000 LR. The weather affected the performance of SSNM during the rice growth period (Fig. 12.2). Of the six crops grown, only the 1999 and 2000 LR seasons were very favorable in terms of climatic conditions. Heavy rains in mid-June during the flowering period of 1998 and 1999 ER caused low grain-filling percentages (73% on average, <60% on 13 farms in 1998 ER), which reduced grain yield. Rice yields in 1999 ER were also low because rainy and cloudy weather and low temperature in April (Fig. 12.2) caused slow seedling growth. Some farmers had to postpone transplanting so that the tillering period became too short. Because of heavy rains in July 1999, harvest of some ER fields and the succeeding planting of LR crops were delayed. Climatic conditions were generally good in 2000 and internal efficiencies of N, P, and K indicated efficient transformation of plant nutrients into grain yield (e.g., 68 and 63 kg grain kg^{-1} plant N with SSNM in ER and LR, respectively). At Jinhua, insect pests, diseases, and weeds did not cause serious yield losses and no water shortages occurred. However, the improper conduct of the mid-season drainage practice caused

N losses in some rice fields and reduced grain yields in both SSNM and FFP plots. Nearly all farmers did mid-season drainage for both SSNM and the FFP, but some farmers prolonged the drainage so that PI and panicle development were negatively affected. Plant density was a major problem for late rice. To save on hybrid rice seed and labor costs, some farmers used 30–35-day-old hybrid rice seedlings and transplanted at low density (18 hills m^{-2} or lower) with only one seedling per hill.

12.3 Conclusions

Current average rice yield in Zhejiang is about 5.5 to 6 t ha^{-1} or only 60% of the yield potential. Our ongoing research project in Jinhua District identified some key problems for increasing rice production in Zhejiang Province. First, the indigenous nutrient supply, rice production, and nitrogen efficiency varied among farms and seasons. However, farmers only receive blanket fertilizer recommendations from the government-owned agricultural extension stations that do not take differences in soil nutrient levels into account. Identifying causes for the variation in soil nutrient supply is a subject for further research. Second, the current N fertilization strategy used by most farmers results in low N-use efficiency. Nitrogen fertilizer is mainly used to promote tillering during the very early growth stage. Applications are not fine-tuned to achieve better congruence of N supply and plant N demand, particularly at later growth stages. Nitrogen losses into the environment are large. Third, wide plant spacing appears to be a constraint to achieving higher yields in late rice.

The new SSNM approach increased average grain yield and nutrient uptake by 6% to 9%, but these increases were achieved with large reductions in fertilizer use of all three macronutrients, particularly N. As a result, large increases in N-use efficiency were observed and profits increased by 10%. Increases in N-use efficiency and profit with SSNM were larger in 1999 and 2000 than in 1998, suggesting that a gradual improvement in the SSNM approach had occurred. Improvements introduced over time included (1) the availability of better estimates of indigenous nutrient supply, (2) improved N management and the use of the SPAD for decision making, (3) an improved fertilizer recommendation model and P and K management strategy, and (4) greater planting density. It appears that the currently existing extension systems for agriculture in China offer good opportunities for extending the SSNM approach to larger rice areas in the near future. However, to consistently achieve average yields of about 7 t ha^{-1} for early rice and 8 t ha^{-1} for late rice demands further optimization of N management, water management, planting density, and transplanting date. More research is needed to quantify the effects of the commonly practiced mid-season drainage on water savings, N losses, fertilizer-N recovery efficiency, and yield.

References

Dobermann A, White PF. 1999. Strategies for nutrient management in irrigated and rainfed lowland rice systems. Nutr. Cycl. Agroecosyst. 53:1-18.

National Soil and Fertilizer Station. 1993. Symposium of national soil fertility monitoring (1984-1991). Beijing: Labor Press.

Soil Survey Office of Zhejiang. 1994. Zhejiang soils (in Chinese, English summary). Hangzhou: Science and Technology Press of Zhejiang.

ten Berge HFM, Shi QH, Zheng Z, Rao KS, Riethoven JJM, Zhong XH. 1997. Numerical optimization of nitrogen application to rice. Part II. Field evaluations. Field Crops Res. 51:43-54.

Wang GF, Lu JX, Zhang CS. 1994. Nutrient-supplying characteristics and response to fertilizer in middle- and low-lying rice fields in North Zhejiang (in Chinese, English summary). J. Zhejiang Agric. 2:55-58.

Witt C, Dobermann A, Abdulrachman S, Gines HC, Wang GH, Nagarajan R, Satawathananont S, Son TT, Tan PS, Tiem LV, Simbahan GC, Olk DC. 1999. Internal nutrient efficiencies of irrigated lowland rice in tropical and subtropical Asia. Field Crops Res. 63:113-138.

Zhejiang Statistical Bureau. 1998. Statistical yearbook of Zhejiang. Beijing: Chinese Statistics Press.

Zhejiang Statistical Bureau. 1999. Statistical yearbook for rural areas of Zhejiang Province. Beijing: Chinese Statistics Press.

Zheng SX, Xiao QY. 1992. Nutritional characteristics and fertilizer technique in high-yielding hybrid rice. In: The third international high yield and high efficiency conference. Beijing, China. p 1-11.

Zheng Z, Yan L, Wang ZQ. 1997a. Simulation of potential production and studies on optimum quantitative population indices of hybrid rice. J. Zhejiang Agric. Univ. 23:59-64.

Zheng Z, Yan L, Wang ZQ. 1997b. Simulation of nitrogen uptake and numerical optimization of fertilizer N management in rice. J. Zhejiang Agric. Univ. 23:211-216.

Notes

Authors' addresses: Wang Guanghuo and Y. He, College of Environmental and Natural Resources Sciences, Zhejiang University, Huajiachi, Hangzhou 310029, e-mail: ghwang@mail.hz.zj.cn; Q. Sun and J. Wu, Agicultural Research Station, Jinhua, Zhejiang; R. Fu and X. Ding, Agricultural Technical Extension Station, Jinhua, Zhejiang; X. Huang, Jinhua Agricultural School, Zhejiang, China; A. Dobermann and C. Witt, International Rice Research Institute, Los Baños, Philippines; A. Dobermann, University of Nebraska, Lincoln, Nebraska, USA.

Acknowledgments: We are particularly grateful to the farmers in Jinhua County and on Shimen farm for their patience and excellent cooperation in conducting the on-farm experiments since 1997. We wish to thank Piedad Moya, Don Pabale, and Marites Tiongco (IRRI) for help with the socioeconomic farm monitoring, Ma. Arlene Adviento (IRRI) for help with the soil and plant analysis work, and Gregorio C. Simbahan, Rico Pamplona, Olivyn Angeles, Julie Mae Criste Cabrera-Pasuquin, and Edsel Moscoso (all IRRI) for help with data management and statistical analysis. The authors are much obliged to Prof. Chang Yong Huang, Dean of the College of Natural Resources and Environmental Sciences, Zhejiang University, and INMNet steering committee member, for his steady support of our research. We acknowledge contributions made by Dr. David Dawe of IRRI and many other scientists involved in this research.
Citation: Dobermann A, Witt C, Dawe D, editors. 2004. Increasing productivity of intensive rice systems through site-specific nutrient management. Enfield, N.H. (USA) and Los Baños (Philippines): Science Publishers, Inc., and International Rice Research Institute (IRRI). 410 p.

13 Combining field and simulation studies to improve fertilizer recommendations for irrigated rice in the Senegal River Valley

S.M. Haefele and M.C.S. Wopereis

13.1 Introduction

Colonial powers introduced irrigated rice (*Oryza sativa* L.) in sub-Saharan Africa in the early 1900s. It became more widespread in the Sahel region after the severe droughts of the 1970s. The technologies for irrigated rice production that were introduced were based on those of Asia, but farmers and researchers are gradually adapting them to fit the African biophysical and economic environment.

The high production potential in the Senegal River Valley and elsewhere in Sahelian West Africa (8 to 12 t ha^{-1} season^{-1}) is often not realized because of a range of biophysical and socioeconomic constraints such as crop, soil, and water management factors, the cost of labor and machinery, and uncertain input and output markets (Miézan and Sié 1997). Suboptimal nutrient management and fertilizer use (especially N, P, and K) contributes substantially to the existing gaps between potential and actual paddy yields. Major constraints were the large differences in the amount of N applied, the extremely variable timing of N-fertilizer split applications, the lack of P fertilization in P-deficient soils, and severe weed pressure resulting in poor rice recovery efficiencies of N, with average losses of 70% of applied N (Wopereis et al 1999, Haefele et al 2001).

Fertilizer recommendations in sub-Saharan Africa have often remained unchanged since the introduction of irrigated rice and are presently uniform over large areas and cut across diverse climatic and edaphic environments. Ideally, fertilizer recommendations should be developed at the field level. The detailed information that is needed for this type of recommendations, however, is not easily available in sub-Saharan Africa. As a first step in the right direction, we propose to develop recommendations for agroecological zones using crop growth and phenology simulation models. These models can help to advise farmers on fertilizer application strategies for a given agroecological zone × season × rice cultivar combination. Using this approach, we developed new fertilizer and integrated crop management recommendations for the Senegal River Valley. The major aim was to improve fertilizer-N recovery efficiency, yield, and productivity. This paper illustrates the concept

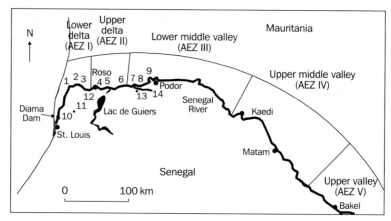

Fig. 13.1. Agroecological zones (AEZ) and study sites in Senegal and Mauritania: Keur Macène (1), Awlicq (2), Gouer (3), Garack (4), Fleuve (5), Gani (6), N'Diawane (7), Koundi (8), Leixeiba (9), WARDA research farm in Ndiaye (10), Pont Gendarme (11), Thiagar (12), WARDA's research farm in Fanaye (13), and Guédé (14).

and reports on the evaluation and adaptation of the newly developed approach at the farm level.

13.2 Site description and general characteristics of the irrigated rice cropping system in the region

The Senegal River Valley is situated in the Sahel savanna vegetation zone (Keay 1959). Five agroecological zones (AEZ) are distinguished in the region based on climate, soil type, and geography (e.g., Boivin et al 1995a)—AEZ I, lower delta; AEZ II, upper delta; AEZ III, lower middle valley; AEZ IV, upper middle valley; and AEZ V, upper valley (Fig. 13.1). Average rainfall is approximately 200 mm year^{-1} in one rainy season (July to September), minimum air temperatures of about 13 to 15 °C occur in December and January, and maximum air temperatures of about 38 to 40 °C or higher occur in May. Solar radiation is high and for a large part of the year is from 20 to 30 MJ m^{-2} day^{-1}.

Soils suitable for irrigated rice in the West African Sahel are invariably linked to valley systems and lowland areas. For the Senegal River delta and valley, the FAO soil map (FAO-UNESCO 1977) shows Eutric Fluvisols with level to gently undulating relief as the dominant soil type from the river mouth near St. Louis to Bakel (Fig. 13.1), coarse- to medium-textured soils in the delta, and medium- to fine-textured soils for the middle and upper river valley. For the Senegal River delta, associated soils (>20% cover) are Thionic Fluvisols and Orthic Solonchaks, whereas Eutric Gleysols and Chromic Vertisols are associated soil types in the middle and upper river valley. The soils are developed in recent deposits of fluviatile origin, but marine influences (salt deposits and/or sulfidic materials) in the subsoil are common in the delta and can be found up to 350 km upstream (Deckers et al 1997) because of a series of sea transgressions and regressions during the last 30,000 years (Deckers et al 1997). For irrigated rice cropping, mostly fine-textured soils with low percolation rates situated in basins behind natural river levees are used. De Poitevin (1993) found that clay soils (>45% clay) occupy around 75% of the cultivated area in the middle valley. About 60% of the clay fraction is composed of swelling clay minerals (Boivin et al 1995b), which explains soil swelling and shrinkage upon contact with water and the importance of vertic characteristics of soils in depressions. The exchange complex of clay is essentially saturated by calcium and magnesium, often in equal proportions. The pH of soil samples extracted with water is generally medium to slightly acidic, with a considerable reserve of exchangeable acidity (Boivin et al 1995b).

Irrigated rice schemes in the Senegal River Valley cover approximately 60,000 ha in Mauritania and Senegal. Irrigation schemes are relatively small, ranging from about 25 ha for village irrigation schemes to 1,000 ha for formerly state-managed irrigation schemes, now managed by farmers' organizations. The crop is mainly direct-seeded and grown in the wet season (WS, July-November) with about 10% to 20% of farmers growing a second crop in the dry season (DS, February-June) in the same field. Pump irrigation with Senegal River water is used. Land preparation is mostly done on dry soils with tractor-driven disk plows, often resulting in only a shallow-tilled topsoil layer (0.05–0.10 m). The current recommended NPK fertilizer dose given by local extension services for the Senegal River Valley is 120-26-50 kg ha^{-1} for both seasons. Recommended urea splits, if applied in two splits, are 67% at early tillering when direct-seeded (or 7 to 10 d after transplanting) and 33% at panicle initiation (PI), and, if applied in three splits, are 50% at early tillering (or 7 to 10 d after transplanting), 25% at midtillering, and 25% at PI. Harvesting and threshing are done manually or mechanically, depending on farm size and farmers' means. Straw normally remains in the field and is often burned after the season, but is sometimes left for grazing cattle. After the dry season, the straw is sometimes sold to cattle owners. Manure or compost is hardly ever used in the region.

13.3 Materials and methods

The approach presented to improve fertilizer recommendations for irrigated rice in the Senegal River Valley is built on several components. The farmers' practice was studied in field surveys to detect areas where improved management could contribute to increasing the performance of rice production. The surveys included an analysis of indigenous soil nitrogen supply (INS) and recovery efficiency of N fertilizer (REN) in farmers' fields. Based on the three-quadrant approach of De Wit (1953), new nitrogen dose recommendations were developed using survey data and crop modeling. Information from fertilizer trials was used to determine accompanying phosphorus and potassium fertilizer doses. Resulting fertilizer doses in combination with recommendations on application timing and weed management were then evaluated together with farmers.

Fertilizer trials

Results from four fertilizer trials that had been conducted by researchers of WARDA's irrigated rice research station in Senegal were used in this study (Table 13.1). Long-term fertility experiments (LTFE) have been conducted since the 1991 WS at WARDA's research farms in Ndiaye and Fanaye (Fig. 13.1), with two rice crops per year on the same plot. The objective is to evaluate long-term effects of intensive irrigated rice cropping under different N, P, and K fertilizer regimes. Cultivars used were Jaya in the 1991 to 1997 WS, IR50 in the 1991 to 1997 DS, and IR13240-108-2-2-3, released in Senegal as Sahel108, in both seasons since the 1998 DS. Straw was completely removed from the plots after each season. In Thiagar, Senegal (1995 WS), and Fleuve, Mauritania (1997 WS), fertilizer trials were conducted in farmers' fields. Treatments are summarized in Table 13.1 and the location of the trials is depicted in Figure 13.1. In contrast to the LTFE, these trials allowed researchers to evaluate the effect of P and K applications separately. All trials had a completely randomized block design with four (LTFE) or three (Thiagar, Fleuve) replications.

Field surveys on farmers' practices

Farmer and plot-level surveys were conducted in Senegal during the 1995 WS (Thiagar) and 1996 WS (Guédé) and in Mauritania during the 1997 WS (i.e., at Keur Macène, Awlicq, Gouer, Garack, Fleuve, Gani, N'Diawane, Koundi, and Leixeiba) (Fig. 13.1, Table 13.2). Detailed information on these surveys can be found in Wopereis et al (1999), Donovan et al (1999), and Haefele et al (2001). In total, these surveys covered 77 farmers. Crop management practices of survey farmers recorded throughout the season included fertilizers and doses applied, timing of crop management actions, weeding practices, and basic economic data. In each participating farmer's field (FF), a subplot (100 m^2) was established, in which farmers managed the crop as in the rest of the field but did not apply any fertilizer (–F).

Plant measurements

Grain yields were obtained from a 6-m^2 harvest area at maturity and yields (kg ha^{-1}) are reported at a standard moisture content of 14%. Total aboveground N uptake in farmers' fields and in fertilizer trials was determined from N concentrations of grain and straw, grain yield, and straw yield. Straw yield was derived from the harvest index of a subsample (12 hills) and grain yield of the 6-m^2 sample. N concentration of oven-dried (80 °C) plant samples was determined using the Micro-Kjeldahl method (Bremner 1996). N uptake in –F plots was used as an estimation of INS (kg N ha^{-1}). REN (kg plant N kg^{-1} applied fertilizer N) in farmer surveys was based on the difference in N uptake between FF and –F and on fertilizer dose applied. For the LTFE, K concentration in straw and grain was measured for three seasons according to Yoshida et al (1976). Total K uptake was calculated as explained for N.

Soil measurements

Five soil samples were taken at 0–0.2-m depth before the onset of the growing season in –F plots in farmers' fields and in all subplots of the LTFE, and then composited

Table 13.1. Fertilizer treatments and grain yield in two long-term fertilizer experiments at Ndiaye and Fanaye, 1991 to 1998 wet season (WS, $n = 8$) and 1991 to 1998 dry season (DS, $n = 8$), and two fertilizer trials at Thiagar (1995 WS) and Fleuve (1997 WS), Senegal River Valley.

Site	Treatment	N	P	K	Zn	N splits[a]	Yield WS		Yield DS	
				(kg ha⁻¹)			Mean	Min.-Max.	Mean (t ha⁻¹)	Min.-Max.
Ndiaye (lower delta)	T1	120	26	50	0	60/30/30	6.8 a[b]	5.2–8.0	7.1 a	5.2–9.0
	T2	120	52	100	0	60/30/30	6.7 a	5.4–7.7	6.9 a	5.5–9.0
	T3	120	0	0	0	60/30/30	5.6 b	4.2–6.9	5.3 b	4.4–6.5
	T4	180	26	50	0	90/45/45	6.6 a	4.8–8.0	7.4 a	6.2–8.5
Fanaye (lower middle valley)	T1	120	26	50	0	60/30/30	6.9 b	5.4–8.4	6.0 b	4.5–7.3
	T2	120	52	100	0	60/30/30	7.1 ab	5.9–8.2	5.9 b	4.8–6.9
	T3	120	0	0	0	60/30/30	6.0 c	3.5–8.1	4.6 c	3.5–5.7
	T4	180	26	50	0	90/45/45	7.5 a	5.9–8.8	7.0 a	5.9–7.5
Thiagar (upper delta)	T1	0	0	0	0		4.6 b			
	T2	138	26	0	0	55/42/42	6.5 a			
	T3	138	26	50	0	55/42/42	6.3 a			
Fleuve (upper delta)	T1	0	0	0	0		3.2 c			
	T2	120	0	0	0	48/36/36	6.6 b			
	T3	120	26	0	0	48/36/36	9.5 a			
	T4	120	26	50	0	48/36/36	6.1 b			

[a]N was applied topdressed in three splits at early tillering, panicle initiation, and booting; P and K fertilizer were applied before planting. [b]Treatment means followed by a common letter are not significantly different according to Duncan's multiple range test ($P<0.05$).

Table 13.2. Fertilizer management practices, grain yield, and nitrogen efficiency in farmers' fields in the Senegal River delta, based on field surveys 1995 to 1997 and the participatory evaluation trial (only sites 4 to 7, Fig. 13.1) in 1998. WS = wet season.

Parameter			Country: Senegal Site: Thiagar Season: 1995 WS[a] Principal cultivar: Jaya Unit $n = 10$	Senegal Thiagar 1995 WS[b] Jaya $n = 10$	Senegal Guedé 1996 WS Jaya $n = 20$	Mauritania Sites 1–9 1997 WS Jaya $n = 37$	Mauritania Sites 4–7 1998 WS Jaya $n = 17$
Applied N	% of n		100	100	100	100	100
	kg ha^1	Av	101	80	117	115	110
		Min.	68	45	79	37	46
		Max.	138	117	177	251	156
Timing first	DAS[c]	Av	45	39	18[d]	32	31
topdressing		Min.	21	30	9[d]	20	25
		Max.	62	61	34[d]	60	38
Timing second	DAS	Av	71	67	43[d]	63	66
topdressing		Min.	59	61	20[d]	42	55
		Max.	85	80	58[d]	83	86
Applied P	% of n	Cases	90	100	100	14	12
	kg ha^1	Av	22	15	21	20	1.5
		Min.	10	9	17	10	0.5
		Max.	66	26	47	30	2.5
Applied K	% of n	Cases	0	20	0	0	0
	kg ha^1	Av		7			
Indigenous	kg ha^1	Av	72 ($n = 8$)	60 ($n = 8$)	31	32	nd
soil N supply		Min.	54	26	16	6	nd
(INS)		Max.	89	76	56	78	nd
Grain yield	t ha^1	Av	4.9	4.1	5.6	4.4	3.9
		Min.	3.6	0.3	2.3	0.9	0.3
		Max.	6.4	6.7	7.2	8.5	6.0
Internal	kg kg^1	Av	50	56	81	77	76
efficiency of N		Min.	40	38	55	38	37
(IEN)		Max.	60	71	109	117	126
Recovery	kg kg^1	Av	0.30	0.38	0.44	0.33	nd
efficiency of N		Min.	0.10	0.06	0.16	0.03	nd
(REN)		Max.	0.76	0.67	0.69	0.73	nd

[a]Single rice cropping. [b]Double rice cropping. [c]DAS = days after seeding. [d]Transplanted rice cropping and therefore days after transplanting. n.d. = not determined.

(Table 13.3). Soil samples were analyzed for pH_{H_2O}, 1:5 soil-extract electrical conductivity, Bray-1-P, Olsen-P, and 1N ammonium-acetate extractable K as described by van Reeuwijk (1992). To analyze the K soil-supplying capacity and the effect of constant rice cropping without K application, soil samples of the LTFE in Ndiaye, taken before the 1998 WS, were analyzed for exchangeable, nonexchangeable, and total soil-K according to MacLean (1961). The nonexchangeable K fraction was divided into a step-K fraction, a more soluble fraction of boiling 1 M HNO_3 extractable-K (Richards and Bates 1988), and a constant-rate fraction (CRF) according to Haylock (1956). Although different rates govern the release of step-K and CRF-K, K from both fractions is assumed to be derived solely from K held in the wedge posi-

Table 13.3. Soil properties (mean and standard deviation) in the two long-term fertilizer experiments (LTFE) at Ndiaye and Fanaye, and in farmers' fields.

Site	n	pH	EC (mS cm^{-1})	Bray 1-P (mg kg^{-1} soil)	Olsen P	Exch. K (cmol kg^{-1})
Ndiaye, LTFE, 1997	6	5.9 (0.3)	0.4 (0.1)	4.7 (2.5)	8.0 (3.4)	0.30 (0.05)
Fanaye, LTFE, 1997	6	5.9 (0.3)	0.1 (0.0)	5.9 (3.2)	14.0 (9.4)	0.24 (0.04)
Thiagar, survey, 1995	20	5.2 (0.3)	0.9 (0.6)	3.5 (1.6)	11.9 (2.6)	0.53 (0.12)
Guedé, survey, 1995	20	6.0 (0.4)	0.2 (0.3)	4.7 (5.6)	12.2 (2.2)	0.40 (0.11)
Mauritania, survey, 1997	41	5.8 (0.7)	1.6 (2.6)	3.8 (1.2)	–	0.27 (0.14)
Mauritania, survey, 1998	20	5.7 (0.5)	0.4 (0.9)	4.0 (1.7)	–	–

tions of illites and vermiculites (Martin and Sparks 1985). Simple K balances for two treatments of the LTFE were calculated from K exports (K_{exp} in kg ha^{-1}):

$$K_{exp} = NS \times (GY \times K_g + SY \times K_s) \qquad (1)$$

where NS = number of seasons (i.e., 14), GY = average treatment grain yield (kg ha^{-1}), SY = average treatment straw yield (kg ha^{-1}), K_g = mean value of K concentration in grain (i.e., T2 0.0037 and T3 0.0034 kg kg^{-1}), and K_s = mean value of K concentration in straw (i.e., T2 0.0166 and T3 0.015 kg kg^{-1}). K inputs (K_{inp} in kg ha^{-1}) were calculated according to

$$K_{inp} = NS \times K_{fert} \qquad (2)$$

where NS = number of seasons (i.e., 14) and K_{fert} = K fertilizer dosage (100 kg K ha^{-1} season^{-1} for T2 and 0 kg K ha^{-1} season^{-1} for T3). K input from irrigation water and dust deposition and K export through leaching were not taken into account. Soil K supply was estimated by multiplying the extractable K fraction in question by a bulk density of 1.5 t m^{-3} (0–0.2-m depth) and 1.6 t m^{-3} (0.2–0.5-m depth), that is, the average values from field measurements. Chemical analyses of soil and plant material were conducted in WARDA's laboratory in Ndiaye, Senegal, and at the Soil Science Department of the University of Hamburg, Germany.

Derivation of new recommendations for fertilizer application rates

The five agroecological zones defined for the Senegal River Valley (Fig. 13.1) were interpreted for irrigated rice cropping using the ORYZA-S model (Dingkuhn and Sow 1997). ORYZA-S simulates the potential yield (Y_{max}) of irrigated rice. Model inputs are solar radiation, daily minimum and maximum temperatures, latitude, and photothermal constants of cultivars used (Dingkuhn and Miézan 1995). Y_{max} was determined for the DS and WS using 10 years of historical weather data (1970-79) from four weather stations located in zones I, II, III, and V along the Senegal River— St. Louis, Rosso, Podor, and Bakel (Fig. 13.1). Results presented here were simulated for medium-duration rice cultivar Jaya and common planting dates for the respective sites for the DS and WS.

To develop N fertilizer recommendations for the different zones, we developed a simple static spreadsheet model based on the three-quadrant approach of De Wit (1953). The model allows us to construct three graphs: the relation between grain yield and total N uptake (internal efficiency of N; IEN), the relation between total N uptake and rate of N application (REN), and the relation between grain yield and rate of N application (PFPN). In the model, yield initially increases proportionally with N uptake but levels off at higher N uptake values. The initial slope of the IEN function (ε_i) for rice was assumed to be 70 kg grain kg^{-1} crop N (M.J. Kropff, personal communication). In the model, e remains constant at low crop N levels, that is, if yields are from 0 to 0.5 Y_{max}. For yields > 0.5 Y_{max}, ε declines linearly from 70 to 0 kg grain kg^{-1} crop N at Y_{max}. IEN reached at Y_{max} was assumed to be constant and was set at $\varepsilon_i/1.5$ to make sure that the uptake yield curve reached the maximum yield level. The resulting value of IEN, 47 kg grain kg^{-1} N uptake at Y_{max}, corresponds to an average crop N concentration at maturity of about 1% with a harvest index of 0.5. It is close to the minimum value of 42 kg grain kg^{-1} N uptake at maturity reported by Cassman et al (1997) for Asian rice-growing conditions.

For the REN function, we assumed an average soil N supply of 39 kg N ha^{-1}, a value that was based on the results of our field surveys. Calculations were performed for the average REN determined in our field surveys (0.35 kg crop N kg^{-1} fertilizer N applied), a lower value (0.25 kg kg^{-1}) representing the case of farmers that face many constraints and a higher value (0.45 kg kg^{-1}) representing the best farmers' practice. To calculate the optimal N application rate, we used the higher value (0.45 kg kg^{-1}) to avoid overfertilization. The PFPN function can then be constructed easily as it is a combination of the first two graphs. The optimal N-fertilizer dose for the different agroecological zones was calculated for a target yield set at 80% of potential yield, that is, $0.8 \times Y_{max}$, as potential yields are rarely reached in farmers' fields because of a range of socioeconomic and biophysical constraints. The slope of the PFPN curve was also determined to obtain insight into the profitability of N application by comparing this slope with a price ratio between 1 kg urea-N and 1 kg rice of about 4 as reported by Donovan et al (1999) for the Senegal River Valley.

The approach described above does not take into account uncertainty in soil and management parameters such as soil N supply, sowing date, or fertilizer recovery and is based on average simulated potential yields, ignoring variability between years. We also ignored the effect of pest and disease pressure or competition between rice and weeds on the N uptake-yield curve. The effect of uncertainty in model input parameters on model outcome can be determined using the approach of Bouman (1994). This, however, is beyond the scope of this paper. We used the three-quadrant model to evaluate the effect of the new fertilizer recommendations on yield for different soil-N supply rates, measured in 75 farmers' fields in our regional surveys. We also calculated site-specific N-fertilizer recommendations for these 75 farmers to reach 80% of potential yield.

Farmer participatory evaluation of integrated crop management recommendations

Discussions with farmers during earlier surveys reported by Wopereis et al (1999) and Haefele et al (2001) revealed a lack of farmers' knowledge on

- Threshold dates for sowing in the WS and DS, as a function of cultivar and location along the Senegal River. The harsh climatic conditions of the Sahel can cause partial or complete crop failure as a result of a delayed start of the growing season.
- Timing, mode, and rate of N, P, and K fertilizer applications.
- Timing of drainage before harvest and timing of harvest. Proper timing helps to reduce irrigation costs, harvest losses, and harvest delays.
- Timing, mode, and rate of herbicide applications.
- Relative importance of N versus P and K.

Integrated crop management recommendations were therefore developed addressing these gaps in farmer knowledge and incorporating the optimal fertilizer application rates derived as explained before. Timing of management actions can be simulated with RIDEV (Dingkuhn 1997). RIDEV simulates rice phenology on the basis of cultivar, sowing date, and weather data and can be used to give recommendations on optimal sowing dates, timing of N applications, timing of weeding, timing of last drainage, and timing of harvest. Model inputs are minimum and maximum daily temperature, photothermal constants of the rice cultivar grown, sowing date, and establishment method (Dingkuhn 1997, Dingkuhn and Miézan 1997). Optimal timing, mode, and dosage of herbicide applications were based on Diallo and Johnson (1997) and on existing recommendations of national agricultural research and extension systems in Senegal and Mauritania.

To evaluate the integrated crop management recommendations, farmer participatory field trials were conducted in rice irrigation schemes located at five sites along the Senegal River, one in Senegal (Pont Gendarme) and four in Mauritania (Garack, Fleuve, Gani, and N'Diawane), during the 1998 WS (Fig. 13.1). All sites were located in either the lower delta (zone I) or the upper delta (zone II). Ten farmers participated in Pont Gendarme and 20 in Mauritania. At each farm, four weed and soil fertility management options were evaluated in four adjacent fields:

FP: farmers' practice
FPF: farmers' practice but with new recommended fertilizer management
FPW: farmers' practice but with recommended weed management
FPFW: new recommended fertilizer and weed management

Treatments differed in weed and fertilizer management, combining the farmers' practice and a recommended practice. Recommended weed management consisted of an overall application of 8 L ha^{-1} propanil and 1.0 L ha^{-1} 2,4-D in Senegal and 6 L ha^{-1} propanil and 2.0 L ha^{-1} 2,4-D in Mauritania, applied at the 2–3-leaf stage of rice. Recommended fertilizer management consisted of an early (basal or topdressed) application of 100 kg ha^{-1} diammonium phosphate (DAP, 18% N, 20%

P) before direct seeding and 300 kg ha^{-1} urea (46% N) in three split topdressings. The first application of 120 kg urea ha^{-1} was made at early tillering (21 d after sowing), the second application of 120 kg urea ha^{-1} at PI, and a third application of 60 kg ha^{-1} at booting (21 days after PI). All farmers used the local variety Jaya. Sowing date was determined by extension officers and it was the same for all four treatments. Dates of fertilizer application were determined using RIDEV and 10 years of weather data from WARDA's weather station in Ndiaye for the Pont Gendarme farmers and 30 years of weather data from Rosso for the farmers in Mauritania. Crop management practices of each farmer were recorded throughout the season and grain yield and yield components were measured. Extension officers supervised the recommended practices at all sites.

13.4 Results and discussion

Soils
Table 13.3 shows the results of some soil properties for farmers' fields and trial sites. The pH values found in the delta and in the middle valley ranged from 4.5 to 7.5, with average values in the medium acid to slightly acid range. The possible danger of alkalization, especially for the middle valley, had been recognized earlier but until now no clear trends have been observed (Wopereis et al 1998). Electrical conductivity (EC) was in general below 0.9 mS cm^{-1}, but higher in the delta because of the salt concentration in the topsoil through capillary rise from the marine salt deposits in the subsoil. In the middle valley and further upstream, EC values were normally below 0.5 mS cm^{-1}. Even in highly saline soils in the delta, rice yields of more than 6 t ha^{-1} can be obtained with good irrigation and drainage facilities (Ceuppens and Wopereis 1999). Bray-1 plant available P was around 3.5 to 4.5 mg kg^{-1} soil, which overlaps with critical levels reported in the literature (Sanchez 1976). This is also true for Olsen-P, for which the critical level ranges from 4 to 29 mg kg^{-1} (Doberman et al 1995). The range of critical levels reported in the literature for K availability in rice soils is from 0.1 to 0.4 cmol kg^{-1} soil (Doberman et al 1995), which is similar to the range of values found in our surveys. INS was low, with average values from 31 to 72 kg N ha^{-1} and the lowest values in Mauritania (Table 13.2). The average value for all survey farmers was 39 kg N ha^{-1}.

Farmers' fertilizer management practices
Urea (46% N) and diammonium phosphate (18% N, 20% P) are the most common mineral fertilizers in the region and they are mostly applied topdressed. Rarely used fertilizers are potassium chloride (50% K), triple superphosphate (20% P), and compound fertilizers distributed by the United Nations High Commission for Refugees (10% N, 4% P, 15% K). Mineral fertilizers are mostly applied in two splits, in which, for the second topdressing in general, only urea is used. Table 13.2 shows the farmers' fertilizer strategy for four sites in four different years. All farmers applied N, but the timing of fertilizer application was extremely variable and often did not coincide

Table 13.4. Simulated potential yield (Y_{max}) of rice for four weather stations in the Senegal River Valley. Potential yields were simulated for the cultivar Jaya with sowing dates February for the dry season (DS) and July for the wet season (WS) using ORYZA-S (Dingkuhn and Sow 1997). Results are based on weather data from 1970 to 1979. For location of weather stations, see Figure 13.1.

Y_{max}	St. Louis		Rosso		Podor		Matam		Bakel	
	WS	DS	WS	DS	WS	DS	WS	DS	WS	DS
					(t ha^{-1})					
Average	9.2	10.0	8.6	9.0	8.0	7.8	–	–	8.2	4.8
Minimum	8.0	8.6	7.6	7.6	7.0	6.0	–	–	7.1	3.4
Maximum	10.2	12.0	10.0	9.8	8.4	8.4	–	–	9.6	5.8

with the critical growth stages of the rice plant. The range for the first topdressing of fertilizer in direct-seeded rice was 20 to 62 d after sowing (DAS) and 42 to 86 DAS for the second topdressing. Average N doses were close to the currently recommended level of 120 kg ha^{-1} at all sites but ranged from 37 to 251 kg ha^{-1}. The average P application rate was about 20 kg P ha^{-1}, but not all farmers applied P fertilizer and, especially in Mauritania, P fertilizer use was rare. Some farmers applied extreme rates of P fertilizer of up to 66 kg ha^{-1}. Only two farmers used K fertilizer at low rates. Farmers' yields were on average from 3.9 to 5.6 t ha^{-1} but were highly variable from season to season and among farmers. Other agronomic constraints included delayed start of the growing season, causing yield loss because of cold sterility, weed problems, and late harvesting (Wopereis et al 1999, Haefele et al 2001). As a consequence, the average REN per site ranged from 0.30 to 0.44 kg crop N kg^{-1} fertilizer N (Table 13.2) and almost total losses were observed at all sites.

Potential yields

Table 13.4 presents simulated potential rice yields along the Senegal River. For the WS, these averaged 9 t ha^{-1} in the Senegal delta and 8 t ha^{-1} in the middle and upper valley, fluctuating within a range of 2 t ha^{-1} for the 10 weather scenarios. In the DS, potential rice yields were 10 t ha^{-1} in St. Louis, 9 t ha^{-1} in Rosso, 8 t ha^{-1} in Podor, and 5 t ha^{-1} in Bakel, with a slightly higher fluctuation. The drop in potential yields in the upper valley and in the dry season is caused by high average air temperatures around flowering, which induce spikelet sterility.

On the basis of these results, we distinguished four potential yield zones in the DS: the lower delta with Y_{max} = 10 t ha^{-1} (AEZ I), the upper delta with Y_{max} = 9 t ha^{-1} (AEZ II), the lower middle valley with Y_{max} = 8 t ha^{-1} (AEZ III), and the upper valley with Y_{max} = 5 t ha^{-1} (AEZ V). We were not able to simulate Y_{max} for the upper middle valley because of missing weather data. Because of the high costs of irrigated rice cropping in the region and the low average yield potential, rice cropping in the dry season in the upper valley (AEZ V) is extremely risky and not recommended. For the WS, we distinguished two potential yield zones, one with a potential of 9 t ha^{-1}

Table 13.5. Optimal N-fertilizer application rates (Nopt, in kg ha^{-1}) and slope of the partial factor productivity function of N at Nopt for target yields (80% of yield potential) and three different fertilizer-N recovery efficiencies (REN) as derived for four agroecological zones (AEZ) in the wet and dry seasons (WS and DS) using the three-quadrant model.

AEZ	Season	Target yield (t ha^{-1})	REN (0.25 kg kg^{-1})		REN (0.35 kg kg^{-1})		REN (0.45 kg kg^{-1})	
			Nopt (kg ha^{-1})	Slope (kg kg^{-1})	Nopt (kg ha^{-1})	Slope (kg kg^{-1})	Nopt (kg ha^{-1})	Slope (kg kg^{-1})
I, II	WS	7.2	290	11	207	15	161	20
III, IV, V	WS	6.4	240	11	172	15	135	20
I	DS	8.0	340	11	242	15	187	20
II	DS	7.2	290	11	207	15	161	20
III	DS	6.4	240	11	172	15	135	20

(overlapping AEZ I and II) and the other with a potential of 8 t ha^{-1} (overlapping AEZ III, IV, and V). In this case, we assumed Y_{max} in zone IV to be similar to Y_{max} in AEZ III and V, that is, 8 t ha^{-1}.

N application rates

The three-quadrant model was used to derive the optimal N application rates for the different agroecological zones and seasons. Table 13.5 presents the results. Taking the best recovery rate (0.45 kg crop N kg^{-1} N applied), the optimal fertilizer application rates are 135 kg N ha^{-1} (AEZ III, IV, V) and 161 kg N ha^{-1} (AEZ I, II) in the wet season and 187 kg N ha^{-1} (AEZ I), 161 kg N ha^{-1} (AEZ II), and 135 kg N ha^{-1} (AEZ III) in the dry season. No recommendations were derived for AEZ IV because of missing weather data and for AEZ V because of the low yield potential, which presents too much risk for farmers to grow irrigated rice. Optimal N application rates are expected to be profitable as indicated by the slope of the fertilization rate versus yield curve (about 20 kg grain kg^{-1} N applied) compared to the price ratio of about 4 between 1 kg urea-N and 1 kg rice in the Senegal River Valley (Donovan et al 1999).

Figure 13.2 shows simulated yield versus N uptake for the three survey sites (Thiagar, 1995 WS, Guédé, 1996 WS, and Mauritania, 1997 WS). At all sites, observed data corresponded well with model predictions. Only a few farmers reached the target yield.

We used the three-quadrant model, with an REN of 0.45 kg kg^{-1}, to calculate grain yield for 73 farmers based on measured INS and the new N fertilizer recommendations (Table 13.6). Compared with that of the farmers' practice (assuming the same REN), average yield increased by 1.1 t ha^{-1} using an additional 51 kg N ha^{-1}. Field-specific recommendations reach the target yield on average with a 10 kg ha^{-1} lower N dose if the calculated recommendation for the lowest INS class is excluded. This reflects the fact that, with field-specific recommendations, the target yield would be adjusted to such a low INS.

Grain yield (t ha⁻¹)

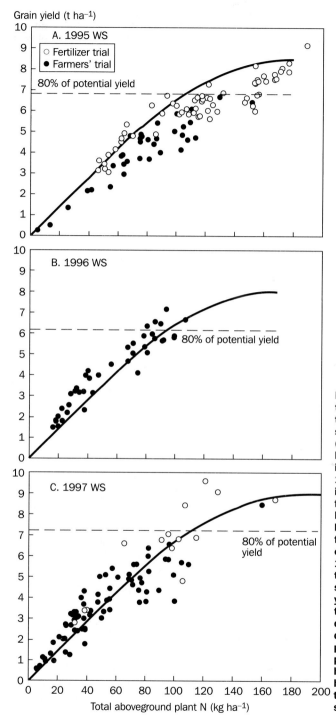

Fig. 13.2. Relation between grain yield and total N uptake at the survey sites Thiagar (A), Guédé (B), and Mauritania (C). The last included sites 1–9 (Fig. 13.1). Closed symbols indicate data from farmers' fields (–F and FP plots). Open symbols indicate data from fertilizer trials in farmers' fields (see Table 13.1). The solid line is the simulated relationship between grain yield and plant N uptake using the three-quadrant model. The dotted line is 80% of potential yield. Yield potentials were 8.5 t ha⁻¹ at Thiagar, 8 t ha⁻¹ at Guédé, and 9.0 t ha⁻¹ at the Mauritania sites.

Table 13.6. Comparison of different N management strategies using actual indigenous N supply (INS) in 73 farmers' fields with rice in agroecological zones I and II during the wet season. Assumed recovery efficiency of applied fertilizer N was 0.45 kg kg^{-1}. Yield was calculated using the model presented in the text.

INS class (kg ha^{-1})	Cases	Mean soil N supply (kg ha^{-1})	Fertilizer N applied (kg ha^{-1})	Recovered fertilizer (kg ha^{-1})	Plant N (kg ha^{-1})	Yield (t ha^{-1})
Farmers' practice						
0<INS<20	12	14	111[a]	50	64	4.5
20<INS<40	35	31	116[a]	52	83	5.7
40<INS<60	12	48	111[a]	50	97	6.5
60<INS<80	11	70	98[a]	44	114	7.3
INS>80	3	83	91[a]	41	124	7.7
Mean		39	110			6.0
Regional recommendation						
0<INS<20	12	14	161	72	87	5.9
20<INS<40	35	31	161	72	103	6.8
40<INS<60	12	48	161	72	120	7.6
60<INS<80	11	70	161	72	143	8.3
INS>80	3	83	161	72	156	8.6
Mean		39	161			7.1
Field-specific recommendation						
0<INS<20	12	14	217	97	112	7.2
20<INS<40	35	31	180	81	112	7.2
40<INS<60	12	48	143	64	112	7.2
60<INS<80	11	70	92	41	112	7.2
INS>80	3	83	63	28	112	7.2
Mean		39	151[b]			7.2

[a]Average N doses used by the survey farmers. [b]Does not include the recommended N dose for the lowest INS class, as this recommendation is not realistic. Further explanations are given in the text.

P and K application rates for the different zones

K and especially P are less mobile and not easily lost from the root zone in comparison with N, which is affected by several biological and chemical processes (nitrification, denitrification, NH_3 volatilization, biological N_2 fixation). Therefore, the issue of maximizing recovery efficiency is less important for fertilizer P and K than for N. Nevertheless, unnecessarily high application rates will not be economical and low application rates may decrease the available soil nutrient supply, limit crop growth, and thus reduce N-use efficiency.

Table 13.1 summarizes data from fertilizer trials used to estimate P and K fertilizer needs for irrigated rice cropping in the region. The LTFEs in Ndiaye and Fanaye show a clear effect of PK applications. At both sites and in both seasons, grain yields of treatments without P and K application were at least 1 t ha^{-1} lower than grain yields in treatments with P and K application. Treatment effects on grain yield among the treatments with NPK 120-26-50 (T1) and NPK 120-52-100 (T2)

Table 13.7. Exchangeable, nonexchangeable, and total K soil supply after 14 seasons of continuous cropping in treatments receiving 200 kg K ha^{-1} y^{-1} and 0 kg K ha^{-1} y^{-1} of the LTFE in Ndiaye. Soil K supply is estimated by multiplying the K fraction in question by the bulk density of 1.5 t m^{-3} (0–0.2-m depth) and 1.6 t m^{-3} (0.2–0.5-m depth).

Item	200 kg K ha^{-1} y^{-1}			0 kg K ha^{-1} y^{-1}	
	Depth (m)	Soil K (mg kg^{-1})	Soil K (t ha^{-1})	Soil K (mg kg^{-1})	Soil K (t ha^{-1})
Exchangeable K	0–0.2	301 a[b]	0.9	260 a	0.8
	0.2–0.5	321 a	1.5	320 a	1.5
Step K[a]	0–0.2	1,243 a	3.7	1,095 a	3.3
	0.2–0.5	1,134 a	5.4	1,139 a	5.5
Constant rate K[a]	0–0.2	1,018 a	3.1	1,052 a	3.2
	0.2–0.5	1,050 a	5.0	977 a	4.7
Total soil K	0–0.2	11,586 a	34.8	11,748 a	35.2
	0.2–0.5	11,049 a	49.7	11,254 a	50.6

[a]Nonexchangeable K fractions, step K, and constant rate K according to MacLean (1961). [b]Values in the same row followed by a common letter are not significantly different according to Duncan's multiple range test ($P<0.05$).

were insignificant when analyzing average seasonal values (Table 13.1) or individual seasons (data not shown). We concluded that an application rate of 26 kg P ha^{-1} and 50 kg K ha^{-1} is high enough to reach yields of 8 t ha^{-1} at both sites and in both seasons. Unfortunately, the design of the LTFE does not allow us to separate the relative importance of P and K. The trials in Thiagar and Fleuve are currently the only available data sources for which the effect of both elements can be evaluated separately. In both trials, yield increased with N and P application, whereas K fertilizer did not affect yield. Treatments that differed only in K application rate resulted in even lower yields when K was applied. Cy-Chain Chen et al (1989) observed similar effects in rice soils and attributed them to NH_4^+ entrapment in clay minerals caused by added K. The high yield difference between treatments with and without K application (T3 vs T4) in Fleuve might also be due to other unknown yield-determining factors.

To analyze (1) the K soil-supplying capacity, (2) the partitioning of soil K in different pools, and (3) the effect of constant rice cropping without K application, soil samples of the LTFEs were analyzed for exchangeable, nonexchangeable, and total soil K according to MacLean (1961). Table 13.7 presents the results for Ndiaye only, but similar results were obtained for Fanaye. Although treatment T3 never received any K fertilizer and the K balance difference between the two treatments was estimated to be 0.9 t ha^{-1} (0.5 t ha^{-1} loss for T2 and 1.4 t ha^{-1} loss for T3), no significant difference in any soil K fraction was detectable. The estimated losses of treatment T3 even equal the complete exchangeable soil K supply in the topsoil (0 to 0.2 m). We concluded that the exchangeable soil K pool has been replenished by nonexchangeable soil K and/or by inputs from dust deposition, irrigation water, and capillary rise during the noncropped time of the year. Table 13.7 also shows the immense plant-available soil-K stock, since plant K uptake is served from exchangeable as well as nonexchangeable K. The application of K fertilizer in the region therefore seems to be currently unnecessary and uneconomical, especially if rice straw is not exported from the fields. Low K fertilizer application is recommended in the case of double cropping, high yields, and straw removal.

We cannot decide at what fertilizer P level crop growth would be affected based on the data that are currently available. Most probably, fertilizer application rates can be lower than 26 kg P ha^{-1} since with 26 kg P ha^{-1} response saturation was achieved for yields up to 8 t ha^{-1} and higher in the LTFE (Table 13.1). We decided that an application of 20 kg P ha^{-1} would be sufficient, except for relatively high N application rates (i.e., 180 kg N ha^{-1}). The existing uniform NPK recommendation of 120-26-50 kg ha^{-1} regardless of location along the Senegal River and season was therefore changed to 161-20-0 kg ha^{-1} (AEZ I, II) and 135-20-0 kg ha^{-1} (AEZ III, IV, V) in the WS. For the DS, we recommend 187-25-0 kg ha^{-1} (AEZ I), 161-20-0 kg ha^{-1} (AEZ II), and 135-20-0 kg ha^{-1} (AEZ III).

Farmer participatory evaluations

Two of the 20 farmers that participated in Mauritania (1998 WS) abandoned their fields (one because of flooding and the other because of late sowing) and another field became inaccessible because of flooding. Observations on farmers' crop management practices confirmed observations made in earlier seasons (Table 13.2) even if some management parameters were influenced by recommended treatments (detailed results in Wopereis et al 1999 and Haefele et al 2001).

The farmers' practice (FP) resulted in a mean yield of 3.9 t ha^{-1} in Mauritania and 3.8 t ha^{-1} in Senegal. For Mauritania, this corresponds well to average yields of 3.8 t ha^{-1} reported by SONADER (1998) and 4.1 t ha^{-1} measured by AGETA in the 1998 wet season (AGETA 1998). In Senegal, mean yields in the delta region ranged from 4.2 to 5.0 t ha^{-1} in the 1990 to 1996 WS (SAED 1998), indicating that the yields found in the present study were not exceptional. At all sites, grain yields of FP were significantly lower than for the other treatments. Grain yield increased by 0.9 t ha^{-1} in Mauritania and by 0.8 t ha^{-1} in Senegal when applying the recommended fertilizer management (FPF). The recommended weed management (FPW) resulted in a yield increase of 0.9 t ha^{-1} in Mauritania and of 1.1 t ha^{-1} in Senegal. No significant differences were apparent between FPF and FPW, but the effects were additive and, compared with FP, FPFW gave a mean yield increase of 1.8 t ha^{-1} in Mauritania and 1.7 t ha^{-1} in Senegal. Improving either weed or fertilizer management had a similar effect on yield, and only by improving both could higher yields be obtained.

Table 13.8 shows the economic results. Partial budgeting shows that the highest treatment costs were encountered in FPFW, followed by FPF, whereas costs were lowest in the treatment with improved weed management (FPW). The gross added product and treatment net benefit increased in the order FP > FPF > FPW > FPFW. The value-cost ratio in partial budgeting indicates the relationship between total increased value and amount spent to get this value for a treatment compared with a

Table 13.8. Economic evaluation of four crop management practices in the farmer participatory testing of the new recommendations in the 1998 wet season. FP = farmers' practice, FPF = farmers' practice but recommended fertilizer management, FPW = farmers' practice but recommended weed management, FPFW = recommended fertilizer and weed management.

Item	FP	FPF	FPW	FPFW
			(Euro ha^{-1})	
Partial budgeting				
Senegal, site 12				
Additional costs per treatment[a]	–	61	43	99
Gross added product[b]	–	129	179	283
Treatment net benefit[c]	–	68	137	184
Value-cost ratio[d]	–	2.1	4.2	2.8
Mauritania, sites 2 to 5				
Additional costs per treatment[a]	–	50	36	82
Gross added product[b]	–	162	165	323
Treatment net benefit[c]	–	112	129	241
Value-cost ratio[d]	–	3.2	4.6	3.9
Total budgeting				
Senegal, site 12				
Net benefit[e]	215	283	352	399
Net benefit increase in % of TP	–	32	64	86
Mauritania, sites 2 to 5				
Net benefit[e]	284	396	413	525
Net benefit increase in % of TP	–	39	45	85

[a]Additional costs = costs of the treatment – costs for TP (total product). [b]Gross added product = paddy price × (yield increase – costs caused by the higher yield). [c]Treatment net benefit = gross added product – additional costs. [d]Value-cost ratio = gross added product/additional cost. [e]Net benefit = paddy price × yield – total costs.

reference treatment. The partial value-cost ratio was highest for FPW and lowest for FPF. These trends were similar in both countries, but in Senegal costs were generally higher and net benefits were lower. Average net benefits increased in the order FP > FPF > FPW > FPFW and ranged from 284 to 525 Euro ha^{-1} in Mauritania and from 215 to 399 Euro ha^{-1} in Senegal. In Mauritania, the net benefit increase above FP was similar for FPF and FPW (40%), whereas, in Senegal, FPW showed a much higher increase (60%) than FPF (30%). At all sites, FPFW gave the greatest total net benefit, with an 85% increase compared with FP.

A simple decision tool for fertilizer management

Farmers often asked advice about what to do under nonoptimal situations, for example, in the case of financial constraints at the time that fertilizer needs to be bought. They also often do not know the relative importance of N, P, and K for rice growth. We therefore developed a simple decision tool to guide farmers in fertilizer management (Table 13.9). The optimum quantity of fertilizer to apply in the WS or DS is given for each AEZ (calculated using an REN of 0.45 kg crop N kg^{-1} fertilizer-N

Table 13.9. A simple decision tool developed to help farmers calculate required fertilizer application rates (50-kg bags ha^{-1}). Optimal fertilizer application rates are given for the dry season (DS) and wet season (WS) depending on agroecological zones (AEZ). Farmers can choose lower rates in the case of financial and/or general crop management constraints. No recommendations are given for the dry season in AEZ IV and V (see text). Fertilizer sources most commonly used in the region are urea (46% N), diammonium-phosphate (DAP; 18% N and 20% P), and potassium chloride (KCl; 50% K). nd = not determined.

WS AEZ	DS AEZ	Target yield	Total	Recommended number of urea, DAP, or KCl fertilizer bags ha^{-1}												
				Urea	Urea	Urea	DAP	Urea	Urea	DAP	Urea	Urea	DAP	Urea	DAP	KCl
	nd	3	1	1												
	nd	4	1	1												
	nd	5	1	1												
	nd	6	1	1												
III,IV,V	III	6.4	7	1	1	1	1	1	1	1						
I,II	II	7.2	8	1	1	1	1	1	1	1	1					
I	I	8.0	9.5	1	1	1	1	1	1	1	1			1	0.5	
I	I	8.0[a]	10.5	1	1	1	1	1	1	1	1			1	0.5	1

[a] = recommendation for rice double cropping including straw removal.

applied) in bags of 50 kg for a 1-ha rice field (a very common size). Lower application rates are given for suboptimal situations, but, below three bags of urea ha^{-1}, the probability of negative returns to irrigated rice cropping becomes high because of the high total production costs. From these rates, farmers can choose according to their means. At low rates, only nitrogen is limiting and, with increasing N dose, P application becomes more important. The diagram is based on urea (46% N) and diammonium phosphate (18% N, 20% P) inputs because they are the N and P fertilizers used almost exclusively in the region. The decision tool is given to farmers together with the integrated crop management recommendations described above, aimed at increasing fertilizer recovery.

13.5 Conclusions

We developed improved fertilizer recommendations for irrigated rice in the Senegal River Valley based on a simple three-quadrant model, farmer surveys, on-station experiments, and simulated potential yield.

Potential yield (Y_{max}) in both the wet and dry season was determined for all five agroecological zones that have been defined for the Senegal River Valley using the ORYZA-S rice growth simulation model. A simple spreadsheet-based three-quadrant model was developed to derive a partial factor productivity function for N (PFPN), based on Y_{max}, a function for internal efficiency of N (IEN), and survey data on recovery efficiency of N (REN) and indigenous soil N supply (INS). The best farmer REN (0.45 kg crop N kg^{-1} N applied) and average INS (39 kg N ha^{-1}) from farmer surveys were used to estimate the optimum N application rate to reach 80% of Y_{max} (assumed to be equivalent to the maximum attainable yield in farmers' fields). Long-term fertility trials and fertilizer trials conducted in farmers' fields were interpreted to derive accompanying recommendations for phosphorus and potassium. The existing uniform NPK recommendation of 120-26-50 kg ha^{-1}, regardless of location along the Senegal River and season, was changed to 161-20-0 kg ha^{-1} (AEZ I, II) and 135-20-0 kg ha^{-1} (AEZ III, IV, V) for the WS. For the DS, we recommend 187-25-0 kg ha^{-1} (AEZ I), 161-20-0 kg ha^{-1} (AEZ II), and 135-20-0 kg ha^{-1} (AEZ III). No recommendations were derived for AEZ IV because of a lack of weather data and for AEZ V because the low yield potential presents too much risk for farmers to grow irrigated rice. Application of K is recommended only for intensive double-cropped systems with high yields and where straw is exported from the field. A simple diagram was developed to help farmers decide what kind of and how much fertilizer to apply in optimal situations as well as in the case of financial constraints.

Recommendations were embedded into integrated crop management recommendations, with emphasis on weed and soil fertility management, and these were evaluated with farmers and extension agents in the field. These evaluations showed the possible productivity gains by improving the timing, mode, and rate of fertilizer and herbicide applications. The integrated crop management recommendations resulted in an increase in net revenues of from 30% to 85% compared with the farmers' practice. Further improvement in fertilizer-use efficiency can be expected from nutrient management recommendations at the field or farm level. However, this is not yet a realistic option for irrigated rice farmers in the region.

The next steps in the optimization of fertilizer recommendations will include calibration and validation of soil tests to get a better handle on soil nutrient-supplying capacity. We will also analyze differences in the N-use efficiency of rice cultivars for further adjustment of the recommendations. Nutrient balances are determined at key sites, taking into account inputs from dust deposition and irrigation water. We hope that this work will result in more precise recommendations that can be applied at the irrigation scheme or field level.

References

AGETA. 1998. Les recommandation de l'AGETA; Objective: 6 t ha^{-1}. Rosso (République Islamique de Mauritanie): Association Général d'Exploitants et Eleveurs pour l'Etude et l'Emploi des Techniques Améliorées Agricoles et Animales. 10 p.

Boivin P, Brunet D, Cascuel C, Zante P, Ndiaye JP. 1995b. Les sols argileux de la région de Nianga-Podor: répartition, caracteristiques, aptitudes et risques de dégradation sous irrigation. In: Boivin P, Dia I, Lericollais A, Poussin JC, Santoir C, Seck SM, editors. Nianga, laboratoire de l'agriculture irriguée dans la moyenne vallée du Sénégal. Paris (France): ORSTOM Collection Colloques et Séminaires. p 67-81.

Boivin P, Dia I, Lericollais A, Poussin JC, Santoir C, Seck SM. 1995a. Nianga, laboratoire de l'agriculture irriguée en moyenne vallée du Sénégal. Paris (France): ORSTOM Collection Colloques et Séminaires. p 39.

Bouman BAM. 1994. A framework to deal with uncertainty in soil and management parameters in crop yield simulation: a case study for rice. Agric. Syst. 46:1-17.

Bremner JM. 1996. Nitrogen-total. In: Sparks DL, editor. Methods of soil analysis. Part 3. Chemical methods. Madison, Wis. (USA): Soil Science Society of America Book Series. p 1085-1122.

Cassman KG, Peng S, Dobermann A. 1997. Nutritional physiology of the rice plant and productivity decline of irrigated lowland rice systems in the tropics. Soil Sci. Plant Nutr. 43:1111-1116.

Ceuppens J, Wopereis MCS. 1999. Impact of non-drained irrigated rice cropping on soil salinization in the Senegal River Delta. Geoderma 92:125-140.

Cy-Chain Chen, Turner FT, Dixon JB. 1989. Ammonium fixation by high-charge smectite in selected Texas Gulf coast soils. Soil Sci. Soc. Am. J. 53:1035-1040.

Deckers J, Dondeyne L, Vandekerckhoven L, Raes D. 1997. Major soils and their formation in the West African Sahel. In: Miézan KM, Wopereis MCS, Dingkuhn M, Deckers J, Randolph TF, editors. Irrigated rice in the Sahel: prospects for sustainable development. Bouaké (Côte d'Ivoire): West Africa Rice Development Association. p 23-35.

De Poitevin F. 1993. Etude d'impact des techniques culturales sur les aménagements hydroagricoles dans la région de Podor (Sénégal). Mémoire de quatrième année de l'ESAP, Multig. Dakar (Sénégal): ORSTOM. 53 p.

De Wit CT. 1953. A physical theory on the placement of fertilizers. Versl. Landbouwk. Onderz. 59(4):1-71.

Diallo S, Johnson DE. 1997. Les adventices du riz irrigué au Sahel et leur contrôle. In: Miézan KM, Wopereis MCS, Dingkuhn M, Deckers J, Randolph TF, editors. Irrigated rice in the Sahel: prospects for sustainable development. Bouaké (Côte d'Ivoire): West Africa Rice Development Association. p 311-323.

Dingkuhn M. 1997. Characterizing irrigated rice environments using the rice phenology model RIDEV. In: Miézan KM, Wopereis MCS, Dingkuhn M, Deckers J, Randolph TF, editors. Irrigated rice in the Sahel: prospects for sustainable development. Bouaké (Côte d'Ivoire): West Africa Rice Development Association. p 343-360.

Dingkuhn M, Miézan KM. 1995. Climatic determinants of irrigated rice performance in the Sahel. II. Validation of photothermal constants and characterization of genotypes. Agric. Syst. 48:411-434.

Dingkuhn M, Sow A. 1997. Potential yields of irrigated rice in the Sahel. In: Miézan KM, Wopereis MCS, Dingkuhn M, Deckers J, Randolph TF, editors. Irrigated rice in the Sahel: prospects for sustainable development. Bouaké (Côte d'Ivoire): West Africa Rice Development Association. p 361-380.

Doberman A, Cassman KG, Sta. Cruz PC, Neue HU, Skogley EO, Pampolino MF, Adviento MAA. 1995. Dynamic soil tests for rice. In: Fragile lives in fragile ecosystems. Proceedings of the International Rice Research Conference, 13-17 February 1995. Manila (Philippines): International Rice Research Institute. p 343-365.

Donovan C, Wopereis MCS, Guindo D, Nebié B. 1999. Soil fertility management in irrigated rice systems in Sahel and Savanna regions of West Africa. Part II. Profitability and risk analysis. Field Crops Res. 61:147-162.

FAO-UNESCO. 1977. Soil map of the world, scale 1/5 000 000. Volume VI, Africa. Paris (France): UNESCO.

Haefele SM, Johnson DE, Diallo S, Wopereis MCS, Janin I. 2000. Improved soil fertility and weed management is profitable for irrigated rice farmers in Sahelian West Africa. Field Crops Res. 66:101-113.

Haefele SM, Wopereis MCS, Donovan C, Maubuisson J. 2001. Improving productivity and profitability of irrigated rice production in Mauritania. Eur. J. Agron. 14(3):181-196.

Haylock OJ. 1956. A method for estimating the availability of non-exchangeable potassium. Paris (France): Trans. Int. Congr. Soil Sci. 6th Congr. II. 1:403-408.

Keay RWJ. 1959. Vegetation map of Africa south of the tropic of cancer. London (UK): O.U.P.

MacLean AJ. 1961. Potassium-supplying power of some Canadian soils. Can. J. Soil Sci. 41:196-206.

Martin HW, Sparks DL. 1985. On the behaviour of non-exchangeable potassium in soils. Commun. Soil Sci. Plant Anal. 16:133-162.

Miézan KM, Sié M. 1997. Varietal improvement for irrigated rice in the Sahel. In: Miézan KM, Wopereis MCS, Dingkuhn M, Deckers J, Randolph TF, editors. Irrigated rice in the Sahel: prospects for sustainable development. Bouaké (Côte d'Ivoire): West Africa Rice Development Association. p 443-456.

Richards JE, Bates TE. 1988. Studies on the potassium-supplying capacities of Southern Ontario soils. II. Nitric acid extraction of nonexchangeable K and its availability to crops. Can. J. Soil Sci. 68:199-208.

Sanchez PA. 1976. Properties and management of soils in the tropics. New York (USA): Wiley and Sons. 618 p.

SAED. 1998. Recuil des statistiques de la vallée du fleuve Sénégal, annuaire 1995/1996, version détaillée. St. Louis (Sénégal): Societé Nationale d'Aménagement et d'Exploitation des terres du Delta du fleuve Sénégal et des vallées du fleuve Sénégal et de la Falémé (SAED). 142 p.

SONADER. 1998. Etude de la filière riz en Mauritanie. Rapport definitif de la Cabinet GLG Consultants et la SONADER. Nouakchott (Republique Islamique de Mauritanie) et Paris (France): GLG Consultants. 132 p.

van Reeuwijk LP. 1992. Procedures for soil analysis. 3rd ed. Wageningen (Netherlands): ISRIC. 60 p.

Wopereis MCS, Ceuppens J, Boivin P, Ndiaye AM, Kane A. 1998. Preserving soil quality under irrigation in the Senegal river valley. Neth. J. Agric. Sci. 46:97-107.

Wopereis MCS, Donovan C, Nebié B, Guindo D, N'Diaye MK. 1999. Soil fertility management in irrigated rice systems in the Sahel and Savanna regions of West Africa. Part I. Agronomic analysis. Field Crops Res. 61:125-145.

Yoshida S, Forno DA, Cock JH, Gomez KA. 1976. Laboratory manual for physiological studies of rice. Third edition. Los Baños (Philippines): International Rice Research Institute. 83 p.

Notes

Authors' addresses: S.M. Haefele, West Africa Rice Development Association, St. Louis, Senegal, e-mail: warda-sahel@metissacana.sn; M.C.S. Wopereis, West Africa Rice Development Association, Bouaké, Côte d'Ivoire.

Acknowledgments: We thank the farmers of Mauritania and Senegal and the staff of the local extension services (SAED, AGETA). This study was partly financed by the Federal Ministry for Economic Cooperation (BMZ) through the German Agency for Technical Cooperation (GTZ).

Citation: Dobermann A, Witt C, Dawe D, editors. 2004. Increasing productivity of intensive rice systems through site-specific nutrient management. Enfield, N.H. (USA) and Los Baños (Philippines): Science Publishers, Inc., and International Rice Research Institute (IRRI). 410 p.

Part 3

4 Yield formation analysis of irrigated rice: characteristics of cultivars and on-farm crop diagnosis

P. Siband, C. Witt, H.C. Gines, G.C. Simbahan, and R.T. Cruz

14.1 Introduction

Yield analysis using several yield components was probably first applied to irrigated rice (Matsushima 1966). These yield components, number of plants per square meter, number of panicles per plant, number of spikelets per panicle, etc., are formed sequentially at different stages of the plant growth cycle (Yoshida 1981). The first yield component, the plant population density, is mainly determined early during the vegetative stage and largely depends on farmers' decision making (e.g., seeding rate for direct seeding or hill density for transplanted rice). The number of panicles per plant is achieved around the panicle initiation stage. In contrast, the percentage of filled spikelets is fixed around or after flowering.

The maximal or potential values of these yield components depend on the cultivar (Siband et al 1999). For example, the yield structure of widely adopted conventional modern cultivars such as IR72 and that of the new generation of new plant types (NPT) are very different (Khush and Peng 1996). The latter have fewer panicles per plant but significantly more spikelets per panicle than the former. These characteristics contribute to determine the varietal potential, which we define as the potential yield of a variety including its plasticity in the formation of yield components depending on competition relationships for resources. Yield components have also been used to design new plant types.

Each yield component has a variety-specific maximum value, which can be achieved only under optimal climatic conditions as the duration and rate of yield component formation during the growing season depend on climatic conditions, that is, solar radiation and temperature (Yoshida 1981). The formation of some yield components is highly variable and may depend on the previously formed yield components. For example, if panicles per unit of area are more numerous, they are generally also smaller. This arises from competition pressure within the plant population, that is, competition among plants regarding the formation of tillers, spikelets, or grains. The competition status or phenotypic plasticity in the formation of yield components is variety-specific and can be defined as the ability of the plant to react to the conditions of the cropping environment through yield component variation.

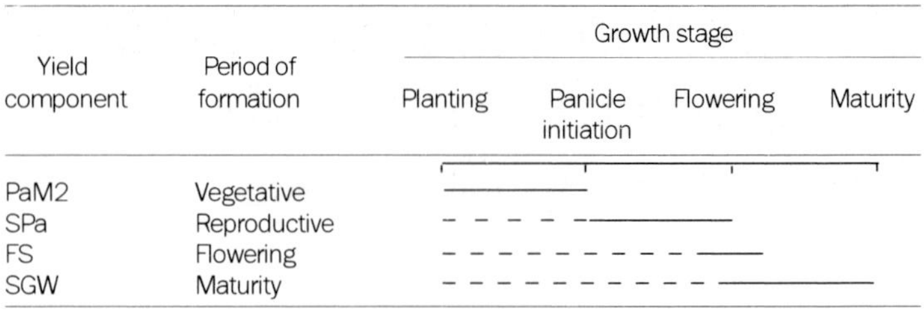

Yield component	Period of formation	Growth stage			
		Planting	Panicle initiation	Flowering	Maturity
PaM2	Vegetative				
SPa	Reproductive				
FS	Flowering				
SGW	Maturity				

Fig. 14.1. Period of potential (broken line) and actual (straight line) yield component formation at certain growth stages of the rice crop. PaM2 = number of panicles per m^2, SPa = number of spikelets per panicle, FS = filled spikelets in %, SGW = mean unit grain weight in mg.

Yield components are also affected by local stress conditions, that is, stresses related to nutrient availability, cultivation practices, salinity, acidity, or pests and diseases (Wey et al 1999). In some cases, the effect of a stress on one yield component can be compensated for partially or totally by a greater value of the succeeding yield component. The ability of a crop to stabilize yield in case of localized stress largely depends on the phenotypic plasticity of a crop, or its compensation ability. A good indicator of the conditions prevailing during the formation of yield components is the stage realization index relating the yield component actually formed during a particular growth stage to its potential value at the beginning of the respective growth stage (Fig. 14.1) (Wey et al 1999).

In this paper, we describe a simple empirical model that has been used to (1) characterize groups of varieties (Fleury 1990, Siband et al 1999), (2) evaluate the yield potential of varieties and their ability to compensate for the encountered environmental variability (Siband et al 1999), (3) compare cultivation methods (Wey et al 1999), and (4) diagnose on-farm situations (Doré et al 1997).

After introducing the theoretical framework of the model, we describe the steps involved in model calibration using on-farm data of four rice cultivars grown in Maligaya, the Philippines. Finally, we use the calibrated model in a yield formation analysis (YFA) to evaluate stress intensities during yield formation in farmers' fields and their implications for improving yields.

14.2 Theory and model calibration

Yield components

The yield formation analysis requires the identification of measurable yield components that are formed during characteristic growth stages. As a first step, rice yield can be expressed according to the following equation (for abbreviations, see Table 14.1):

$$Y = PM2 \times TP \times FT \times SFT \times FS \times SGW \tag{1}$$

Table 14.1. List of abbreviations used in the yield formation analysis.

Parameter	Description
C_S	Surface-based component (PaM2, SM2, GM2, Y)
C_F	Component in formation
FS	Filled spikelets (in % of total number of spikelets)
FT	Fertile tillers (in % of total tiller number)
I_V, I_R, I_F, I_M	Stage realization index for vegetative (I_V), reproductive (I_R), flowering (I_F), and maturity (I_M) stage of crop growth
I_Y	Yield realization index for the entire cropping period
LC	Limit of competition among C_S and C_F
GM2	Number of grains per m²
PM2	Plant population density ($=$ number of plants per m²)
PaM2	Number of panicles per m²
SM2	Number of spikelets per m²
TP	Number of tillers per plant
SFT	Number of spikelets per fertile tiller
SPa	Number of spikelets per panicle ($=$ SFT)
SGW	Single-grain weight (mg)
Y	Grain yield (kg ha[1])
Y_{max}/PaM2	Maximum possible yield at a given PaM2 value
Y_{max}	Maximum Y_{max}/PaM2 (i.e., potential yield)
Y_V, Y_R, Y_F, Y_M	Updated potential yield at the end of vegetative (Y_V), reproductive (Y_R), flowering (Y_F), and maturity (Y_M) stage of crop growth

It is difficult, however, to separate the formation periods of TP and FT. Also, PM2 and TP are difficult to measure at harvest and thus the data are seldom available from previous on-farm experiments or surveys. Hence, it is simpler to express yield as a function of four yield components:

$$Y = PaM2 \times SPa \times FS \times SGW \qquad (2)$$

To analyze competition relations, it is necessary to express yield components on a surface or unit area basis. On the basis of equation 2, two important surface-based secondary yield components can be derived by multiplying PaM2, the only primary surface-based yield component, by other yield components: SM2 is the result of PaM2 × SPa and GM2 is the result of PaM2 × SPa × FS.

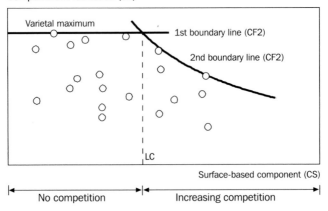

Component in formation (CF)

Varietal maximum

1st boundary line (CF2)

2nd boundary line (CF2)

LC

Surface-based component (CS)

|← No competition →|← Increasing competition →|

Fig. 14.2. Relationship between the value of a surface-based yield component (CS) after its formation and the potential value of a subsequently formed yield component (CF). LC = limit of competition.

Periods of formation

The formation periods of selected yield components during a rice cropping cycle are schematically presented in Figure 14.1. The potential value of each yield component is genetically limited and can be reached only under optimal conditions. Any limitation in the formation of a yield component during the cropping season (Fig. 14.1, broken line) may reduce the potential value of a subsequently formed yield component. On the basis of the still possible potential value of a yield component at the beginning of the formation period, the actual value of this yield component could be reduced further depending on cropping conditions (Fig. 14.1, straight line). The quantity of the yield component elements formed can be further reduced after their formation because of stresses or unfavorable conditions. For example, the first yield component that is formed, PaM2, is a surface-based component. Its period of formation begins at planting and ends at the panicle initiation stage. However, PaM2 may be reduced at later stages because of stem borers.

Relationships among yield components: compensation effects

The working hypothesis of the yield formation analysis is that an already formed, surface-based yield component (CS) determines the potential level of another, subsequently formed yield component (CF). Figure 14.2 presents the schematic relationship between CS and CF. If the assimilate source is not limiting, the value of a developing yield component will depend on the cultivar. Otherwise, the assimilate source will limit the formation of the yield component. If a yield component is never source-limited during formation, the limit of competition will never be passed (Fig. 14.3). In this case, CF is formed without any limitation and should reach the potential value of the respective cultivar. The assimilate resource is limiting if the value of the previously formed surface-based yield component is so large that it creates a competition

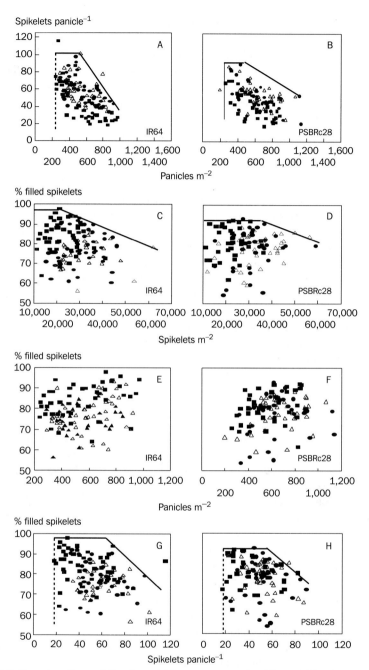

Fig. 14.3. Boundary lines of the four yield component pairs SPa × PM2, FS × SM2, FS × PM2, and FS × SPa for the rice varieties IR64 and PSBRc28. Closed square = control without fertilizer N, closed circle = farmers' fertilizer practice, open triangle = site-specific nutrient management.

pressure among the growing yield component elements, that is, among panicles, spikelets, or grains in formation. Therefore, if the previous surface-based yield component value CS is low, the potential value of the component in formation is genetically determined (first boundary line, Fig. 14.2) and independent of CS. But if CS exceeds a certain limit of competition, the yield component in formation will depend on CS. As the entire assimilate resource is used as soon as the limit of competition is reached, CF values will decrease with increasing CS values, which is described in the model by a second boundary line with negative slope (Fig. 14.2).

Model calibration

The model requires the calibration of the relationships between the following three pairs of yield components: PaM2 vs SPa, SM2 vs FS, and GM2 vs SGW. The two boundary lines describe the relationship between CS and CF depending on the level of competition as depicted in Figure 14.2 and can be derived using experimental data. As yield component characteristics vary among cultivars, boundary lines have to be determined for individual cultivars. This obviously requires a sufficiently large data set covering a wide range of experimental conditions not only to adequately describe the boundary lines but also to be able to exclude potential outliers caused by errors in sampling or plant analysis. Once the model is calibrated, the boundary lines can be used to exclude outliers from data sets under evaluation. Since our data contained replicated measurements, we used the following two complementary methods to eliminate outliers (Siband et al 1999):

1. The statistical analysis requires that the differences between two replicates within plots should be normally distributed across the entire sample. Where this was not the case, data points with the greatest differences between replicates were eliminated.

2. Outliers were further excluded under the assumption that the plot average values are normally distributed. Values of the entire data set were not always normally distributed. This may be because our data sets contained fewer data points, as yield components reached their upper limits close to potential values. Thus, any error in sampling or plant analysis had a greater effect on the maximum than on average values of a certain yield component. Therefore, when the distribution of a particular yield component did not follow a Gaussian distribution, the ten highest values were repeatedly checked and data from fields with the highest differences between replicates were excluded until the condition was met that data are normally distributed.

The first boundary line, describing the genetically determined, cultivar-specific maximum value of a yield component, was determined using the greatest value in the data set. The second boundary line was derived using the two greatest values in the right part of each graph (Siband et al 1999) after outliers had been excluded as described above (see example in Fig. 14.2).

Maximum yield as a function of panicle population density

As a consequence of the relationship between potential values of yield component and the previous surface-based yield component, the first yield component (PaM2) determines the potential values of succeeding yield components (max indexed) by a recursive process:

$$\{ \ [\ (PaM2 \times SPa_{max}) \times FS_{max} \ ] \times SGW_{max} \ \}$$
$$(\quad SM2_{max} \quad)$$
$$[\quad\quad GM2_{max} \quad\quad]$$
$$\{ \quad\quad\quad Y_{max}/PaM2 \quad\quad\quad\quad \}$$

Thus, a maximum yield value (Y_{max}/PaM2) can be associated with a particular panicle density. By varying the value of PaM2, it is possible to determine the panicle density associated with the maximum yield (Y_{max}) of the respective cultivar:

$$Y_{max} = max \ (Y_{max}/PaM2) \tag{3}$$

Stage updated maximum yield

In the previous section, we discussed the concept of potential values for yield components as determined by boundary lines. Under on-farm conditions, yield components do not always reach their potential values because of local limitations. In such cases, points representing CS × CF relationships (Fig. 14.2) would be located below the boundary line indicating a stress period during CF formation. The vertical distance from an observed point to the envelope curve can be used as an indicator of intensity. Production constraints can then be diagnosed by relating these stress indicators to additional information such as environmental and cultural parameters.

On the basis of the yield component relationships calibrated in the model, it is possible to calculate an updated potential yield at each stage of the cropping cycle and for each case in a survey. For example, the yield potential Y_{max} may not be reached in a particular season in which PaM2 was low at panicle initiation, so that the updated potential yield would now be equal to Y_{max}/PaM2, and so forth. Based on the established model relationships between successive yield components, the updated intermediate potential yield at the end of a certain stage is calculated retrospectively by correcting the potential values of the yield components to be formed next depending on the actual values of the already formed yield components. In summary, one preplanting and four successive, intermediate potential yields are calculated at the end of vegetative (V), reproductive (R), flowering (F), and maturity (M) growth stages (see also Fig. 14.1) :

$$Y_{max} \geq Y_V \geq Y_R \geq Y_F \geq Y_M \tag{4}$$

A local yield analysis: stage and yield indices

A stress during the growth stages can be described using a stage realization index I_S:

$$I_S = Y_S / Y_{S-1} \qquad (5)$$

where Y_S is the updated intermediate yield at stage s and Y_{S-1} equals the updated intermediate yield of the previous growth stage. Four stage indices (I_V, I_R, I_F, I_M) can be defined referring to the four main stages of crop growth: the vegetative (V), reproductive (R), flowering (F), and maturity (M) stages. Note that the stage realization indices are affected by phenotypic plasticity, that is, the ability of the plant to compensate for a smaller number of yield components with a larger number of the subsequently formed yield components. The actual yield can then be expressed as

$$Y = Y_{max} \times I_V \times I_R \times I_F \times I_M = Y_{max} \times I_Y \qquad (6)$$

where I_Y is the yield realization index for the entire cropping period, that is, the product of all intermediate indices. Thus, yield is expressed as the product of (1) the genetically determined potential yield (Y_{max}) and (2) the yield realization index (I_Y), which depends on local limitations. The advantage of using stage indices is that compensation effects are taken into account. Stage indices also allow comparisons across sites, seasons, and/or cultivars.

A summary of crop history: stress profiles

Yield realization indices allow a more detailed analysis of on-farm situations than yield or intermediate potential yield. However, they are more difficult to relate to other local information. A synthetic expression is needed that preserves timing information. A binary note may be attributed to each stage depending on whether stress was identified (1) or not (0). We suggest applying a threshold index value of 0.85 for the relationship given in equation 8 to decide between stressed and unstressed situations as suggested by Wey et al (1999). Association of the different binary notes with the four growth stages (vegetative, reproductive, flowering, maturity) allows a rapid classification of stress profiles. For example, 0100 would indicate that a stress occurred in the formation of yield components during the reproductive stage but not during other growth stages. There are 16 possible stress profiles. Profile types represent categories of situations. They can be compared to environment parameters by multiple correspondence analysis.

The data set

The model was calibrated using existing on-farm data obtained in the project on Reversing Trends in Declining Productivity (RTDP) at Maligaya, the Philippines. Full details of the data set are given elsewhere (Dobermann et al, Chapter 2, this volume, Gines et al, Chapter 8, this volume). The two most widely used rice cultivars, IR64 and PSBRc28, were used for the model calibration. The cultivars were grown in the 1995, 1997, and 1998 dry seasons (DS) and in the 1997 and 1998 wet seasons (WS). Briefly, three treatments were implemented in 27 on-farm experiments with two replicates: a control without N, P, and K fertilization (CTRL), a farmers' fertilizer practice (FFP) with fertilizer NPK applied by the farmer, and a site-specific nutrient management treatment aiming at a balanced NPK nutrition of the rice plant (Gines et al, this volume). Two replicate plant samples per plot (500–1,000 m^2) were taken from two sampling areas each of 0.25 m^2 (direct seeding) or from two 12-hill samples (transplanting). The following parameters were determined: number of panicles per m^2 (PaM2), number of spikelets per m^2 (SM2), percentage of filled spikelets (FS), single-grain weight in mg (SGW), total grain yield in kg ha^{-1} (Y), and grain moisture. Grain weights (Y, SGW) were adjusted to 0.14 g H$_2$O g^{-1} fresh weight. Number of spikelets per panicle (SPa) and number of grains per m^2 (GM2) were calculated as SM2/PaM2 and Y/SGW, respectively.

14.3 Results and discussion

Relationships between two yield components

Yield component data from all three treatments (CTRL, FFP, and SSNM) were used to calibrate the model. The boundary lines of yield component pairs are depicted for each variety in Figures 14.3 and 14.4 and the linear equations of the boundary lines are given in Table 14.2. Variety-specific maximal values for spikelets per panicle, percentage of filled spikelets, and single grain weight are given in Table 14.3.

Depending on the cultivar, spikelets per panicle strongly decreased as the number of panicles per m^2 increased beyond 500–600 and the boundary line was delineated with two straight lines (Fig. 14.3A,B). The maximum value of spikelet filling was successfully linked to the number of spikelets per m^2 with the limit of competition determined at 20,000 to 30,000 spikelets per m^2 (Fig. 14.3C,D). As the number of spikelets per m^2 is the product of panicles per m^2 and spikelets per panicle, the relationship could be analyzed with each of these two components. It appeared that maximum spikelet filling depends on the number of spikelets per panicle rather than on the number of panicles per m^2 (Figs. 14.3E,F vs 14.3G,H). An explanation could be that grain filling was limited by the size of the panicles. When the number of spikelets per m^2 (and the number of spikelets per panicle) increased, the reduction in maximum spikelet filling seemed to be greater for IR64 than for PSBRc28, but IR64 reached greater maximum values for spikelet filling than PSBRc28 below the limit of competition (Table 14.3).

Maximum single-grain weight diminished when the number of grains per m^2 increased beyond 30,000 to 35,000 grains per m^2 (Fig. 14.4A,B), but the trend was not as evident as the previous relationships. Two opposite relationships could be observed when analyzing the relationship between single-grain weight and each component of number of grains per m^2: maximum single-grain weight decreased with an increase in spikelets per panicle (Fig. 14.4C,D) but increased with increasing spikelet filling until 70% to 75% (Fig. 14.3E,F). We hypothesize that the relationship between maximum single-grain weight and spikelets per panicle (like the relation between maximum spikelet filling and spikelets per panicle) could be a competition

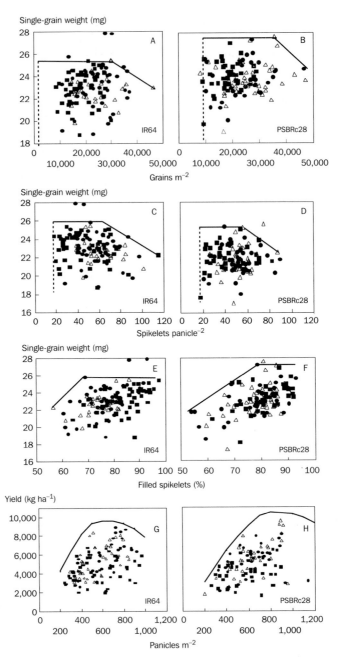

Fig. 14.4. Boundary lines of the three yield component pairs SGW × GM2, SGW × SPa, and SGW × FS and boundary line for the relationship between actual (points) and potential (line) grain yield and panicles m⁻² of rice varieties IR64 and PSBRc28. Closed square = control without fertilizer N, closed circle = farmers' fertilizer practice, open triangle = site-specific nutrient management.

Table 14.2. Regression equations including constants (a, b) for pairs of yield components of four rice varieties. Equations describe the second boundary line used in the yield formation analysis, that is, the linear relationship between a component in formation (y-axis) and a surface-based component (x-axis) beyond the limit of competition as depicted in Figures 3–9.

Equation	IR64		PSBRc28	
	a	b	a	b
SPa = a PaM2 + b	−0.123	157	−0.073	122
FS = a SM2 + b	-5.9×10^{-4}	110	-3.0×10^{-4}	102
FS = a SPa + b	−0.13	102	−0.31	107
SGW = a GM2 + b	-1.8×10^{-4}	31.2	-2.1×10^{-4}	33
SGW = a FS + b	0.25	8	0.19	10.4
SGW = a SPa + b	−0.053	29	−0.083	29.9

Table 14.3. Genetically determined maximum values of spikelets per panicle, filled spikelets, and single-grain weight used in the yield formation analysis of two rice varieties.

Item	IR64	PSBRc28
Spikelets per panicle	100	90
Filled spikelets (%)	0.98	0.93
Single-grain weight (mg)	25.8	25.6

relationship for grain filling. In contrast, the relationship between maximum single-grain weight and spikelet filling cannot simply be another expression of the two previous ones, that is, spikelets per panicle × spikelet filling (Fig. 14.3G,H) and spikelets per panicle × single-grain weight (Fig. 14.4C,D). Indeed, spikelets per panicle × spikelet filling involved maximum spikelet-filling values, whereas spikelet filling × single-grain weight implied low values of spikelet filling. This may be because the hull size, defining the potential single-grain weight, is formed in about the same period as spikelet filling occurs. Thus, these two parameters would be affected by the same stress situations. This aspect needs further investigation.

Maximum yield

The parameters and relationships obtained for the four rice cultivars were used to calculate the maximum yield of each cultivar depending on planting densities (Y_{max}/PaM2). According to the model, the maximum yield potential of all varieties is reached at a density of about 500 to 800 panicles per m^2 with yields ranging from about 9.5 to 10.5 t ha^{-1} (Fig. 14.4G,H). At low planting densities, model-predicted maximum yields of IR64 were about 1 t ha^{-1} higher than the observed yields (Fig. 14.4G). Likewise, the number of spikelets per panicle (Fig. 14.3A) and spikelet filling (Fig. 14.3C) of

Yield variation

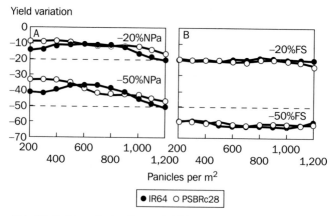

Fig. 14.5. Expected effect of different degrees of transient stresses, that is, the reduction in the number of spikelets per panicle (SPa) by 20% or 50% at panicle initiation (A) and flowering (B) on the grain yield variation (DY) of IR64 and PSBRc28.

IR64 were also low at low planting densities. One explanation may be that a substantial limitation during the formation of a certain surface-based yield component such as panicle density (PaM2) or spikelet density (SM2) also limits the potential value of the next yield component, that is, spikelets per panicle (SPa) or spikelet filling (FS). Adjusting the boundary line accordingly may improve the ability to correctly predict the maximum yield of IR64. However, more data points may be needed to justify such an adjustment. This phenomenon was not observed for the other varieties.

Effects of transient stresses

The model allows the simulation of transient stresses on yield formation, that is, temporary stresses during the growing season limiting the formation of a single yield component. As an example, two separate stress scenarios were simulated for each cultivar: reducing (1) the number of spikelets per panicle before panicle initiation (PI) and (2) the percentage of filled spikelets at flowering. Both parameters were reduced by 20% and 50%, respectively. Results are presented as the predicted change in yield in relation to the potential yield depending on planting density (Fig. 14.5).

The predicted change in yield (DY) because of a reduction in SPa at panicle initiation was always less than the percentage by which the number of spikelets per panicle was reduced except for very high panicle densities (Fig. 14.5A). Thus, both IR64 and PSBRc28 were in part able to compensate for a lower number of spikelets per panicle by increasing subsequently formed yield components. At elevated panicle densities, stress because of a reduction in SPa was less successfully compensated for by both cultivars and the phenotypic plasticity in response to panicle density was similar. Planting density had no effect on the change in yield caused by a reduction in the percentage of spikelet filling (Fig. 14.5B). The reduction in yield was proportional to a moderate reduction in spikelet filling (−20%) for both cultivars. The consistent effect of a reduction in spikelet filling on yield reduction regardless of panicle density is probably because of the link between single-grain weight and percentage of spikelet filling. Single-grain weight is the last yield component formed and its reduction would have the same proportional effect on yield for all cultivars and at all panicle densities.

In summary, the effect of a stress at a particular growth stage on the formation of yield appears to depend on the stress intensity, the crop status (population below or above optimum), and the phenotypic plasticity of the respective cultivar, that is, the possible interactions and compensations among successively formed yield components until maturity. Crops in farmers' fields may suffer from one or more stresses of different intensities at different growth stages. The overall effect of these stresses on yield and yield components can be measured at harvest, but more detailed information on yield component formation during the cropping season would be required to identify stress situations and contributing factors and to develop mitigation options for removing yield constraints.

Yield variability in farmers' fields

On-farm yields generally show a very large variability because of many factors, such as indigenous nutrient supplies or crop management practices, that vary in small domains or even differ in adjacent fields. We do not know a priori which factors control yield in a given field. We are not even sure whether we capture the most decisive factors influencing yield when deciding on parameters to monitor in farmers' fields. Therefore, it is extremely difficult if not impossible to analyze a survey data set according to the collected factors (i.e., according to cause), as is done in designed experiments. It is possible, however, to analyze data from farmers' fields according to yield results (i.e., according to effect). Yield is the most relevant information as it determines profit. Furthermore, yield components are relatively easy to collect and they allow a one-way arrangement of the data set, as they are formed in sequence. However, an absolute yield value has no direct significance. Detecting the existence of limiting factors during the cropping season and evaluating their importance require the comparison of actual yield with potential yield. In the following, we will elaborate on the possibilities of using stage and yield realization indices as well as stress profiles to identify crop growth stages with stress depending on variety and nutrient management treatment.

The frequency distribution of stress profiles in farmers' fields in Muñoz, the Philippines, is depicted in Figure 14.6. Four stress profiles included about 75% of all observed cases for IR64. Stress-free conditions were observed in 17.5% of all cases (0000), whereas, in 28% of all cases, stresses occurred only during the reproductive period (0100). The other major categories were 0110 with prolonged stress from panicle initiation up to flowering (15% of all cases) and 1110 with stress lasting from planting to maturity (12.5% of all cases). Four kinds of stress profiles were never observed for IR64 and these were profiles with stresses mainly during the vegetative and maturity stages (1000, 1001, 1011, 1101). It appears that stress situations during the vegetative stage (24% of cases) always resulted in stress during flowering. This

Frequency

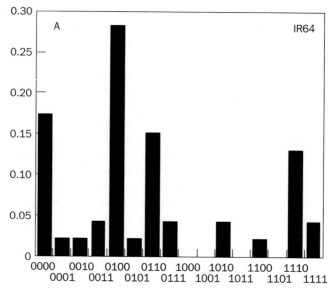

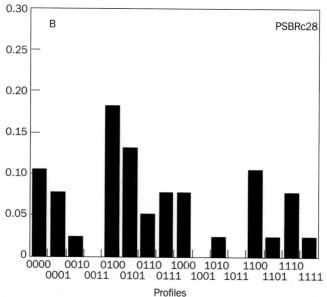

Profiles

Fig. 14.6. Observed cumulative frequency distribution of the stage realization indices for the vegetative (I_V), reproductive (I_R), flowering (I_F), and maturity (I_M) stage of crop growth and the yield realization index for the entire cropping period (IM) for IR64 (n = 117) and PSBRc28 (n = 110) in plots with the farmers' fertilizer practice, 1995-98.

Frequency

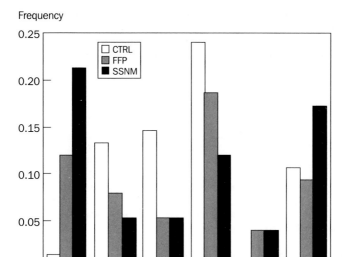

Fig. 14.7. Frequency distributions of main stress profile classes according to fertilizer treatments (*n* = 68) including both varieties IR64 and PSBRc28. A stress profile consists of 4 binary notes (0 = no stress/1 = stress), each indicating the absence or occurrence of stress at a particular growth stage (vegetative, reproductive, flowering, maturity). CTRL = control without fertilizer N, FFP = farmers' fertilizer practice, SSNM = site-specific nutrient management.

suggests that early stress might affect both the number of panicles and the strength of fertile tillers, making them more susceptible to subsequent stresses. Although there were similarities in the distribution of profiles among both varieties, PSBRc28 showed a more uniform distribution of stress profiles. The number of stress-free profiles decreased and stresses were observed more frequently at maturity (0001, 0101, 1101). Note, however, that the two varieties were not grown by the same farmers so that differences among varieties were probably also affected by differences in crop management among farms.

Data from the two varieties were pooled to obtain a larger data set for investigating the effect of nutrient management treatments on stress profiles (Fig. 14.7). Cases without stress (0000) increased in the order CTRL > FFP > SSNM. Applying fertilizer nutrients and balanced fertilization increased the frequency of nonstressed fields from 1% in the control to 12% in FFP and 21% in SSNM. Unfertilized control treatments faced the greatest stress during the vegetative and reproductive growth stages (1000, 1100, and 0100) presumably because of N deficiency at early growth stages. These stress profiles were less frequent in FFP and SSNM because of N fertilization. The greatest stress in FFP and SSNM was observed during the reproduc-

Table 14.4. The stage realization indices during vegetative (I_V), reproductive (I_R), flowering (I_F), and maturity stage (I_M) resulting in the overall yield realization index (I_Y) for two cultivars in control (CTRL), farmers' fertilizer practice (FFP), and site-specific nutrient management treatments, Maligaya, Philippines, 1995-98.

Variety	n	Treatment	I_V	I_R	I_F	I_M	I_Y
IR64	30	CTRL	0.84 b*	0.65 b	0.86 a	0.90 a	0.41 b
IR64	30	FFP	0.91 a	0.78 a	0.82 b	0.91 a	0.54 a
IR64	30	SSNM	0.93 a	0.80 a	0.83 b	0.91 a	0.56 a
PSBRc28	38	CTRL	0.80 b	0.64 b	0.88 a	0.89 a	0.39 c
PSBRc28	38	FFP	0.85 a	0.79 a	0.85 b	0.89 a	0.52 b
PSBRc28	38	SSNM	0.86 a	0.83 a	0.86 ab	0.89 a	0.55 a

*Means within treatments of each variety followed by the same letter in each column are not significantly different at $P<0.05$ (t test).

tive (0100) growth stages. This stress sometimes continued, thus causing stress at flowering (0110).

In the following, we investigate the effect of nutrient management treatments on the stage realization indices at individual growth stages and the overall yield realization index (Table 14.4). The yield realization index was significantly greater in treatments receiving different amounts of fertilizer N, P, and K (FFP and SSNM) compared with the unfertilized control. Differences in stage realization were already significant during the vegetative growth stage and, as mentioned above, this was probably because of nitrogen deficiency in the control treatments (see Gines et al, this volume). Differences in stage realization indices between FFP and SSNM were insignificant for both varieties, but effects during the season were accumulative for PSBRc28, so that the site-specific nutrient management treatment aiming at balanced N, P, and K fertilization had a significantly greater yield realization index than the FFP treatment. This was mainly related to less stress during the vegetative and reproductive growth stages.

The stage realization index at flowering was greater for the unfertilized treatments than for the fertilized treatments and this difference was significant in the case of IR64 (Table 14.4). Note that the stage realization index refers to the updated intermediate potential yield based on the yield components actually formed during the previous crop stage. It is therefore possible that an unfertilized crop made efficient use of its (limited) potential during the flowering stage, whereas the full potential was not achieved in the fertilized treatments.

Across varieties, the yield gap between actual and potential yield as predicted by the model was about 60% in unfertilized control treatments and about 44–48% in treatments receiving fertilizer N, P, and K (Table 14.4). Although stress profiles and realization indices can indicate the growth stages that most likely encounter stresses, it is much more difficult to identify the limiting factors. Growth-stage-specific information on pest problems and crop management practices may help to understand the effect of stress on yield components under formation and to formulate and test hypotheses. Yield analysis separating varietal potential and local realization would also allow comparisons across sites and cultivars.

14.4 Summary

A yield component model is proposed for assessing limitations in the formation of yield components at various stages of crop growth. The model is based on the sequential formation of yield components and on competition relationships between successively formed components. The analysis requires the estimation of four yield components: number of panicles per m^2, number of spikelets per panicle, percentage of filled spikelets, and single-grain weight. Competition relationships between yield components are used to define the maximum value of each yield component depending on the components previously formed. By replacing maximum with actual yield component values, it is possible to calculate stage updated potential yields (Y_S) during the cropping season. The ratio of the potential yield updated for a particular stage divided by the potential yield updated for the previous stage ($Y_S/Y_S - 1$) is used as an indicator of yield loss during this stage because of stress. The model was calibrated for the two rice cultivars IR64 and PSBRc28 using on-farm data from Maligaya, the Philippines, collected from 1995 to 1999. In general, yield formation was mainly stressed in the middle of the crop cycle for IR64 and during the entire cropping cycle for PSBRc28. Stresses were frequently observed during the reproductive growth stage. Yield formation analysis was also used to compare the performance of different nutrient management treatments in farmers' fields. It appeared that the increase in yield observed in fields with the farmers' fertilizer practice compared with a control without fertilizer application was mainly caused by eliminating stresses during vegetative and reproductive growth. Compared with the FFP, an improved site-specific nutrient management approach positively affected the formation of all yield components rather than the formation of certain components during particular growth stages. The proposed yield formation analysis allows us to separate varietal potential from local realization, thus providing a powerful tool to analyze large data sets covering a wide range of biophysical and socioeconomic conditions. Identifying periods of stress during the formation of yield components and evaluating the stress intensity may help to develop mitigation options. Future work will focus on analyzing a possible association of differences in yield formation across sites with cultivation practices and/or environmental parameters.

References

Doré T, Sébillotte M, Maynard JM. 1997. A diagnostic method for assessing regional variations in crop yield. Agric. Syst. 54:169-188.

Fleury A. 1990. Méthodologie de l'analyse de l'élaboration du rendement. In: Physiologie et production du mais. Paris: Inra. p 279-290.

Khush GS, Peng S. 1996. Breaking the yield frontier of rice. In: Reynolds MP, Rajaram S, McNab A, editors. Increasing yield potential in wheat: breaking the barriers. Proceedings of a Workshop held on 26-28 March 1996 in Ciudad Obregón, Sonora, Mexico. El Batán (Mexico): International Maize and Wheat Improvement Center. p 36-51.

Matsushima S. 1966. Crop science in rice: theory of yield determination and its application. Japan: Fuji Publ. Co. Ltd. p 1-365.

Siband P, Wey J, Oliver R, Letourmy P, Manichon H. 1999. Analysis of the yield of two groups of tropical maize cultivars: varietal characteristics, yield potentials, optimum densities. Agronomie 18:545-561.

Wey J, Oliver R, Mandal R, Siband P. 1999. Analysis of local limitations to maize yield under tropical conditions. Agronomie 18:545-561.

Yoshida S. 1981. Fundamentals of rice crop science. Los Baños (Philippines): International Rice Research Institute. p 1-269.

Notes

Authors' addresses: P. Siband, C. Witt, G.C. Simbahan, International Rice Research Institute, Los Baños, Philippines; P. Siband, Centre de Coopération International de Recherche Agronomique pour le Développement, (CIRAD) BP 5032, 34031, Montpellier Cedex 1, France, e-mail: pierre-lucien.siband@cirad.fr; H.C. Gines, R.T. Cruz, Philippine Rice Research Institute, Maligaya, Nueva Ecija, Philippines; G.C. Simbahan, University of Nebraska, Lincoln, Nebraska, USA.

Acknowledgments: The authors gratefully acknowledge the assistance of Edsel Moscoso in performing statistical analysis.

Citation: Dobermann A, Witt C, Dawe D, editors. 2004. Increasing productivity of intensive rice systems through site-specific nutrient management. Enfield, N.H. (USA) and Los Baños (Philippines): Science Publishers, Inc., and International Rice Research Institute (IRRI). 410 p.

5 | Agronomic performance of site-specific nutrient management in intensive rice-cropping systems of Asia

A. Dobermann, S. Abdulrachman, H.C. Gines, R. Nagarajan, S. Satawathananont,
T.T. Son, P.S. Tan, G.H. Wang, G.C. Simbahan, M.A.A. Adviento, and C. Witt

We defined site-specific nutrient management (SSNM) as the field-specific adjust-ment of fertilizer rates of N, P, and K combined with a field- and season-specific timing of N applications. This form of SSNM attempts to account for (1) regional and seasonal differences in yield potential and crop nutrient demand, (2) between-field spatial variability in indigenous nutrient supply, (3) within-season dynamics of crop N demand, and (4) location-specific cropping systems and crop management practices. It adds important regional and real-time components to more commonly used definitions of SSNM, which mainly focus on managing spatial variability of nutrients within large production fields (Pierce and Nowak 1999).

Applied at different sites, SSNM involved (1) simultaneously varying rates of the three nutrients N, P, and K, (2) variation in the timing of N applications at the same or at different total N rates, and (3) variations in other crop management prac-tices that affect the response to macronutrient supply. This differs from traditional soil fertility research, which mostly evaluates crop response to nutrient supply on the basis of a single-nutrient response curve with only one factor being varied and others held constant (Bray 1954, Havlin et al 1999). Agronomists have traditionally plotted yield response versus the supply of nutrient from soil and fertilizer to derive an opti-mal rate of fertilizer, above which profit does not increase anymore (Colwell 1994). However, such "optimal rates" vary widely among sites and from year to year (Dawe and Moya 1999) and they contain no information about different aspects of the system's performance, factors affecting resource-use efficiency (de Wit 1992), and sustainability of the cropping system.

In this chapter, we summarize the agronomic performance of SSNM across eight geographical domains. Increased yield and profit are the ultimate short-term indicators of a successful new technology. However, the concept employed in our on-farm studies was one that aimed at optimizing the short- and long-term perfor-mance of SSNM in intensive rice systems. Therefore, a more complete agronomic evaluation was performed that took into account short-term as well as potential long-term effects and how they interact with other crop management factors.

Table 15.1. Agronomic and economic performance indicators to assess gains in productivity, resource-use efficiency, and sustainability resulting from the introduction of a new nutrient management technology.

Indicator	Unit	Interpretation
Increase in grain yield	t ha^{-1}	Gross productivity
Increase in straw yield	t ha^{-1}	Useful by-product
Probability of yield increase	%	Achievement of yield increases in different geographical environments and under favorable and unfavorable environmental conditions
Achievement of yield goal	% of yield goal	Climatic variability and quality of crop management
Yield gap 1	t ha^{-1}	Yield reduction mainly because of insufficient uptake of N, P, and K
Yield gap 2	t ha^{-1}	Yield loss mainly because of other stresses (water, pests, micronutrients)
Internal efficiency of N, P, K	kg kg^{-1}	Balanced nutrition within the plant, occurrence of other stresses
Nitrogen-use efficiency (AEN, REN)	kg kg^{-1}	Congruence of N supply and crop N demand. Negative effects on the environment
Partial factor productivity (PFP) of N, P, and K fertilizer	kg kg^{-1}	Grain yield increase per unit of fertilizer applied
Input-output balance of P and K	kg ha^{-1} crop^{-1}	Medium- and long-term sustainability of soil productivity

15.1 Methodological considerations for performance assessment

Types of comparisons

We propose to use the agronomic indicators shown in Table 15.1 to assess the performance of SSNM relative to current farmers' fertilizer practices (FFP) as well as in absolute terms. In addition to this, Chapter 16 provides an economic assessment. To assess performance in relative terms, we will use two types of comparisons:

- Compare SSNM (four crops in 1997-99) with a baseline data set of the FFP for two consecutive rice crops sampled in 1996-97 before an SSNM treatment was established. This is reasonable because both data sets include the same farmers and the two major climatic seasons at each site. Because of the large sample size and wide geographical distribution, we assume that the comparison of sample means or medians is not significantly biased by climatic effects on yield. The main advantage of comparing SSNM with a baseline data set is that the latter is not affected by possible changes in crop management that farmers may have adopted in their fields from the SSNM treatment established in 1997 or 1998.

- Compare SSNM in 1997-99 with the FFP sampled in the same seasons, that is, summarize and assess pairwise treatment differences. The main advantage of comparing SSNM with the FFP in the same season is that the comparison is not affected by climatic differences in yield potential. However, this comparison may be somewhat biased toward smaller treatment differences. There is a chance that the FFP treatment is not an original farmers' practice anymore because some farmers may have changed their crop management based on what they have observed in an attempt to compete with the SSNM plot. In on-farm experiments, such subtle changes cannot be controlled and are difficult to quantify. In our experiments, the overall effect of such changes in FFP appeared to be small, but there was evidence for changes in fertilizer rates and timing of N applications in some cases.

For both comparisons, we will mainly discuss results that comprise everything from very good to very poor conditions, not separated based on differences of crop growth and crop management. These results indicate the *average achieved* increases in yield and nutrient efficiency through changing fertilizer management, but still constrained by other factors such as pests or climate at many sites. In contrast, upper percentiles of treatment differences between SSNM and FFP can be used as an indicator of the *achievable* increases if stress factors can be better controlled, although control of these factors may be difficult in actual practice. Our studies did not include an unaffected control group of farmers, which would have been another alternative for assessing SSNM.

Evaluating the performance and sustainability of a new technology cannot be based on the difference from an inferior practice alone. In on-farm studies, relative treatment comparisons of the type described above are associated with uncertainties because of the lack of control over key inputs and crop management operations. For example, uncertainties may be caused by nonuniform, biased crop management in the two treatments compared. Crop management in SSNM plots was discussed in Chapter 5, but uncontrollable factors in FFP plots have to be considered too. Perhaps the greatest uncertainty is associated with the rates of N, P, and K applied by the farmers, in terms of both the accuracy of weights reported and the uniform distribution of fertilizer within different parcels managed by the same farmer. Therefore, we will also assess absolute values of yields and nutrient efficiency in the SSNM compared to what is known to be attainable under controlled experimental conditions.

Data analysis
The data set included 179 farms with rice-rice, rice-rice-rice, or rice-rice-maize cropping systems (see Table 2.1 in Chapter 2). Sites with rice-wheat cropping at Pantnagar were excluded as well as three farms at Thanjavur (SWMRI) for which data sets were incomplete.

Mostly descriptive statistics (e.g., median and quartile) were used to characterize and compare whole populations such as the FFP baseline (1995-97) and SSNM or FFP in 1997-99. Pairwise treatment differences between SSNM and FFP measured from 1997 to 1999 were subjected to analysis of variance (ANOVA) and will be discussed mostly based on means. PROC GLM of SAS (SAS Institute Inc. 1988) was used to perform ANOVA on the differences (D) between SSNM and FFP measured at each farm for four consecutive rice crops grown (D = SSNM – FFP). To test the hypothesis that the performance of SSNM improves with time, we divided the data set into year 1 (= crop 1 and 2) and year 2 (= crop 3 and 4). To test the hypothesis that the performance of SSNM differed between climatic seasons, we classified the data obtained from each site into low-yielding seasons (LYS, seasons with low climatic yield potential such as a wet season in the tropics) and high-yielding seasons (HYS, seasons with high climatic yield potential such as a dry season in the tropics).

To analyze yield gaps, the yield potential (Y_{max}) was defined as grain yield limited by climate and genotype only. Y_{max} fluctuates from year to year and among sites because of climatic factors, different planting dates, and varietal differences. We used an estimated average Y_{max} for irrigated rice in Asia of 8.3 t ha^{-1} (Matthews et al 1995), assuming that our large sample represents a range similar to the one used for estimating this average regional Y_{max} through crop simulation[1]. We defined the average attainable yield (Ya) as the yield that can be achieved with the best management practices at the measured actual levels of plant N, P, and K accumulation, assuming that water was not a limiting factor in our experiments. Ya was estimated using a modified version of the QUEFTS model (Janssen et al 1990, Witt et al 1999) with the average Y_{max} and the median N, P, and K uptake as model inputs. Two yield gaps were then defined as

Yield gap 1 = Y_{max} – Ya
Yield gap 2 = Ya – Y

where Y is the actual yield measured (t ha^{-1}).

15.2 The baseline: current agronomic characteristics of irrigated rice farms in Asia

Tables 15.2 and 15.3 show key parameters of productivity, nutrient supply, and nutrient-use efficiency in 179 farmers' fields of Asia before any intervention by SSNM took place. Although other diagnostic surveys have been conducted in rice systems of Asia and Africa, none of them matches the data set compiled here in terms of wide geographical coverage, measurement detail, and standardized methodology. This sample is probably representative for most of the irrigated rice area in Asia and provides the best overview currently available. The numbers discussed below provide an update of earlier reports in which both on-farm and research-station data were reviewed (Bouldin 1986, Cassman et al 1998, De Datta and Buresh 1989, Dobermann et al 1998, Greenland 1997, Olk et al 1999, Vlek and Byrnes 1986, Witt et al 1999).

---

[1]Average rice yield potential in the main planting season. Regional geographic information systems analysis based on climate data from 68 sites in 15 countries of Asia. Average Y_{max} was estimated at 8.1 t ha^{-1} using the ORYZA1 model (Kropff et al 1993) or 8.5 t ha^{-1} using SIMRIW (Horie et al 1995).

Table 15.2. Baseline agronomic measurements in irrigated rice fields of Asia before the introduction of site-specific nutrient management. The data set consists of 179 farm sites in China, India, Indonesia, the Philippines, Thailand, and Vietnam with at least two rice crops grown per year. Values shown are summary statistics of two successive rice crops grown on each farm in 1995-96 or 1997.

Factor[a]	Mean	Min.	25% quartile	Median	75% quartile	Max.
Grain yield (GY, t ha^{-1})	5.01	1.91	3.93	5.12	5.91	9.93
Harvest index	0.48	0.27	0.44	0.48	0.52	0.63
No. of panicles m^{-2}	457	146	307	436	585	1,136
Total no. of spikelets panicle^{-1}	69	19	46	64	91	152
Total no. of spikelets m^{-2} ($\times$ 1,000)	28	9	22	27	33	80
Filled spikelets (%)	82	57	76	83	88	99
Fertilizer N use (FN, kg ha^{-1})	116	31	86	110	137	270
Fertilizer P use (FP, kg ha^{-1})	18	0	11	18	25	51
Fertilizer K use (FK, kg ha^{-1})	27	0	0	18	48	184
N uptake (UN, kg ha^{-1})	88	27	67	86	107	169
P uptake (UP, kg ha^{-1})	16	6	12	16	19	30
K uptake (UK, kg ha^{-1})	84	27	62	80	100	219
Input-output N balance (kg ha^{-1} crop^{-1})	3	−53	−10	3	16	73
Input-output P balance (kg ha^{-1} crop^{-1})	7	−18	−2	5	15	46
Input-output K balance (kg ha^{-1} crop^{-1})	−21	−100	−42	−26	−3	152
Partial productivity of N (PFPN, kg kg^{-1})	46.7	17.2	32.1	44.3	58.0	112.0
Agronomic efficiency of N (AEN, kg kg^{-1})	11.4	0.0	4.0	9.4	17.0	49.0
Recovery efficiency of N (REN, kg kg^{-1})	0.30	0.00	0.15	0.27	0.43	0.91
Internal efficiency of N (IEN, kg kg^{-1})	58.7	35.6	51.1	56.9	65.7	100.2
Physiological efficiency of N (PEN, kg kg^{-1})	34.0	0.0	21.4	34.1	48.0	80.7

[a]PFPN = GY_N/FN, AEN = $(GY_N - GY_O)$/FN, REN = $(UN_N - UN_O)$/FN, IEN = GY_N/UN_N, PEN = $(GY_N - GY_O)$/$(UN_N - UN_O)$, all as kg ha^{-1}. Symbols with subscript $_N$ refer to measurements made in the farmers' fertilizer practice. Symbols with subscript $_O$ refer to measurements made in replicated 0-N plots embedded within the farmers' fields.

Table 15.3. Estimates of the indigenous N, P, and K supply in irrigated rice fields of Asia. The data set consists of 179 farm sites in China, India, Indonesia, the Philippines, Thailand, and Vietnam with at least two rice crops grown per year. Values shown are based on nutrient omission plots established in three to four successive rice crops grown on each farm in 1997 to 1998, including the low- and high-yielding growing season.

Parameters	Minimum	25% quartile	Median	75% quartile	Maximum
Average indigenous nutrient supply[a]					
Plant N uptake in 0-N plot (kg ha^{-1})	25.1	43.0	55.2	66.4	90.2
Plant P uptake in 0-P plot (kg ha^{-1})	6.7	12.6	14.5	17.1	26.9
Plant K uptake in 0-K plot (kg ha^{-1})	40.6	66.1	77.6	90.4	156.4
Grain yield 0 N (t ha^{-1})	1.45	3.31	3.86	4.65	6.09
Grain yield 0 P (t ha^{-1})	2.59	4.08	5.07	5.89	7.51
Grain yield 0 K (t ha^{-1})	2.57	4.11	4.92	5.74	7.62
Potential indigenous nutrient supply[b]					
Plant N uptake in 0-N plot (kg ha^{-1})	29.0	51.5	63.6	76.0	107.2
Plant P uptake in 0-P plot (kg ha^{-1})	6.7	13.8	16.5	19.6	31.5
Plant K uptake in 0-K plot (kg ha^{-1})	42.6	74.0	89.6	108.6	197.8
Grain yield 0 N (t ha^{-1})	1.83	3.81	4.51	5.21	6.54
Grain yield 0 P (t ha^{-1})	2.68	4.50	5.73	6.67	8.23
Grain yield 0 K (t ha^{-1})	2.57	4.59	5.56	6.57	8.75

[a]Average of 3 or 4 successive rice crops grown in each field in 1997-98. [b]Average of the two largest values of 3 or 4 successive rice crops grown in each field in 1997-98.

Grain yield

Average rice yield was 5.01 t ha^{-1}, which is roughly identical to the global average yield of irrigated rice assuming 1991 production figures (IRRI 1993) updated by yield growth occurring since then. Despite irrigation, yields were below <4 t ha^{-1} in 26% of all cases. Very high yields of ≥8 t ha^{-1} were recorded in only eight cases (2%), illustrating how difficult it is to achieve such yields under actual production conditions. The typical irrigated rice farmer in Asia achieves roughly 60% of the average climatic yield potential of current modern rice varieties. The average grain yield was 5.62 t ha^{-1} in HYS versus 4.37 t ha^{-1} in LYS. The wide range in the number of panicles m^{-2} (150 to >1,100) reflects the variation in crop establishment methods, ranging from transplanting at 25 × 25-cm hill spacing to direct seeding with seed rates of +200 kg ha^{-1}.

Nutrient uptake

Because of unbalanced NPK nutrition, inefficient N use, and other constraints to grain yield, the average actual nutrient uptake with grain and straw to produce 1 t of rice yield is currently about 12–20% larger than what would be required under conditions of balanced nutrition at the current average yield. Median actual nutrient uptake of rice was 17.6 kg N, 3.1 kg P, and 16.3 kg K t^{-1} of grain yield produced. For comparison, under optimal growth conditions and within a range of about 60–75% of the maximum yield, optimal nutrient uptake requirements of rice are only 14.7 kg N, 2.6 kg P, and 14.5 kg K t^{-1} of grain yield produced (Witt et al 1999). Therefore, using acquired nutrients more efficiently by eliminating other constraints such as pests or poor water supply is as important for improving nutrient-use efficiencies as the balanced application of fertilizers themselves.

Nitrogen

Fertilizer rates varied widely among farms, but were typically 90 to 140 kg N ha^{-1} per crop, with a median of 110 kg N ha^{-1}. Median rates in HYS (115 kg N ha^{-1}) were only slightly greater than those in LYS (102 kg N ha^{-1}), suggesting insufficient adjustment of N rates according to the climatic yield potential and the actual yield achieved. Among countries, N use was highest in China, where farmers usually apply 150–180 kg N ha^{-1}. At present, fertilizer-N-use efficiency is low in intensive rice systems (Table 15.2). The medians of the agronomic efficiency of N (AEN, 9.4 kg kg^{-1}) and of the recovery efficiency of applied N (REN, 0.27 kg kg^{-1}) were below half of what is typically achieved in well-managed on-farm or research-station field experiments and there was no difference between HYS and LYS. The low median physiological efficiency of N (PEN, 34 kg kg^{-1}) indicates that numerous stresses other than N supply affect the use of fertilizer-N in farmers' fields. For comparison, REN of 0.50–0.70 kg kg^{-1}, AEN of >20–25 kg kg^{-1}, and PEN of about 50 kg kg^{-1} can be achieved in a well-managed rice crop by adjusting the amount and time of N application to the indigenous N supply (INS) and to the actual plant N status (Cassman et al 1998, Dobermann et al 2000, Peng et al 1996, Peng and Cassman 1998). Moreover, there appears to be little consistency in N management, which is reflected by large differences among sites, among farmers within each site, and among different cropping seasons for the same farmer.

Phosphorus

The median P use of 18 kg P ha^{-1} $crop^{-1}$ appears to be sufficient for current average yields and for sustaining a neutral or slightly positive P balance on most farms. However, site variations need to be considered. Phosphorus use was close to 20 kg P ha^{-1} at most sites but low in West Java and Central Luzon (median of 10–12 kg P ha^{-1}). The latter confirms recent concerns about low soil P status in large areas of Central Luzon (Dobermann and Oberthür 1997). Overall, the P budget was positive on 75% of all farms and averaged 5 kg P ha^{-1} $crop^{-1}$.

Potassium

Because of low K fertilizer use (median 18 kg K ha^{-1} $crop^{-1}$) and significant straw removal at many sites, the K input-output balance was estimated to be negative for about 80% of all rice farms, with an average of about -25 kg K ha^{-1} $crop^{-1}$. However, K input-output budgets can vary from -100 to $+150$ kg K ha^{-1} $crop^{-1}$, depending on factors such as fertilizer-K use, the source of irrigation water, straw management, and yield (Dobermann et al 1998). Almost all farmers in West Java and Central Thailand did not apply any K fertilizer, whereas median use was as high as 52 kg K ha^{-1}

in North Vietnam and 64 kg K ha^{-1} in Zhejiang. At some sites, particularly in direct-seeded rice (Philippines, Thailand, South Vietnam) or transplanted hybrid rice (China), visible plant symptoms of K deficiency were observed during our studies, a new phenomenon not observed before.

Indigenous nutrient supply

Two- to threefold ranges of the indigenous supply of N (INS), P (IPS), and K (IKS) among fields were found within all rice-growing domains of South and Southeast Asia monitored in this study (see Chapters 6–12). Table 15.3 provides summary values across all sites, divided into average and potential indigenous supplies. Potential supplies typically exceeded the average supplies by 15%, providing evidence that seasonal fluctuations in growth conditions affect such plant-based estimates of the indigenous nutrient supply. Potential supplies as defined in Table 15.3, however, are more relevant for field-specific adjustment of fertilizer rates as done in our SSNM approach. Rice yield without applying N ranged from <2 to >6 t ha^{-1}. Similar ranges were found in other studies (Bouldin 1986, Cassman et al 1996, Wopereis et al 1999). Median potential indigenous supplies were 64 kg N, 17 kg P, and 90 kg K ha^{-1} crop^{-1}. Median grain yields in nutrient omission plots used to estimate potential indigenous supplies increased in the order 0-N (4.5) < 0-K (5.6) < 0-P (5.7 t ha^{-1}). Clearly, N deficiency is a general feature of all irrigated rice systems, although, because of the large N input from indigenous sources, these systems are able to sustain unfertilized yields at 2–4 t ha^{-1} for decades (Cassman et al 1998, Dobermann et al 2000). Our data also provide new evidence that, after 30 years of intensive cropping, potassium and phosphorus are equally limiting in many irrigated rice areas of South and Southeast Asia. Depletion of the median potential IKS to levels below the average current farm yields has not yet occurred, but, considering the almost universal character of negative K-input balances in these systems (Dobermann et al 1998), this is probably only a matter of a few more years.

15.3 Effects of site-specific nutrient management on rice productivity and nutrient-use efficiency

Grain yield and nutrient uptake

Across all sites and crops, average grain yields in the SSNM increased by 0.36 t ha^{-1} (7%, $P<0.001$) compared with the FFP measured in the same year (Table 15.4) or by 0.54 t ha^{-1} (11%) compared with the baseline yield before intervention (Fig. 15.1). Mean FFP yields during 1997-99 (Fig. 15.1) also increased by 0.18 t ha^{-1} (3.5%) compared with the baseline (1996-97). Yield increases over FFP were significant in all crops grown after the first crop (Fig. 15.1) and the yield difference increased from 0.31 t ha^{-1} in year 1 to 0.41 t ha^{-1} in year 2 (crop-year effect, $P = 0.016$). This is evidence for an improvement in the performance of SSNM over time as the concept and its practical application evolved (see Chapter 5). Yield increases were similar in HYS (0.39 t ha^{-1}) and LYS (0.34 t ha^{-1}; crop-season effect nonsignificant).

Table 15.4. Effect of site-specific nutrient management on grain yield and plant nutrient uptake in irrigated rice fields of Asia. Values shown are means of 179 farm sites in China, India, Indonesia, the Philippines, Thailand, and Vietnam (1997-99).

Parameters	Levels[a]	Treatment[b] SSNM	Treatment[b] FFP	Δ[c]	$P > \|t\|$[c]	Effects[d]	$P > \|F\|$[d]
Grain yield (GY)	All	5.54	5.18	0.36	<0.001	Village	<0.001
(t ha^{-1})	Year 1	5.56	5.25	0.31	0.002	Year[e]	0.016
	Year 2	5.53	5.12	0.41	<0.001	Season[e]	0.466
	HYS	6.02	5.63	0.39	<0.001	Year × season[e]	0.217
	LYS	5.08	4.74	0.34	0.001	Village × crop	<0.001
Plant N uptake (UN)	All	99.2	89.8	9.4	<0.001	Village	<0.001
(kg ha^{-1})	Year 1	99.2	91.2	8.0	<0.001	Year	0.029
	Year 2	99.0	88.3	10.7	<0.001	Season	0.854
	HYS	106.4	97.1	9.3	<0.001	Year × season	0.596
	LYS	91.9	82.4	9.5	<0.001	Village × crop	<0.001
Plant P uptake (UP)	All	19.4	17.4	2.0	<0.001	Village	<0.001
(kg ha^{-1})	Year 1	19.8	18.0	1.8	<0.001	Year	0.002
	Year 2	19.1	16.9	2.2	<0.001	Season	<0.001
	HYS	20.7	19.0	1.7	<0.001	Year × season	<0.001
	LYS	18.1	15.7	2.4	<0.001	Village × crop	<0.001
Plant K uptake (UK)	All	102.4	94.0	8.4	<0.001	Village	<0.001
(kg ha^{-1})	Year 1	100.8	93.7	7.1	0.006	Year	0.092
	Year 2	103.7	94.2	9.5	<0.001	Season	0.801
	HYS	110.6	102.0	8.6	<0.001	Year × season	0.807
	LYS	94.0	85.9	8.1	<0.001	Village × crop	0.611

[a]All = all four successive rice crops grown from 1997 to 1999; year 1 = crops 1 and 2; year 2 = crops 3 and 4; HYS = high-yielding season; LYS = low-yielding season. [b]FFP = farmers' fertilizer practice; SSNM = site-specific nutrient management. [c]Δ = SSNM – FFP; $P > \|t\|$ = probability of a significant mean difference between SSNM and FFP. [d]Source of variation of ANOVA of the difference between SSNM and FFP by site; $P > \|F\|$ = probability of a significant F-value. DF: site 7, year 1, season 1, year-season 1, site-crop 21. [e]Year refers to two consecutive cropping seasons.

The probability of a yield increase occurring because of SSNM was 73% (Fig. 15.2), with no difference between HYS and LYS crops, thus demonstrating a relatively small risk associated with this new technology. Yield increases of >1 t ha^{-1} were observed in 13% of all cases. In 1996-97, only 22% of the farmers achieved yields greater than 6 t ha^{-1}. This proportion increased to 42% in the SSNM plots in 1997-99 (27% in FFP). The proportion of yields >8 t ha^{-1} increased to 12% versus 2% in the baseline data. On 41 out of the 179 farms, average actual yields of four successive rice crops grown were within ±10% of the yield goal.

Increases in grain yield were mainly associated with an increase in the sink size. Compared to the FFP in the same year, the average number of spikelets m^{-2} increased by 6.5%, whereas differences in harvest index and other yield components were not significant (data not shown). The average straw yield increased by 0.28 t ha^{-1} (6%) compared to the FFP in the same season, suggesting that improved nutrient supply had a slightly greater effect on grain yield than on straw yield. Nevertheless,

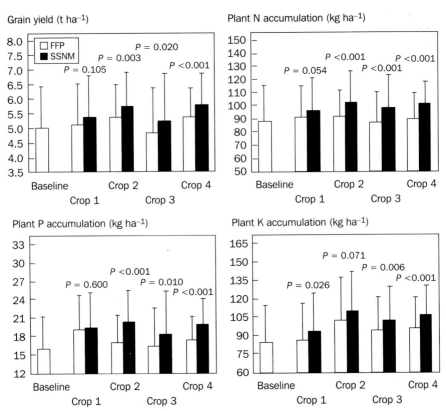

Fig. 15.1. Grain yield and plant N, P, and K accumulation in the farmers' fertilizer practice (FFP) and site-specific nutrient management (SSNM) plots at eight sites. Baseline data refer to two successive rice crops monitored before intervention (1995-97), whereas crop 1 to crop 4 indicates the four rice crops grown thereafter (1997-99). Bars: mean of 179 farms; error bars: standard deviaton of 179 farms.

the increase in straw yield represents another potential benefit in areas where farmers use straw as fuel, forage, or bedding material or for soil improvement.

Increases in total uptake of N, P, and K were larger than increases in grain yield. Average nutrient uptake under SSNM increased by 13% (N) and 21% (P, K) compared with the baseline or by 9–11% compared with the FFP in the same years (Table 15.4, Fig. 15.1). Increases in plant uptake were significant in most crops grown (Fig. 15.1) and they also improved significantly from year 1 to year 2 for all three nutrients. The latter is particularly remarkable because rates of N, P, and K fertilizer declined from year 1 to year 2 (see below). Only in the case of P was the increase in plant uptake in HYS larger than in LYS (crop-season effect, $P<0.001$).

Fertilizer use

SSNM and FFP treatments differed in the amounts of N, P, and K applied. On average, 4.6 kg ha^{-1} less N was applied in SSNM plots than in FFP (Table 15.5; 4% less,

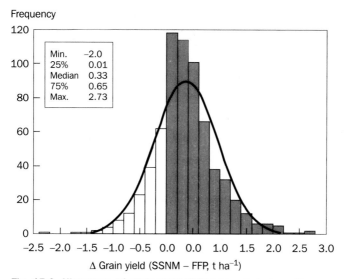

Fig. 15.2. Histogram of grain yield differences between site-specific nutrient management (SSNM) and the farmers' fertilizer practice (FFP). Frequencies shown are based on four consecutive rice crops grown on 179 farms in 1997-99.

$P = 0.013$). Moreover, this difference increased from 1.7 kg N ha^{-1} crop^{-1} in the first year (nonsignificant) to 7.4 kg N ha^{-1} in the second year (6% less, $P = 0.004$). Similarly, the average P use in SSNM exceeded that in FFP by 4.7 kg P ha^{-1} crop^{-1} in year 1, but this difference was reversed to -2 kg P ha^{-1} in year 2. Although the farmers differentiated little between HYS and LYS in their P use, P rates under SSNM were about 3 kg ha^{-1} higher in HYS than in LYS to account for the differences in the climatic yield potential and yield goals. Average fertilizer-K rates in the SSNM (58 kg K ha^{-1} crop^{-1}) were generally about 90% higher than in the FFP (31 kg K ha^{-1} crop^{-1}). However, the difference between the two treatments declined from 35 kg K ha^{-1} in year 1 to 21 kg K ha^{-1} in year 2 (crop-year, $P < 0.001$).

Greater P and K rates in the first SSNM year were mainly caused by insufficient levels of IPS and IKS at many sites. The decreasing rates in the second year resulted from (1) more accurately measured values of IPS and IKS that had become available, (2) increases in IPS and IKS estimated from the actual input-output balance of the previous crops (see Chapter 5), and (3) setting more realistic yield goals. Median P use in the FFP remained the same (18 kg P ha^{-1} crop^{-1}) as for the period before the introduction of SSNM. Interestingly, the farmers increased their K use in FFP plots from a median of 18 kg K ha^{-1} crop^{-1} in 1995-97 to 24 kg K ha^{-1} crop^{-1} in 1997-99 (33%), which may be partly a response to the higher K rates observed in the SSNM plots. However, an increase in K use occurred mainly at three sites (Thanjavur, Maligaya, and Omon). The latter two sites represent direct-seeded rice systems in which plant symptoms of K deficiency were first observed in 1997. For example, at Omon (Mekong Delta), average K use in the FFP increased from 9 to 21 kg K ha^{-1}

Table 15.5. Effect of site-specific nutrient management on fertilizer use and nitrogen-use efficiency in irrigated rice fields of Asia. Values shown are means of 179 farm sites in China, India, Indonesia, the Philippines, Thailand, and Vietnam (1997-99).

Parameters	Levels[a]	Treatment[b] SSNM	Treatment[b] FFP	Δ[c]	$P > \lvert t \rvert$[c]	Effects[d]	$P > \lvert F \rvert$[d]
N fertilizer (FN)	All	112.3	116.9	−4.6	0.013	Village	<0.001
(kg ha ^{1})	Year 1	115.1	116.8	−1.7	0.520	Year[e]	0.017
	Year 2	109.5	116.9	−7.4	0.004	Season[e]	0.298
	HYS	116.0	121.7	−5.7	0.025	Year × season[e]	0.857
	LYS	108.6	112.1	−3.5	0.182	Village × crop	<0.001
P fertilizer (FP)	All	19.3	17.9	1.4	0.015	Village	<0.001
(kg ha ^{1})	Year 1	22.3	17.6	4.7	<0.001	Year	<0.001
	Year 2	16.4	18.3	−1.9	0.009	Season	0.148
	HYS	20.7	18.5	2.2	0.005	Year × season	0.180
	LYS	17.9	17.3	0.6	0.497	Village × crop	<0.001
K fertilizer (FK)	All	58.2	30.5	27.7	<0.001	Village	<0.001
(kg ha ^{1})	Year 1	65.6	30.5	35.1	<0.001	Year	<0.001
	Year 2	51.0	30.5	20.5	<0.001	Season	0.035
	HYS	59.9	33.1	26.8	<0.001	Year × season	<0.001
	LYS	56.5	27.9	28.6	<0.001	Village × crop	<0.001
Agronomic efficiency	All	14.8	11.5	3.3	<0.001	Village	0.003
of N (AEN)	Year 1	14.6	11.8	2.8	<0.001	Year	0.063
(kg grain kg ^{1} N)	Year 2	15.0	11.2	3.8	<0.001	Season	0.552
	HYS	16.2	12.7	3.5	<0.001	Year × season	0.217
	LYS	13.4	10.3	3.1	<0.001	Village × crop	<0.001
Recovery efficiency	All	0.40	0.31	0.09	<0.001	Village	<0.001
of N (REN)	Year 1	0.40	0.33	0.07	<0.001	Year	0.028
(kg plant N kg ^{1} N)	Year 2	0.40	0.29	0.11	<0.001	Season	0.975
	HYS	0.44	0.36	0.08	<0.001	Year × season	0.940
	LYS	0.37	0.28	0.09	<0.001	Village × crop	<0.001
Physiologic efficiency	All	37.2	36.3	0.9	0.320	Village	0.028
of N (PEN)	Year 1	35.9	35.6	0.3	0.804	Year	0.614
(kg plant N kg ^{1} N)	Year 2	38.6	37.2	1.4	0.247	Season	0.778
	HYS	37.6	36.1	1.5	0.226	Year × season	0.015
	LYS	36.9	36.6	0.3	0.809	Village × crop	0.134
Partial productivity	All	52.2	49.2	3.0	0.006	Village	<0.001
of N (PFPN)	Year 1	51.8	49.2	2.6	0.109	Year	0.679
(kg grain kg ^{1} N)	Year 2	52.6	49.3	3.3	0.022	Season	0.535
	HYS	54.5	50.7	3.8	0.004	Year × season	0.166
	LYS	49.8	47.6	2.2	0.207	Village × crop	<0.001

[a]All = all four successive rice crops grown from 1997 to 1999; year 1 = crops 1 and 2; year 2 = crops 3 and 4; HYS = high-yielding season; LYS = low-yielding season. [b]FFP = farmers' fertilizer practice; SSNM = site-specific nutrient management. [c]Δ = SSNM − FFP; $P > \lvert t \rvert$ = probability of a significant mean difference between SSNM and FFP. [d]Source of variation of ANOVA of the difference between SSNM and FFP by site; $P > \lvert F \rvert$ = probability of a significant F-value. DF: site 7, year 1, season 1, year-season 1, site-crop 21. [e]Year refers to two consecutive cropping seasons.

crop^{-1} from 1995-96 to 1997-99. On the other hand, in the same time periods, farmers around Sukamandi (West Java) responded to the economic crisis in their country by cutting their K use from an average of 16 to less than 5 kg K ha^{-1} crop^{-1}.

Earlier studies have shown no relationship between N use and INS in farmers' fields (Cassman et al 1996, Olk et al 1999). Our results suggest that, with SSNM, fertilizer use can be adjusted to account for indigenous nutrient supplies, whereas farmers do not in general appear to be following this practice. The correlation between potential IPS (see Table 15.3) and the average fertilizer-P applied to all four crops on each farm was –0.36 in the SSNM treatment ($P<0.001$) versus 0.10 in the FFP ($P = 0.175$). Similarly, K use was negatively correlated with potential IKS (r = –0.27, $P<0.001$) in the SSNM, but not in the FFP (r = 0.14, $P = 0.047$). Many farmers tended to apply more N on soils with high potential INS (r = 0.25, $P = 0.025$) in contrast to a weak tendency in the SSNM to apply less N on such soils (r = –0.15, $P = 0.043$). Clearly, these are weak correlations and further research will have to clarify the general validity of these preliminary findings[2]. For example, we hypothesize that farmers' PK applications show little relationship with IPS and IKS because farmers are less familiar with the management of such nutrients.

Nitrogen-use efficiency

Site-specific nutrient management led to large gains in N-use efficiency. Average AEN under SSNM rose to 15 kg kg^{-1}, REN to 0.40 kg kg^{-1}, and partial factor productivity of N (PFPN) to 52 kg kg^{-1}. Compared to the FFP in the same season, average AEN and REN increased by almost 30% and PFPN by 6% (Table 15.5). Comparing the medians of the SSNM sample with the FFP baseline data, SSNM increased AEN by 56%, REN by 44%, and PFPN by 14%. Despite differences among sites, increases in AEN and REN were consistent and occurred in all crops (Fig. 15.3). The probability of an increase occurring in AEN and REN because of SSNM was 72% and 74%, respectively (Fig. 15.4). Increases in N-use efficiency by SSNM over FFP were similar in high-yielding and low-yielding climatic seasons (nonsignificant crop-season effects, Table 15.5). However, treatment differences in AEN and REN increased significantly from 21–24% in year 1 to 34–38% in year 2 (crop-year effect), reflecting the gradual improvements in dynamic N management algorithms made over time (see Chapter 5). SSNM had no significant effect on PEN. In both SSNM and FFP treatments, average PEN remained at suboptimal levels of 36–37 kg kg^{-1}, although this represents a 6–9% increase over the baseline data set.

Large increases in AEN of >5 kg kg^{-1} were observed in 38% of all cases. In 15% of all cases, AEN increases exceeded 10 kg kg^{-1}, which is equivalent to doubling the current average AEN achieved by rice farmers in Asia (Table 15.2). The proportion of fields with a high AEN (>20 kg kg^{-1}) increased from 19% in 1996-97

---

[2]This correlation analysis was done using the potential indigenous supplies measured as the average of the largest two values of three to four crops. Correlations between indigenous supply and fertilizer rate in SSNM would have been larger if the actual model input values in each year had been used.

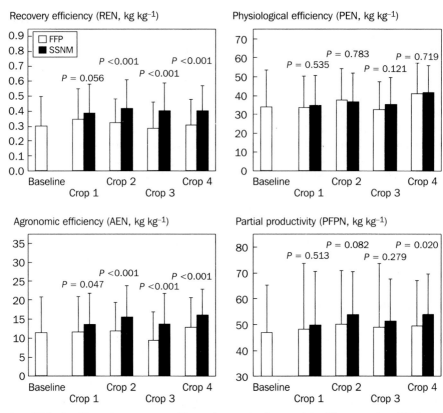

Fig. 15.3. Fertilizer nitrogen-use efficiencies in the farmers' fertilizer practice (FFP) and site-specific nutrient management (SSNM) plots at eight sites. Baseline data refer to two successive rice crops monitored before intervention (1995-97), whereas crop 1 to crop 4 indicates the four rice crops grown thereafter (1997-99). Bars: mean of 179 farms; error bars: standard deviation of 179 farms.

(FFP baseline) to 30% in 1997-99 (SSNM), but decreased to 14% in the FFP during the same period. Before the introduction of SSNM, 22% of the farmers achieved a high REN of >0.5 kg kg^{-1}. This number decreased to 15% in the FFP during 1997-99, but increased to 35% under SSNM. A very high REN of >0.7 kg kg^{-1}, which is comparable with that at maximum rates of N uptake (Peng and Cassman 1998), was measured in 16% of all cases.

Differences in N management between SSNM and FFP are further illustrated by analyzing the individual split applications of N (Fig. 15.5). Most farmers applied high doses early in the season when the capacity for crop uptake is small (Peng et al 1996). Under SSNM, N applications were typically delayed by 5 to 6 d compared with the farmers' practice and individual doses of preplant or topdressed N were commonly about 10 kg N ha^{-1} smaller than in FFP. Did some of the farmers adopt these practices during the course of our experiments? The median AEN in the FFP increased from 9.4 kg kg^{-1} in 1996-97 to 10.3 kg kg^{-1} in 1997-99. Similarly, REN

Frequency

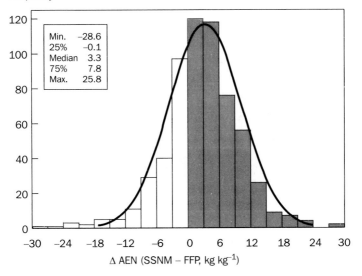

Min.   −28.6
25%   −0.1
Median   3.3
75%   7.8
Max.   25.8

Δ AEN (SSNM − FFP, kg kg⁻¹)

Frequency

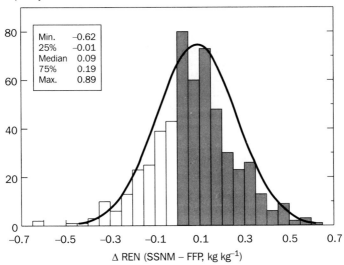

Min.   −0.62
25%   −0.01
Median   0.09
75%   0.19
Max.   0.89

Δ REN (SSNM − FFP, kg kg⁻¹)

Fig. 15.4. Histogram of nitrogen-use efficiency differences between site-specific nutrient management (SSNM) and the farmers' fertilizer practice (FFP). Frequencies shown are based on four consecutive rice crops grown on 179 farms in 1997-99. AEN = agronomic efficiency of N, REN = recovery efficiency of applied N.

N rate (kg N ha^{-1})

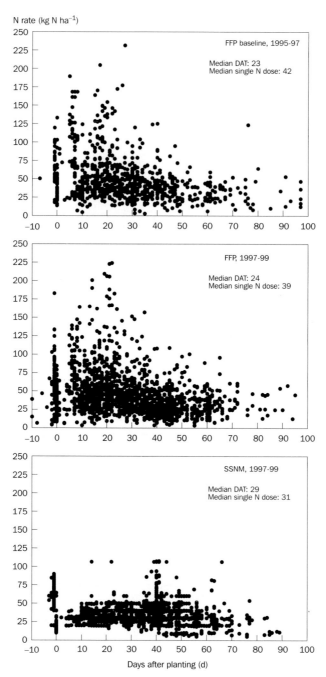

Fig. 15.5. Nitrogen applications in the farmers' fertilizer practice (FFP) and with site-specific nutrient management (SSNM). Values shown represent all individual preplant or topdressed N applications made on 179 farms at different dates.

increased only slightly from 0.27 kg kg^{-1} (1996-97) to 0.29 kg kg^{-1} (1997-99). Fertilizer-N rates in FFP and SSNM were positively correlated (r = 0.34, $P<0.001$). These numbers and the slight decrease in the amount of N applied with single split applications in the FFP (Fig. 15.5) suggest that such an adoption may have partially occurred, but was probably not widespread.

Our experimental design did not allow us to statistically assess separate effects of amount of N, timing of N, or balanced NPK nutrition on AEN or REN. Presumably, all three factors played a role and our data confirm that, when grown under optimal crop management, rice is capable of using fertilizer-N very efficiently (Peng and Cassman 1998). It is possible to achieve high N-use efficiency under on-farm conditions by using common fertilizers in a more appropriate way. The gains described above were all achieved with topdressed applications of prilled urea and no major changes in other cropping practices, that is, without using expensive slow-release fertilizers or labor-intensive deep-placement techniques. Spreading N applications more evenly throughout the growing season was probably the major reason for the increases in N-use efficiency and this also reduces the risk of environmental pollution associated with gaseous N losses or losses through runoff or leaching after a heavy fertilizer application. A more balanced NPK nutrition practiced in the SSNM may have contributed to increases in AEN and REN through more vigorous plant growth and greater resistance to diseases.

Phosphorus and potassium input-output budgets

Positive or neutral nutrient budgets are one indicator of the sustainability of cropping systems, except for some locations at which very large native nutrient resources can be mined for a limited period of time. Historically, natural components of the nutrient balance such as sedimentation, nutrient inflow by irrigation, organic residues, biological N$_2$ fixation (BNF), and carbon assimilation by floodwater flora and fauna played an important role in securing the sustainability of irrigated rice systems. Since the advent of the Green Revolution, the contribution of natural nutrient sources has declined and crop intensification has increased the annual plant nutrient accumulation five- to sevenfold compared with the pre-Green Revolution period (Dobermann et al 1998, Greenland 1997).

Using the assumptions described in Chapter 2, we estimated P and K input balances for each individual data set. We acknowledge that these estimates represent only a partial nutrient balance at best because several important components such as nutrient input from irrigation and rainfall or losses from leaching or straw burning were not measured. However, within each farm, the same assumptions were used for both FFP and SSNM treatments so that the balance estimates mainly show differences among these treatments as an indicator of how sustainable they are in maintaining soil nutrient reserves. Leaching of mineral P or K is probably a minor pathway of losses at the sites included in our study because of the generally high clay content and low percolation rates. Perhaps the only exception to this are the coarse-textured soils at Thanjavur and a few sites in the Sukamandi domain.

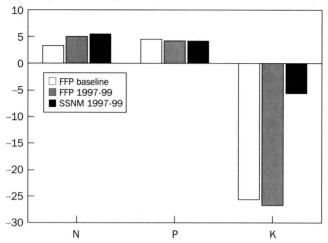

Input-output balance (kg ha⁻¹ crop⁻¹)

Fig. 15.6. Estimated average input-output balances of nitrogen (N), phosphorus (P), and potassium (K) in the farmers' fertilizer practice (FFP) and under site-specific nutrient management (SSNM). Values shown are averages of four rice crops grown in 179 farmers' fields (1997-99) compared with the baseline data set (two rice crops) collected before the introduction of an SSNM plot.

There was no difference between FFP and SSNM treatments in the average P input-output balance, which remained positive at about 4 kg P ha⁻¹ crop⁻¹. Note that in both treatments the P input-output balance was negative in 30–35% of all cases. Although the average fertilizer-P input in the SSNM was greater than in the FFP (Table 15.5), at about one-third of all sites high levels of IPS justified P rates that were slightly below the net crop removal. Applying such a strategy can save money over the short term, but this requires monitoring of IPS levels every 4 to 5 years.

The negative K balance was reversed on many rice farms. Site-specific application of K reduced the average K balance to –6 kg ha⁻¹ crop⁻¹, whereas about 20 kg K ha⁻¹ more were lost with each crop grown in the FFP (Fig. 15.6). Because of greater K use, the K balance was 3 kg K ha⁻¹ crop⁻¹ in the first SSNM year, but dropped to –10 kg K ha⁻¹ crop⁻¹ in year 2 when previous increases in IKS were taken into account for lowering the K rates. The proportion of farms with a negative K balance decreased from 80% in the FFP baseline to 50% in SSNM plots. Further improvements in K balances should probably focus on both the increased use of K fertilizer and better straw management. Many farmers currently use more expensive compound fertilizers as a source of K for rice. At issue is whether cheaper single-nutrient K fertilizers such as muriate of potash will play a greater role in rice systems in the future. Rice straw is one of the largest components of the K input-output balance and typically contains 12–17 kg K t⁻¹ dry matter. However, its incorporation may be associated with problems such as higher cost and increased pest pressure.

Table 15.6. Estimated average yield gaps in irrigated rice fields of Asia before and after the introduction of site-specific nutrient management. The data set consists of 179 farm sites in China, India, Indonesia, the Philippines, Thailand, and Vietnam with at least two rice crops grown per year.

Data set[a]	Y_{max}[b] (t ha^{-1})	Y_a[c] t ha^{-1}	Y_a[c] % of Y_{max}	Y[d] t ha^{-1}	Y[d] % of Y_{max}	Yield gap 1: $Y_{max} - Y_a$[e] (t ha^{-1})	Yield gap 2: $Y_a - Y^f$ (t ha^{-1})
FFP baseline (1995-97)	8.30	5.78	70	5.12	62	2.52	0.62
FFP year 1 (1997-98)	8.30	6.14	74	5.21	63	2.16	0.88
FFP year 2 (1998-99)	8.30	6.03	73	5.10	61	2.27	0.88
SSNM year 1 (1997-98)	8.30	6.54	79	5.54	67	1.76	0.94
SSNM year 2 (1998-99)	8.30	6.51	78	5.55	67	1.79	0.90

[a]FFP = farmers' fertilizer practice; SSNM = site-specific nutrient management. [b]Y_{max} = estimated average climatic and genetic yield potential of irrigated rice in Asia using the ORYZA 1 and SIMRIW models (Matthews et al 1995). [c]Y_a = attainable yield estimated by using the QUEFTS model (Witt et al 1999) with the median actual uptake of N, P, and K in FFP and SSNM plots at 179 farm locations. [d]Y = actual median grain yield at 179 farm locations. [e]Yield loss mainly because of insufficient NPK supply. [f]Yield loss mainly because of factors other than NPK supply (water, other mineral disorders, pests, unexpected events).

Yield gaps

On average, actual yield (Y) in the SSNM was about 67% of the potential yield versus 62% in the FFP. We analyzed the average yield gaps across all sites to quantify to what extent yield limitations caused by macronutrient supply were alleviated through field-specific application of fertilizers (Table 15.6).

Yield gap 1 is an indicator of the unrealized yield potential because of insufficient uptake of N, P, or K, that is, a reduction in nutrient uptake caused by limitations in the availability of nutrients or caused by crop damage. Possible causes are nutrient deficiencies (e.g., micronutrient deficiencies, nutrient losses from leaching, N fertilizer losses from unfavorable splitting and timing of applications), inadequate planting density, poor seed quality, competition with weeds, root damage from nematodes, and rat or pest damage during vegetative growth. Because of increased NPK uptake, we reduced the gap between potential (Y_{max}) and nutrient uptake-limited (Y_a) yield to about 1.8 t ha^{-1} under the SSNM (yield gap 1) versus 2.2 t ha^{-1} in the FFP.

Yield gap 2 is an indicator of the unrealized yield potential because of inefficient conversion of accumulated plant nutrients into grain yield, that is, factors other than N, P, or K supply. It mainly involves location-specific occurrences of drought stress, mineral disorders with a direct impact on physiological processes, unfavorable weather conditions during grain filling, lodging, or insect and disease incidence during reproductive growth. Yield gap 2 remained at about 0.9 t ha^{-1} crop^{-1} in both the FFP and SSNM, indicating similar constraints to the conversion of accumulated plant nutrients to yield. Yield gaps and nutrient uptake-limited yield in the SSNM were similar in the two years compared.

Internal nutrient efficiencies remain high until yields reach about 80% of Y_{max} (Witt et al 1999). The average nutrient uptake in the SSNM treatment was sufficient to attain yields of almost 80% of Y_{max}, but the actual internal efficiencies of N, P, and K in the plant (grain yield per unit of nutrient accumulated) remained well below optimal levels. Average internal efficiencies in the SSNM were 57 kg grain kg^{-1} N, 299 kg grain kg^{-1} P, and 57 kg grain kg^{-1} K versus 68 kg grain kg^{-1} N, 385 kg grain kg^{-1} P, and 69 kg grain kg^{-1} K that can be achieved with balanced nutrition and complete control of other biotic and abiotic stresses (Witt et al 1999). It is difficult to separate the effect of individual crop management factors on yield gaps 1 and 2. For example, better plant nutrition or weed control may mainly reduce yield gap 1 rather than gap 2, whereas improved pest control can have an impact on both yield gaps. However, given its size, there is still substantial scope for closing yield gap 1 further by exploiting the synergy that occurs if crop, pest, and nutrient management are improved simultaneously. Closing yield gap 2 appears to be more difficult because only small increments can be achieved.

15.4 Variation in the performance of site-specific nutrient management

Although the SSNM strategy tested followed generic principles, location-specific adjustments were made (see Chapter 5) and the performance was affected by numerous factors other than the supply of N, P, and K. Many significant site and site-crop effects (Tables 15.4 and 15.5) illustrated the large variation among the rice-growing domains with regard to increases in yield, nutrient uptake, fertilizer use, and N-use efficiency achieved through SSNM. Table 15.7 provides a summary of performance differences by sites. Achievement of the yield goal varied from an average of about 95% at Hanoi to less than 75% in Central Luzon and West Java (Fig. 15.7). What are the reasons for these geographical differences and what lessons can be learned from this?

Climatic factors

The discussion of location-specific results in Chapters 6 to 12 has already highlighted that climate affected the performance of SSNM at several sites. In particular, the experimental period included the El Niño–La Niña climatic cycle, but its effect on rice yield probably varied widely among sites. Climatic events such as El Niño directly affect crop growth through solar radiation, temperature, and moisture. Increased solar radiation can increase the yield potential at certain sites, whereas greater than normal temperatures probably cause increased spikelet sterility and degeneration at others. In contrast, greater rainfall and more cloudy conditions associated with the La Niña period would mostly decrease the yield potential because of less solar radiation, but also because of possibly poor spikelet fertility if heavy rains occur at the flowering stage.

It is difficult to quantify such effects without a complete modeling effort done for all sites. However, the crop-modeling analysis conducted for the Philippine site (Chapter 17) indicated a lower-than-normal yield potential in all four crops grown.

Table 15.7. Comparison of the agronomic performance of site-specific nutrient management (SSNM) at eight sites in Asia. Values shown are SSNM treatment means and mean differences between SSNM and the farmers' fertilizer practice (Δ) for four successive rice crops grown at each site from 1997 to 1999.

Site[a]	Grain yield (t ha⁻¹)		Fertilizer N (kg ha⁻¹)		Fertilizer P (kg ha⁻¹)		Fertilizer K		AEN		PEN (kg kg⁻¹)		REN	
	SSNM	Δ	SSNM	Δ	SSNM	Δ	SSNM	Δ	SSNM	Δ	SSNM	Δ	SSNM	Δ
JI	6.35	0.45	133	-35	15	-4	60	5	11	5.0	40	3.1	0.29	0.11
HA	6.24	0.19	93	-11	16	-4	53	-9	18	4.0	46	2.9	0.39	0.06
AD	6.45	0.49	127	15	26	2	70	32	16	2.1	35	2.2	0.43	0.04
TH	5.64	0.63	129	34	18	2	80	45	15	1.4	31	3.1	0.46	0.01
SU	4.52	0.22	103	-21	19	11	53	49	13	3.8	29	-0.1	0.46	0.15
MA	5.26	0.51	111	1	19	4	49	27	15	3.0	34	-2.9	0.46	0.14
SB	4.90	0.10	111	2	18	-3	45	43	9	1.6	33	-2.9	0.29	0.07
OM	4.77	0.33	98	-13	22	3	62	42	20	5.0	46	0.8	0.44	0.10

Site	N uptake (kg ha⁻¹)		P uptake		K uptake		Panicles per m² (no.)		Spikelets per m² (× 1,000)		Filled spikelets (%)		1,000-grain wt. (g)	
	SSNM	Δ	SSNM	Δ	SSNM	Δ	SSNM	Δ	SSNM	Δ	SSNM	Δ	SSNM	Δ
JI	106	7.7	21	2.4	130	11.8	320	12	26	1.4	80	0.3	26.3	0.0
HA	96	1.5	18	-0.6	109	-6.0	364	-3	31	1.0	84	0.1	22.7	-0.2
AD	104	11.6	23	2.5	109	10.5	473	16	34	2.5	87	0.2	19.2	0.0
TH	96	17.2	24	4.9	79	12.5	526	16	44	4.6	82	-0.5	18.0	0.1
SU	102	8.9	14	1.5	93	2.5	226	15	24	2.3	81	0.6	22.3	0.1
MA	107	17.6	19	2.9	101	9.2	554	18	32	1.6	76	-0.5	21.9	0.0
SB	98	5.8	18	1.1	81	10.5	360	-21	21	0.5	72	-1.0	26.1	0.1
OM	82	5.9	18	2.3	102	17.8	506	17	24	1.4	76	0.2	24.6	0.0

[a]JI = Jinhua, Zhejiang (China); HA = Hanoi, Red River Delta (Vietnam); AD = Aduthurai, Old Cauvery Delta, Tamil Nadu (India); TH = Thanjavur, New Cauvery Delta, Tamil Nadu (India); SU = Sukamandi, West Java (Indonesia); MA = Maligaya, Central Luzon (Philippines); SB = Suphan Buri, Central Plain (Thailand); OM = Omon, Mekong Delta (Vietnam).

Actual yield (% of yield goal)

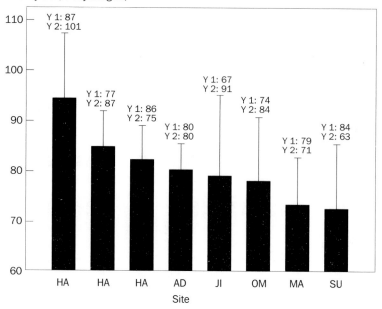

Fig. 15.7. Achievement of the yield goal in the site-specific nutrient management treatment. The columns show the median of the actual yield (% of the yield goal) for all four crops grown at each site. Error bars indicate the 75% quartile. Numbers above the columns show the yield goal (%) in the first and second experimental year at each site. MA = Maligaya, Central Luzon (Philippines), SB = Suphan Buri, Central Plain (Thailand), OM = Omon, Mekong Delta (Vietnam), SU = Sukamandi, West Java (Indonesia), AD = Aduthurai, Old Cauvery Delta, Tamil Nadu (India), TH = Thanjavur, New Cauvery Delta, Tamil Nadu (India), HA = Hanoi, Red River Delta (Vietnam), JI = Jinhua, Zhejiang (China).

Wet-season yield potential in 1997 and 1998 averaged 8. 3 t ha⁻¹, whereas average dry-season yield potentials were 9.8 t ha⁻¹ in 1998 and 8.7 t ha⁻¹ in 1999. Using the same crop model (ORYZA1), yield potential for Central Luzon based on historical climate data was estimated at about 10.8 t ha⁻¹ for WS crops and 13 t ha⁻¹ for DS crops (Matthews et al 1995). Even though we cannot be certain about such estimates, the large difference between earlier estimates and the simulated yields for 1997 to 1999 suggests a significant reduction in yields because of the El Niño–La Niña cycle in Central Luzon. This was probably also the case at a few other sites such as Omon in South Vietnam. Moreover, unusual climatic events indirectly affect crop growth because of suboptimal crop management. This includes issues such as delayed or too early water supply, nonoptimal planting dates, or unusual pest outbreaks. Examples of this were described for several sites, including Sukamandi, Suphan Buri, Omon, and Central Luzon.

In summary, climate was responsible for an unknown proportion of the variation in SSNM performance observed among sites and seasons. However, such events will always be an uncontrollable part of practical farming and every technology has to deal with them. In that sense, the fact remains that SSNM performed well across a wide range of conditions observed from 1997 to 1999.

Differences among geographical domains

Using the criteria shown in Table 15.7 and general knowledge about site differences in cropping systems and practices, we can empirically group sites as follows:

Group 1: High yields of transplanted rice with reduced use of N, P, and K fertilizers. This group consists of 45 rice farms near Hanoi in the Red River Delta of North Vietnam (HA) and at Jinhua (JI) in Zhejiang Province, China (Table 15.7). In both domains, two transplanted rice crops (spring and summer rice) are grown in very small fields (field sizes <0.3 ha) under subtropical climate. Increasing yield is of primary concern for farmers at these sites. All farmers at Hanoi grow an upland winter crop and many at Jinhua used to do this. Labor input ranges from about 100 d ha^{-1} at JI to 230 d ha^{-1} at HA[3] because much time is spent with crop care, such as the intensive use of pesticides (Tables 11.3 and 12.3) and, at Hanoi, farmyard manure. Because of a history of collective farming and the promotion of mineral fertilizers, NPK use was high even before the introduction of SSNM, including average rates of 40–60 kg K ha^{-1} crop^{-1}. Soils at Jinhua also had the largest indigenous nutrient supply among all domains. Site-specific nutrient management on these farms was characterized by (1) high average rice yields of 6.2 to 6.4 t ha^{-1}, (2) a reduction in mineral fertilizer use compared with the FFP, (3) large relative increases in AEN, PEN, and REN because of plant-based N management, and (4) a moderate to excellent quality of crop management. Yield increases over FFP were small at Hanoi because of greater amounts of manure applied in the FFP treatment, whereas large increases in yield were achieved at Jinhua. A greater degree of achieving the predicted yield goal (Fig. 15.7) and site averages that were close to the line describing the optimal relationship between grain yield and plant N accumulation for balanced NPK nutrition (Fig. 15.8) confirm the good cropping conditions and good crop management of these farmers. Except for uncontrollable climatic factors, some yield losses were caused only by diseases and insects, whereas water supply and weed control were generally good. At Jinhua, however, N-use efficiency remained moderate in absolute terms and we hypothesize that it can be further improved by synchronizing N with water management.

Group 2: High yields of transplanted rice with increased use of N, P, and K fertilizers. This group consists of 40 rice farms in the Old (AD) and New (TH) Cauvery Deltas of Tamil Nadu (Table 15.7). Transplanted rice is grown in medium-sized fields (0.5 to 1 ha) under tropical climate. Farmers often prefer fine-grain modern varieties with a 1,000-grain weight of less than 20 g. At both sites, labor input is high (80 to 150 d ha^{-1}) but pesticide use is low (Table 6.3). Hand weeding is the primary weed-

---

[3]Refers to 8-hour person-days.

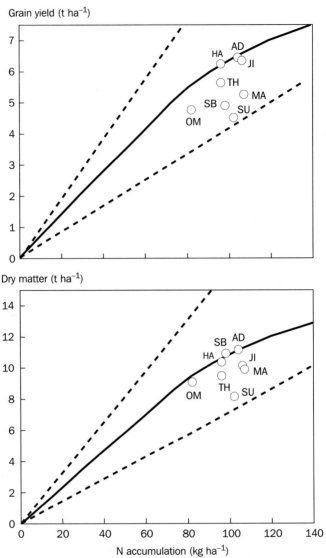

Fig. 15.8. Relationship between grain yield or total aboveground dry matter and N accumulation of rice. The broken lines show the envelope of maximum dilution (YND) and maximum accumulation (YNA) of N in the plant. The solid line shows the relationship for balanced nutrition of N, P, and K as simulated by the QUEFTS model (Witt et al 1999). Circles indicate measured average values in site-specific nutrient management (SSNM) plots at eight sites (1997-99, site means of 4 rice crops). MA = Maligaya, Central Luzon (Philippines); SB = Suphan Buri, Central Plain (Thailand); OM = Omon, Mekong Delta (Vietnam); SU = Sukamandi, West Java (Indonesia); AD = Aduthurai, Old Cauvery Delta, Tamil Nadu (India); TH = Thanjavur, New Cauvery Delta, Tamil Nadu (India); HA = Hanoi, Red River Delta (Vietnam); JI = Jinhua, Zhejiang (China).

control method and farmers attempt to follow IPM guidelines for insect control. This group is characterized by (1) high average rice yields in the SSNM (5.6 to 6.5 t ha^{-1}) and large yield increases over FFP, (2) an increase in the use of N, P, and K fertilizer, (3) high AEN, PEN, and REN because of plant-based N management, and (4) a moderate to excellent quality of crop management. Increased N uptake (13–22%) was probably the major cause of yield increases. Despite relatively small increases in N-use efficiency between SSNM and the FFP, average AEN (16 kg kg^{-1}) and REN (0.45 kg kg^{-1}) were at the upper end of all domains (Table 15.7). At Aduthurai, average yield was high and close to the optimal line describing the relationship between grain yield and plant N accumulation (Fig. 15.8). Yield losses observed there were caused mainly by insects. Average yield increases of 13% were achieved at Thanjavur, but the yield goal achievement was lower (Fig. 15.7) and the internal N-use efficiency was suboptimal because of somewhat larger stresses (water supply, insects).

Group 3: Low yields of transplanted rice because of factors other than N, P, and K supply. This group consists of the 20 rice farms in West Java (SU, Table 15.7). Transplanted rice is grown in the dry and wet seasons in small fields (<0.5 ha) under tropical climate. Labor input averages 150 d ha^{-1} and pesticide use is moderate. The performance of SSNM in this domain was negatively affected by unfavorable climate as well as severe pest infestations in all four crops grown (see Chapter 9) so that yield goals were rarely achieved (Fig. 15.7). Average yields were the lowest among all domains (4.5 t ha^{-1}), although SSNM caused a large increase in N-use efficiency. For example, REN increased by almost 50% from 0.31 under FFP to 0.46 kg kg^{-1} under SSNM, but the extra N taken up was not converted into grain yield. The average internal efficiency of N was close to the line of maximum accumulation on both a grain yield or total dry matter basis (Fig. 15.8), demonstrating that stresses occurred during all stages of growth. Untimely water release into the irrigation system, rats, weeds, and insects (brown planthopper, stem borers) were among the major problems observed (Chapter 9).

Group 4: Direct-seeded rice. This group consists of 74 rice farms in Central Luzon (MA), Central Thailand (SB), and the Mekong Delta (OM, Table 15.7). Broadcast direct-seeded rice is grown in the dry and wet seasons in small to medium-sized fields (<0.5 to 1 ha) under tropical climate. High seed rates (100–200 kg ha^{-1}) are predominant and labor input is 15 to 60 d ha^{-1} only.[4] Pesticide use varies, but farmers generally use herbicides for weed control. Farmers apply little K. At the OM and SB sites, triple cropping of rice is common. Straw is mostly burned in the field. The performance of SSNM in this group was characterized by (1) low to moderate average rice yields of 4.8 to 5.3 t ha^{-1} and widely varying yield increases over FFP, (2) low grain-filling percentage of about 75%, (3) large increases in the use of K fertil-

---

[4] Farms in Central Luzon have both transplanted and direct-seeded rice. Labor use in transplanted rice is typically 60 to 90 d ha^{-1}.

Table 15.8. Selected agronomic performance characteristics in site-specific nutrient management (SSNM) plots that yielded within ±10% of the yield goal. Values shown are medians of 157 cases (1997-99) and include farms in all experimental domains and climatic seasons.

Factor	SSNM	Δ^a	
Grain yield (t ha^{-1})	6.84	0.44	(7%)
Total no. of spikelets m^2 (× 1,000)	32.2	1.9	(6%)
Filled spikelets (%)	84.6	0.4	(0.5%)
Fertilizer N use (kg ha^{-1})	110.0	−9.4	(−8%)
Fertilizer P use (kg ha^{-1})	15.0	−1.7	(−10%)
Fertilizer K use (kg ha^{-1})	53.4	23.2	(77%)
N uptake (kg ha^{-1})	112.9	12.6	(13%)
P uptake (kg ha^{-1})	23.4	3.1	(15%)
K uptake (kg ha^{-1})	113.1	9.1	(9%)
Internal efficiency of N (IEN, kg kg^{-1})	61.2	−2.6	(−4%)
Agronomic efficiency of N (AEN, kg kg^{-1})	18.4	5.2	(39%)
Recovery efficiency of N (REN, kg kg^{-1})	0.45	0.13	(41%)
Partial productivity of N (PFPN, kg kg^{-1})	59.7	8.5	(14%)

[a]Median of the differences between SSNM and the farmers' fertilizer practice (FFP) and percentage difference between SSNM and FFP.

izer compared with the FFP, (4) small to large increases in AEN and REN, and (5) widely varying quality of crop management. At all three sites, the average rice yield per unit of plant N was about 1 to 1.4 t ha^{-1} below that achievable with optimal internal N efficiency (Fig. 15.8A). When total dry matter was plotted versus N uptake, the difference between actual and optimal values diminished (Fig. 15.8B). We interpret this as evidence for yield losses that were mainly caused by abiotic or biotic stresses during reproductive growth. This is also supported by the low grain-filling percentage at all three sites, an empirical modeling approach discussed in Chapter 17, and earlier studies showing an inferior ability of direct-seeded rice to convert high biomass production into grain yield compared with transplanted rice (Schnier et al 1990). Unfavorable climate, water management problems, poor seed quality, weeds, insect pests, and diseases were common problems in these direct-seeded rice areas, particularly in wet-season crops (see Chapters 7, 8, and 10). Considering the large effect of climate during the study period (1997-99) at these sites, it remains to be clarified whether significantly different SSNM strategies are required for direct-seeded rice than for transplanted rice.

Agronomic characteristics on the best farms

What are realistic goals for increasing productivity and resource-use efficiency at the farm level with the currently available germplasm? Table 15.8 lists selected agronomic characteristics for roughly 25% of all cases in which the actual yield in the SSNM treatment was within ±10% of the yield goal. This subset represents situations where the response to N, P, and K was not much limited by other crop management factors and demonstrates what can be achieved under the widely differing environmental conditions across Asia. Note that this data set contained similar numbers of cases for seasons with low and high climatic yield potential. Average rice yield in this subset was 6.84 t ha^{-1}, which exceeded the overall SSNM mean by 23% and was equivalent to about 82% of the climatic yield potential. This elevated yield was associated with a high grain-filling percentage (85%), moderate use of N and P fertilizer, but greatly increased addition of fertilizer-K. The average internal efficiency of N was 61 kg kg^{-1}, which is close to the optimal values specified for rice grown at such yield levels (Witt et al 1999). The agronomic and recovery efficiency of N were 40% greater than with the current farmers' management. Another way to look at this is to select farmers who consistently performed better than average, that is, based on achieving the yield goal in all four crops grown. On 41 out of 179 farms, average yields of four rice crops grown were within ±10% of the yield goal. The average agronomic characteristics of this sample were similar to those shown in Table 15.8, even though this sample of farms was somewhat biased toward a few sites with transplanted rice and a high quality of crop management (HA, JI, AD).

We conclude that increasing average irrigated rice yields from the current 5.1 to 5.3 t ha^{-1} to about 6.8 t ha^{-1} is within reach of the currently available germplasm, which is equivalent to the yield increases required to sustain a sufficient rice supply in Asia until 2020 (Dobermann 2000). However, achieving such average rice yields is not an easy task and currently only a few farmers are able to do so. Growing rice at 70% to 80% of its genetic and climatic yield potential requires more fine-tuned, site-specific crop management than the blanket recommendations that accompanied the steep yield increases during the first phases of the Green Revolution in Asia. Adequate management of N, P, K, water, and pests appears to be a cornerstone for this.

15.5 Conclusions

On-farm monitoring data provided new evidence that current irrigated rice yields and nutrient-use efficiencies are well below levels that can be achieved with good management of existing germplasm. Average grain yields achieved by irrigated rice farmers in Asia were at about 60% of the yield potential. In typical production fields, losses of fertilizer-N appear to be high and there is increasing evidence of growth limitations caused by imbalanced nutrition. An insufficient indigenous K supply has become a yield-limiting factor in some areas. Most rice farmers were not aware of this change and had limited means for adjusting fertilizer rates according to the potential indigenous nutrient supply.

Site-specific management of macronutrients increased nutrient uptake, yields, and nutrient-use efficiency and improved the potassium input-output balance on most farms. Depending on the basis used for comparison, yield increases in diverse environments and climatic seasons averaged 0.4–0.5 t ha^{-1} (7–11%). The performance of SSNM did not differ significantly between HYS and LYS crops. Yield increases were achieved with no increase in the average N rate, but required better N management as well as larger amounts of fertilizer-K at sites where the present K use was low. Compared to current farmers' practices, N losses from fertilizer were typically reduced by 30–40%. Other, subtler positive effects will probably evolve slowly over time because improved plant nutrition will increase biomass production and therefore C sequestration in soil, reduce gaseous and leaching losses of N, and change the chemical composition of crop residues and thereby their decomposition patterns.

If SSNM were implemented on a regional or global scale, average yield increases of 7–11% would be sufficient for matching about 6 to 10 years of annual growth in rice demand in Asia. Yield increases were smaller than the increases in nutrient uptake and N-use efficiency, suggesting that further scope for improvement exists. Water management, crop establishment, and pest management must be fine-tuned to fully exploit the improved plant nutrition potential. If better control of factors other than N, P, and K is achieved, SSNM will probably become a key component for yield increases up to about 80% of the yield potential of currently available varieties, which is equivalent to the average farm yields needed by 2020. Although performance differences were large among sites, SSNM can be considered a low-risk technology for most environments.

The NPK modeling and management concept used in our studies with rice is applicable to other cropping systems, particularly those with irrigation. It can be used for managing plant nutrients at any scale, that is, ranging from a general recommendation for homogeneous management of a larger domain to true site-specific nutrient management of within-field variability. The next steps should involve (1) developing essential tools for implementing SSNM at the farm level, (2) integrating SSNM with other components of crop management to develop integrated site-specific crop management solutions, and (3) studying medium- to long-term agronomic and economic performance in larger areas. Collaboration among research institutions, nongovernment organizations, the public extension sector, and the private sector will be required to achieve a widespread adoption of improved, more knowledge-intensive management technologies.

References

Bouldin DR. 1986. The chemistry and biology of flooded soils in relation to the nitrogen economy in rice fields. Fert. Res. 9:1-14.

Bray RH. 1954. A nutrient mobility concept of soil-plant relationships. Soil Sci. 78:9-22.

Cassman KG, Gines HC, Dizon M, Samson MI, Alcantara JM. 1996. Nitrogen-use efficiency in tropical lowland rice systems: contributions from indigenous and applied nitrogen. Field Crops Res. 47:1-12.

Cassman KG, Peng S, Olk DC, Ladha JK, Reichardt W, Dobermann A, Singh U. 1998. Opportunities for increased nitrogen use efficiency from improved resource management in irrigated rice systems. Field Crops Res. 56:7-38.

Colwell JD. 1994. Estimating fertilizer requirements: a quantitative approach. Wallingford (UK): CAB International.

Dawe D, Moya P. 1999. Variability of optimal nitrogen applications for rice. In: Program report for 1998. Makati City (Philippines): International Rice Research Institute. p 15-17.

De Datta SK, Buresh RJ. 1989. Integrated nitrogen management in irrigated rice. Adv. Soil Sci. 10:143-169.

de Wit CT. 1992. Resource use efficiency in agriculture. Agric. Syst. 40:125-151.

Dobermann A. 2000. Future intensification of irrigated rice systems. In: Sheehy JE, Mitchell PL, Hardy B, editors. Redesigning rice photosynthesis to increase yield. Makati City (Philippines): International Rice Research Institute, Amsterdam (Netherlands): Elsevier Science. p 229-247.

Dobermann A, Cassman KG, Mamaril CP, Sheehy JE. 1998. Management of phosphorus, potassium and sulfur in intensive, irrigated lowland rice. Field Crops Res. 56:113-138.

Dobermann A, Dawe D, Roetter RP, Cassman KG. 2000. Reversal of rice yield decline in a long-term continuous cropping experiment. Agron. J. 92:633-643.

Dobermann A, Oberthür T. 1997. Fuzzy mapping of soil fertility: a case study on irrigated riceland in the Philippines. Geoderma 77:317-339.

Greenland DJ. 1997. The sustainability of rice farming. Oxon (UK) and Manila (Philippines): CAB International, International Rice Research Institute. 273 p.

Havlin JL, Beaton JD, Tisdale SL, Nelson WL. 1999. Soil fertility and fertilizers: an introduction to nutrient management. 6th ed. Upper Saddle River, N.J. (USA): Prentice Hall. 499 p.

Horie T, Nakagawa H, Centeno HGS, Kropff MJ. 1995. The rice crop simulation model SIMRIW and its testing. In: Matthews RB, Kropff MJ, Bachelet D, van Laar HH, editors. Modeling the impact of climate change on rice production in Asia. Wallingford (UK): CAB International, International Rice Research Institute. p 51-66.

IRRI. 1993. Rice almanac. Los Baños (Philippines): International Rice Research Institute (IRRI). 142 p.

Janssen BH, Guiking FCT, Van der Eijk D, Smaling EMA, Wolf J, van Reuler H. 1990. A system for Quantitative Evaluation of the Fertility of Tropical Soils (QUEFTS). Geoderma 46:299-318.

Kropff MJ, van Laar HH, ten Berge HFM. 1993. ORYZA1: A basic model for irrigated lowland rice production. Los Baños (Philippines) and Wageningen (Netherlands): International Rice Research Institute and CABO. 89 p.

Matthews RB, Kropff MJ, Bachelet D, van Laar HH. 1995. Modeling the impact of climate change on rice production in Asia. Wallingford (UK): CAB International and the International Rice Research Institute.

Olk DC, Cassman KG, Simbahan GC, Sta. Cruz PC, Abdulrachman S, Nagarajan R, Tan PS, Satawathananont S. 1999. Interpreting fertilizer-use efficiency in relation to soil nutrient-supplying capacity, factor productivity, and agronomic efficiency. Nutr. Cycl. Agroecosyst. 53:35-41.

Peng S, Cassman KG. 1998. Upper thresholds of nitrogen uptake rates and associated N fertilizer efficiencies in irrigated rice. Agron. J. 90:178-185.

Peng S, Garcia FV, Laza RC, Sanico AL, Visperas RM, Cassman KG. 1996. Increased N-use efficiency using a chlorophyll meter on high-yielding irrigated rice. Field Crops Res. 47:243-252.

Pierce FJ, Nowak P. 1999. Aspects of precision agriculture. Adv. Agron. 67:1-85.

SAS Institute Inc. 1988. SAS/STAT user's guide, release 6.03 edition. Cary, N.C. (USA): SAS Institute Inc. 1,028 p.

Schnier HF, Dingkuhn M, De Datta SK, Mengel K, Faronilo JE. 1990. Nitrogen fertilization of direct-seeded flooded vs. transplanted rice. I. Nitrogen uptake, photosynthesis, growth, and yield. Crop Sci. 30:1276-1284.

Vlek PLG, Byrnes BH. 1986. The efficacy and loss of fertilizer N in lowland rice. Fert. Res. 9:131-147.

Witt C, Dobermann A, Abdulrachman S, Gines HC, Wang GH, Nagarajan R, Satawathananont S, Son TT, Tan PS, Tiem LV, Simbahan GC, Olk DC. 1999. Internal nutrient efficiencies of irrigated lowland rice in tropical and subtropical Asia. Field Crops Res. 63:113-138.

Wopereis MCS, Donovan C, Nebie B, Guindo D, N'Diaye MK. 1999. Soil fertility management in irrigated rice systems in the Sahel and Savanna regions of West Africa. I. Agronomic analysis. Field Crops Res. 61:125-145.

Notes

Authors' addresses: A. Dobermann, G.C. Simbahan, M.A.A. Adviento, C. Witt, International Rice Research Institute, Los Baños, Philippines; A. Dobermann, Department of Agronomy and Horticulture, University of Nebraska, P.O. Box 830915, Lincoln, NE 68583-0915, Nebraska, USA, email: adobermann2@unl.edu; S. Abdulrachman, Research Institute for Rice, Sukamandi, Indonesia; H.C. Gines, Philippine Rice Research Institute, Maligaya, Nueva Ecija, Philippines; R. Nagarajan, Soil and Water Management Research Institute, Thanjavur, India; S. Satawathananont, Pathum Thani Rice Research Center, Pathum Thani, Thailand; T.T. Son, National Institute for Soils and Fertilizers, Hanoi, Vietnam; P.S. Tan, Cuu Long Delta Rice Research Institute, Omon, Cantho; G.H. Wang, Zhejiang University, Hangzhou, China.

Citation: Dobermann A, Witt C, Dawe D, editors. 2003. Increasing productivity of intensive rice systems through site-specific nutrient management. Enfield, N.H. (USA) and Los Baños (Philippines): Science Publishers, Inc., and International Rice Research Institute (IRRI). 410 p.

16 Nutrient management in the rice soils of Asia and the potential of site-specific nutrient management

D. Dawe, A. Dobermann, C. Witt, S. Abdulrachman, H.C. Gines, R. Nagarajan,
S. Satawathananont, T.T. Son, P.S. Tan, and G.H. Wang

16.1 Nutrient management and the Asian rice economy

Nutrient management in rice-growing Asia has changed dramatically in the past 35 years. Before the Green Revolution, organic fertilizers were the primary source of nutrients throughout the region. Because organic fertilizers require relatively large amounts of labor, they were used especially in areas with high labor-to-land ratios, such as many parts of China and India. They were historically less common in much of Southeast Asia, where land was more abundant (Barker et al 1985). In the future, labor costs in Asia will probably continue to rise in line with continued economic development. Thus, the influence of "labor" availability (in terms of both unskilled labor and the time needed to acquire knowledge) on nutrient management decisions will be an important recurring theme in this chapter.

The reliance on organic fertilizers changed rapidly, however, beginning in the mid-1960s (the process began much earlier in Japan). The development of fertilizer-responsive modern varieties increased farmers' demand for nutrients dramatically, and innovations in the production and marketing of inorganic fertilizers resulted in equally dramatic increases in supply. The combination of increased demand and supply generated a tremendous surge in consumption, especially of nitrogen (N). This change in farmer behavior was one of the most significant changes in the natural resource management practices of farmers in Asian history.

N is clearly the most growth-limiting nutrient in nearly all rice-growing areas, so it is rightfully the primary focus of most general discussions on nutrient management. Yet, this focus on N has obscured other substantial changes that have occurred in soil management and farmers' knowledge in the recent past. With the heavy use of N, phosphorus (P) and potassium (K) can also become limiting nutrients, and farmers throughout Asia apply P and K fertilizers to solve this problem, although perhaps in suboptimal amounts. Furthermore, farmers understand, at least implicitly, that these nutrients are more effectively stored in the soil than N, as evidenced by their greater willingness to temporarily abstain from using P and K because of short-term economic factors (see Chapter 3). In many parts of Asia (e.g., Tamil Nadu and Uttar Pradesh in India), zinc deficiency can be a problem and farmers apply doses of zinc sulfate to alleviate this constraint. Sulfur is limiting in other areas, such as Bangladesh, and farmers have responded to this problem as well. In many areas of the Punjab, farmers apply gypsum to guard against alkalinity problems. In Xinjiang, China, farmers grow rice once every three years in order to flood the soil and leach salinity out of the soil, a classic instance of integrated soil and water management. In the Mekong Delta, Vietnamese farmers adjust their use of inorganic fertilizer based on the amount of nutrient-bearing silt deposited during the annual floods. All of these examples show that farmers, while not perfect, have become experts in soil management and are serving well as guardians of the world's natural resources. This speaks eloquently to the many past successes of natural resource management research and extension in agricultural research institutions around the world.

Just as it has in the past, nutrient management will continue to play a key role in the rice economy of Asia during the next 20 years. First, because of continued population growth, yields will need to increase to keep rice prices from rising for poor consumers in both urban and rural areas. Yet, rice yields have not increased during the past decade for many countries (Indonesia and the Philippines) and regions (Punjab) where the Green Revolution was adopted relatively early (Cassman and Dobermann 2000). At the same time, actual farm yields are below what might be achieved with better nutrient management, as shown in Chapter 14. Thus, better nutrient management might be able to restore some momentum to rice yield growth in the coming decades.

Second, reductions in N use may be possible in many areas without any sacrifice in yields. This would improve farm profitability to some extent (lower input costs), perhaps even to a large extent in areas where N fertilizer use is very high, as in China. In general, however, a given percentage increase in yield will do much more for profitability than a similar percentage reduction in N use because the ratio of N costs to gross revenue from paddy is typically 8% or less (Dawe 2000). Reduced use of N fertilizer might also generate off-farm environmental benefits.

16.2 Alternative nutrient management technologies

By far the most common approach to fertilizer management on irrigated rice farms in Asia is simple broadcasting of inorganic NPK fertilizers. This is a viable method because the associated labor costs are low and because these inorganic fertilizers are widely available and much cheaper today than ever before. Figure 16.1 shows that the inflation-adjusted prices of urea, triple superphosphate (TSP), and muriate of potash (MOP) have been declining for many years now. These price declines have occurred because supply increased faster than demand because of technological developments in the production of fertilizer.

Because these traditional application methods often lead to low nutrient-use efficiencies, researchers and governments have promoted many alternative nutrient and soil management technologies. One option promoted by many is urea tablets or briquettes, which release nitrogen more slowly and evenly during the course of plant growth, thus reducing nitrogen losses and improving N-use efficiency (Singh et al

Price (year 2000 US$ t^{-1})

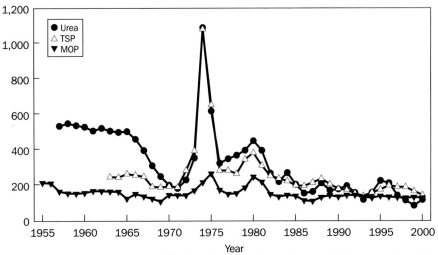

Fig. 16.1. Long-term trends in inflation-adjusted prices of urea, triple superphosphate (TSP), and muriate of potash (MOP).

1995). These have been tried extensively in Indonesia and Bangladesh in recent years. In Bangladesh, some decentralized manufacture of urea briquettes is occurring on a commercial basis in the private sector and adoption by farmers is increasing. Recent figures indicate that this technology is being used on perhaps 30,000 ha, still a small proportion of the total rice area of about 10 million ha (Hossain, personal communication). In Indonesia, many farmers adopted urea tablets (Pasandaran et al 1999), but it is not clear how much of this adoption was voluntary and how much was forced upon farmers through restrictions on the availability of prilled urea (Cohen 1997). Urea tablets were originally promoted in Indonesia by a company controlled by family members of ex-President Suharto, but, with the change in government, urea tablets are now less easily available. One constraint to the wider adoption of tablets is the high labor cost associated with their use, but research is being conducted to design improved applicators and application mechanisms that mitigate this constraint (Savant and Stangel 1995). The high labor cost reduces the applicability in areas with higher wages and broadcast direct seeding, such as the Central Plain (Thailand), the Mekong Delta (Vietnam), and Central Luzon (Philippines). In Indonesia, where farmers used them more commonly a few years ago, it was common practice to grind the tablets first and then broadcast them in the traditional manner, or simply to broadcast the tablets. In Bangladesh, it is an open question as to how long use of this technology will be sustained if economic development raises wages substantially.

Slow-release and controlled-release fertilizers are more expensive versions of urea tablets and briquettes because they are coated with sulfur or polymers that provide more control over the release of nitrogen to the plant. These fertilizers are now quite expensive, and are used only in more developed countries. Even there, they are usually used only on high-value crops such as fruits and vegetables, not on cereal crops.

The use of farmyard manure to increase or maintain soil organic matter (SOM) content is another approach recommended by some researchers to improve soil quality and promote plant growth. Although some farmers use organic manure for rice (including farmers in northern Vietnam and many parts of India), the high labor cost of gathering and application has limited more widespread adoption (or greater use in the cases where it is applied). Use declined sharply in Japan as wages increased because of economic development and farmers changed from full-time farming to part-time farming (Kanazawa 1984). Thus, as wages continue to increase in other Asian countries, the use of organic manure on rice will probably decline even further in the future. Even in areas where wages are relatively low (e.g., India), organic manure is often used as fuel and many farmers prefer this use instead of application to the crop. It is also important to realize that maintenance of SOM content in double- and triple-crop rice systems is usually easy to achieve without adding organic manure because SOM tends to accumulate under wet conditions (Dobermann and Witt 2000, Witt et al 2000). Maintenance of SOM is potentially a larger problem in rice-upland crop systems, however.

Green manure (e.g., azolla or sesbania) is sometimes touted as an alternative source of N that is potentially cheaper than mineral sources and would lead to more sustainable rice production. Ali (1999) shows that green manures are not likely to be economically acceptable to farmers, however, because of the labor cost and the opportunity costs of using land to grow the green manure crop. Ladha and Reddy (2000) and Cassman et al (1996) analyze experiments in which yield trends in plots with green manure are virtually identical to yield trends in plots using only urea as a source of N, thus casting doubt on the hypothesis of enhanced sustainability.

Grain legumes can also supply residual N to a succeeding rice crop. They may have a role in some systems where there is a fallow period with adequate residual moisture and an appropriate temperature regime that makes such cultivation economically attractive to farmers (Ali 1999).

The creation of rice plants with a greater capacity for biological nitrogen fixation (BNF) might also reduce the need to apply mineral sources of N, thus allowing farmers to save on input costs. Many scientists are optimistic that research progress in this area will continue (Ladha and Reddy 2000), but the development of N_2-fixing cereals is still many years away. According to the Kendall report to the World Bank (Kendall et al 1997), "At some point in the future, N fixation may be transferred to crops such as maize and rice, but such an achievement must be seen as a far-off goal." Key issues in this research will be whether such plants carry with them a yield penalty because of the carbohydrates required for BNF, and whether the addition of mineral N will suppress BNF, as occurs in soybeans.

The site-specific nutrient management (SSNM) approach described in this book is an attempt to tailor nutrient management to local conditions while avoiding some of the constraints that hamper adoption of the strategies discussed above (e.g., high labor use, expensive fertilizers). Although SSNM is still evolving, three main guiding principles lie behind this approach. First is an emphasis on balanced nutrition using estimates of nutrients lost through crop removal. Second is the application of fertilizers based on native soil fertility, as determined by nutrient omission plots. Third is the improved timing of N applications based on leaf color (e.g., the leaf color chart or a SPAD meter) at specific growth stages. To some extent, farmers' current nutrient management practices are already site-specific, as the examples given in the first section of this chapter make clear. But the SSNM approach is an attempt to systematize and improve upon these already existing farmers' practices. The use of this approach has generally suggested an increased use of P and K (especially the latter), an extra split application of N, different timing of N applications, and a reduced quantity of N use, although the specifics of SSNM at each site will vary. The need for more balanced nutrition has been stressed by many scientists (e.g., FAO 1998) and may be due to overemphasis on the use of N during the Green Revolution, which led to the mining of key soil nutrients. Some degree of mining of soil P and K may have been economically justified in the past, but farmers may now be reaching a point where replenishment is necessary. Improved splitting of N applications often leads to lower N use by matching the supply of N fertilizer with crop demand (Peng et al 1996).

Because many nonagronomic factors (e.g., availability of labor and time) are important considerations in the decision to adopt improved nutrient management strategies, this chapter will discuss some of the socioeconomic factors that are relevant for the potential adoption of SSNM techniques by farmers.

16.3 Factors affecting technology adoption

In general, many factors influence farmers' decisions regarding the adoption of a new technology or management strategy. Perhaps the most important consideration is financial profitability. If a technology is not financially profitable for farmers, it is not likely to be widely adopted. But assessing the potential for technology adoption is not as simple as measuring financial profitability: profitability is only a necessary, not a sufficient, condition. Other important considerations include the observability of the technology's benefits, risk, credit requirements, the need for coordination with other farmers, the complexity of the technology, and the opportunity cost of farmers' time, all evaluated within the context of farmers' knowledge. This list is by no means exhaustive, but these factors are certainly some of the most important that need to be considered.

A related issue is whether society should encourage the adoption of a new technology or management strategy. This decision depends on more than just the direct effects on farmers. It also depends on the indirect effects of the technology on third parties. In the absence of any third-party effects (externalities in economics jargon) or market distortions, the optimal decision for farmers on adoption is also the optimal decision for society. However, if there are costs to other parties and society, it is possible that the adoption of a technology is optimal from the farmers' point of view but not for the larger society in which the farmers operate. On the other hand, if the farmers do not capture the external benefits of a technology, it is possible that society should encourage the farmers to adopt a technology that is not necessarily in their own short-term private interest. Examples of this are restrictions on N-fertilizer use because of problems with nitrate contamination of groundwater (as has occurred with maize in Nebraska), restrictions on straw burning because of problems with air pollution (rice in California), and restrictions on the use of organic manure in Western Europe.

The objective of this paper is to assess the costs and benefits of SSNM from the viewpoint of both farmers and society at large, and to draw implications for further research that might influence the adoption process.

16.4 Financial profitability

As noted earlier, financial profitability of a technology is a necessary condition for widespread adoption by farmers. If the technology does not generate additional profits for farmers, there is no incentive to adopt the technology and adoption most likely will not occur.

Because the Reversing Trends of Declining Productivity (RTDP) project did not collect data on land rental costs and because of the difficulties in imputing costs to family labor, it is not possible to calculate the absolute level of profit with and without SSNM. This is not a major drawback since the absolute level of profits is less important than the *change* in profits because of the adoption of the technology. The incremental profitability of SSNM (which can be positive or negative) is measured as the difference in gross returns caused by different grain yields in the two treatments (SSNM and farmers' fertilizer practice, or FFP) minus the change in total fertilizer costs because of different fertilizer usage in the two treatments. This is equivalent to gross returns above fertilizer costs in SSNM (GRF_{SSNM}) minus GRF_{FFP} (see equations 2.6 to 2.11 in Chapter 2 for the algebraic expressions).

In this chapter, average regional prices are used for rice (paddy), N, P, and K. Specifically, the prices used are US$0.15 kg^{-1} for rice, $0.35 kg^{-1} for N, $1.10 kg^{-1} for P, and $0.30 kg^{-1} for K. This is in contrast to the individual country chapters, which each use site-specific prices for the profitability analysis. The use of average regional prices in this paper allows us to focus on differences across sites that are due to the technology itself. National policies affecting rice prices and input prices also affect the profitability of a new technology and both factors are in principle important for technology adoption. To focus on the technology itself (SSNM), however, it seems better to avoid a discussion of idiosyncratic national pricing policies, especially since the effect of different national policies on profitability is not large. In interpreting the results, however, readers should be aware that the results in this chapter are not simple aggregations of the results from the individual country chapters. The analysis below includes data from 179 different farms, just as in Chapter 14 (i.e., no data from Pantnagar are included here).

Attributing meaning to the calculation of profit (equation 2.11) implicitly assumes that the only difference in crop management between SSNM and FFP is different quantities of nutrients and different timing of a certain constant number of applications, so that all other management practices and quantities of input use are held constant. Generally speaking, this assumption holds, but three exceptions are worth noting. First, the number of fertilizer applications was different in the SSNM and FFP treatments. Across all sites and crops, the average number of fertilizer applications per crop in the SSNM treatment was 3.83 versus 2.74 in the FFP, a difference of approximately one application per crop (the difference is statistically significant at $P<0.01$). The difference in the number of N applications was smaller at just 0.54 per crop ($P<0.01$). (There were 3.13 N applications per crop on average in the SSNM treatments versus 2.59 in the FFP.) In theory, it should be possible to attribute a labor cost to the time spent in applying an extra dose of fertilizer and incorporate this figure into the profitability calculation. This figure is small, however, since it takes only about 6 h ha^{-1} to apply a dose of fertilizer at most sites. Since this is small relative to the profitability of SSNM, no explicit correction was made for this effect. Nevertheless, farmer surveys need to be carried out to verify whether farmers perceive an extra application of fertilizer as a significant shortcoming of SSNM.

The second exception concerns plant spacing in Zhejiang, China, for the late rice crop in 1999. Plant density was 19 hills m^{-2} in the FFP treatment but 22 hills m^{-2} in the SSNM treatment, a difference of 18%. The higher plant density in the SSNM treatment implies a commensurate percentage increase in seed cost and labor cost for transplanting that amount to approximately \$25 ha^{-1}. This compares to a ΔGRF (gross returns above fertilizer cost) of \$120 ha^{-1} for this particular crop at this site. Because profitability for this specific crop is still substantial even after accounting for the additional costs, and because the average profitability across all 32 crops at eight sites is affected only slightly, this cost adjustment is not incorporated into the estimates reported below.

The third important exception was also unique to one site, northern Vietnam. In these treatments, substantially more farmyard manure (FYM) was applied to the FFP treatments than to the SSNM treatments because of the difficulty researchers had in obtaining supplies of FYM. In fact, no manure was applied at all to the SSNM in the first season (see Chapter 11 for details). It is difficult to adjust the profitability calculations for this effect, however. The financial cost of obtaining FYM in this area is zero since it is obtained on-farm. The opportunity cost of FYM is also zero since there is no market on which farmers can sell FYM. The application of FYM used on average nearly 20 person-days ha^{-1} crop^{-1} compared with total labor use of 260 person-days ha^{-1}. This is a nontrivial amount of labor (nearly 8% of the total), but the opportunity cost of labor at this site is probably close to zero since the FYM is applied using family labor that has no obvious alternative employment for such a short period of time. Thus, no adjustment was made for the additional FYM used in the FFP plots in northern Vietnam. If an adjustment were made, it would increase the profitability of SSNM relative to FFP.

Table 16.1 Effect of site-specific nutrient management on fertilizer costs and gross returns above fertilizer costs from rice production, 1997-99.

Levels[a]		Treatment[b]		Δ[c]	$P > \|T\|$[c]	Effects[d]	$P > \|F\|$[d]
		SSNM	FFP				
Fertilizer cost	All	78.0	69.8	8.2	<0.001	Village	<0.001
(US$ ha^{-1})	Year 1	84.5	69.4	15.1	<0.001	Year[e]	<0.001
	Year 2	71.7	70.2	1.5	0.367	Season[e]	0.636
	HYS	81.3	72.8	8.5	<0.001	Year × season[e]	0.018
	LYS	74.7	66.7	8.0	<0.001	Village × crop	<0.001
Gross returns above	All	754	708	46.2	<0.001	Village	<0.001
fertilizer costs	Year 1	750	718	31.6	0.032	Year[e]	<0.001
(US$ ha^{-1})	Year 2	758	697	60.6	<0.001	Season[e]	0.404
	HYS	821	771	49.9	<0.001	Year × season[e]	0.523
	LYS	686	644	42.5	0.005	Site × crop	<0.001

[a]All = all four crops grown, HYS = high-yielding season, LYS = low-yielding season. [b]SSNM = site-specific nutrient management; FFP = farmers' fertilizer practice. [c]Δ = SSNM − FFP, $P > \|T\|$ = probability of a significant mean difference between SSNM and FFP. [d]Source of variation of analysis of variance of the difference between SSNM and FFP by farm; $P > \|F\|$ = probability of a significant F-value. [e]Year refers to two consecutive cropping years.

The financial profitability calculations also implicitly assume that the FFP treatment is representative of farmers' practices in the area. This was generally true, but not always. For example, in Maligaya and Omon, certified seed was used in both the SSNM and FFP plots based on the judgment of researchers that seed quality was a problem in these areas. Farmers in these areas typically do not use certified seed, primarily because the reliability of the current seed marketing system is suspect. If the financial profitability of SSNM depends on the use of certified seed, then the dissemination of SSNM will encounter difficulties because the creation of a viable seed marketing system is a long and difficult task. Although there is no strong evidence that SSNM performs substantially better with certified seed, future trials of SSNM should take more into consideration farmers' ability to implement the recommended strategy. For other examples of practices that were applied in both SSNM and FFP but were not necessarily standard farmers' practices, see the section titled "Other crop management practices" in Chapter 5.

Nutrient omission plots (−F, +NP, +NK, +PK) were also used in farmers' fields to gather data for implementing SSNM. The combined size of these plots for an individual field was small, however (typically 0.01 ha), and would not be a major constraint to adoption in terms of foregone yield. Construction of these plots may be a burden in terms of complexity, an issue that will be discussed later in the chapter.

On average, across all sites and for all four crops, profitability increased by $46.20 ha^{-1} crop^{-1} (Table 16.1). This increase in profitability compares to an average net return (total value of production minus total costs) of about $400 ha^{-1} crop^{-1} (average annual net return across sites from Table 3.14 divided by two). Thus, on average, SSNM would increase the returns accruing to land and farmers' management by about 12%. This is a respectable improvement in returns to farmers.

The above figure is not a measure of profitability as a percentage return on investment. To make this calculation, it is assumed that fertilizer costs are incurred up front before planting, that returns are realized after harvest 4 months later, and that the first year of SSNM has higher costs and lower returns using the numbers in Table 16.1 (see discussion below for the reasons behind this pattern). Under these assumptions, the internal rate of return (IRR) of investing in SSNM is about 24% per month, or more than 1,300% per year. This clearly indicates a very profitable activity, but the IRR calculation is somewhat misleading. First, the IRR is not defined in cases where the fertilizer costs of SSNM are lower than in FFP, since in such cases no investment is required (assuming that grain yields in SSNM are higher than in FFP). Thus, the IRR calculation can be used only for averages, not for all individual farmers in the sample. More important, it is not possible for farmers to earn this high a rate of return except on very small quantities of money. For example, a farmer that invests $10 ha^{-1} in additional fertilizer costs and generates additional revenues of $50 ha^{-1} 4 months later would earn a monthly rate of return of 400%. Yet, this 400% rate of return can be earned only on an investment of $10 ha^{-1}. It is not possible to scale up the technology *within an individual field* to invest an additional $1,000 ha^{-1} in fertilizer costs and generate $5,000 ha^{-1} in additional revenue. Because of these considerations, it is preferable for our purposes to measure profitability in absolute terms. Thus, the rest of the paper will discuss only changes in absolute levels of financial profitability.

The increase in profitability was different across years (i.e., comparing crops 1 and 2 with crops 3 and 4), as shown by the statistical significance of the crop-year effects for GRF in Table 16.1. Profitability was just $32 ha^{-1} crop^{-1} in the first year, followed by $61 ha^{-1} crop^{-1} in the second year. The increase in profitability over time was due to two main factors in roughly equal measures. First, the technology was more effective at increasing grain yields in the second year, with a yield advantage of 0.31 t ha^{-1} in the first year and 0.41 t ha^{-1} in the second year (difference statistically significant at $P = 0.016$, see Chapter 15, Table 15.4). Second, the SSNM strategy invested more heavily in recapitalizing soil P and K (especially the latter) in the first year. Table 15.5 shows that SSNM plots used on average 35.1 kg K ha^{-1} more than FFP plots in year 1, with the difference falling to 20.5 kg K ha^{-1} in year 2. For P, SSNM used 4.7 kg P ha^{-1} more than FFP in year 1, but used less P in year 2. Because of this initial effort at soil recapitalization, total fertilizer costs (TFC) in the SSNM plots exceeded those in the FFP plots by $15.10 ha^{-1} crop^{-1} in year 1, despite a slightly lower N use in the SSNM treatments (Table 16.1). By the second year, however, fertilizer costs in the two treatments were nearly identical, with SSNM costs exceeding FFP costs by just $1.50 ha^{-1} crop^{-1}.

The increased profitability in the second year because of lower differential fertilizer costs of SSNM indicates that profitability will increase in the longer term as the benefits of soil recapitalization are felt. It is possible to incorporate estimates of these benefits into a long-run IRR, but the IRR even after two years already exceeds 1,000%, and the shortcomings of using the IRR for this technology were discussed

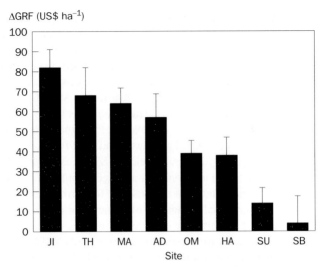

ΔGRF (US$ ha⁻¹)

Fig. 16.2. Difference in financial profitability (ΔGRF, gross returns above fertilizer cost) between SSNM and FFP treatments in US$ ha⁻¹. The height of the columns represents the average profitability over four crops for each site, 1997-99. MA = Maligaya, Central Luzon (Philippines); SB = Suphan Buri, Central Plain (Thailand); OM = Omon, Mekong Delta (Vietnam); SU = Sukamandi, West Java (Indonesia); AD = Aduthurai, Old Cauvery Delta, Tamil Nadu (India); TH = Thanjavur, New Cauvery Delta, Tamil Nadu (India); HA = Hanoi, Red River Delta (Vietnam); JI = Jinhua, Zhejiang (China).

earlier. Thus, we need to keep in mind that the benefits of using SSNM will probably continue to accrue into the future.

Sensitivity analysis was also conducted to investigate the effects of different prices for paddy and N on financial profitability. For example, if paddy prices were to decline, this would lower the value of the higher yields produced by SSNM. Thus, the increase in profitability because of SSNM was simulated with paddy prices 20% lower than the value of $0.15 kg⁻¹ used earlier (i.e., a paddy price of $0.12 kg⁻¹ was used). A 20% drop in paddy prices reduces profits by 30% in the first year to $22.20 ha⁻¹ crop⁻¹. Second-year profits fall to $48.20 ha⁻¹ crop⁻¹, a decline of just 20.5%, consistent with the fact that fertilizer costs under SSNM are virtually identical to those under FFP in the second year. (If fertilizer costs under SSNM and FFP were identical, then a 20% drop in paddy prices would reduce profits by exactly 20%.) Thus, lower paddy prices reduce the incentives for farmers to adopt SSNM, especially in the first year, but the technology remains profitable.

16.5 Differential performance of SSNM among sites

The profitability of SSNM also differed across sites, as shown by the significance of the site effects for GRF (Table 16.1). SSNM was the most profitable at the sites in

China, southern India, and the Philippines (Fig. 16.2). Differential profitability ranged from $57 to $82 ha^{-1} crop^{-1} in these areas. The sites in Vietnam (southern and northern) exhibited intermediate levels of profitability at $38–39 ha^{-1} crop^{-1}. Had the use of FYM been the same in the SSNM and FFP treatments, northern Vietnam would have probably exhibited higher levels of profitability on a par with those in the first group. For example, the FFP treatment was substantially more profitable than SSNM for the first crop in northern Vietnam (when no FYM was applied in the SSNM). If this one crop is excluded from the analysis, then average profitability in northern Vietnam is above that at Aduthurai and almost equal to that at Maligaya. Thus, generally speaking, SSNM was quite profitable at six of the eight sites where on-farm trials were conducted.

The two sites where SSNM generated only small increases in profits were Indonesia ($14 ha^{-1} crop^{-1}) and Thailand ($4 ha^{-1} crop^{-1}). In Indonesia, this may have been due to heavy pest attacks, especially for the first two crops. These crops may also have suffered from El Niño-induced disruptions in water releases in the area, as well as problems in the design of SSNM for this site (unrealistically high yield targets and an initially poor N management scheme; see Chapter 9). For the third and fourth crops, profitability was much improved at $33 ha^{-1} crop^{-1}. In Thailand, profitability of SSNM was consistently low and was negative for two of the four crops. On average across four crops, the average profitability of SSNM was just $4 ha^{-1}. Serious problems occurred with weeds, insects, and disease at this site. Also, farmers at this site used low amounts of labor (meaning perhaps less attention to crop care) and had low levels of education. It is not known to what extent these factors contributed to the poor performance of SSNM.

Site by crop effects were also significant for profitability, indicating that SSNM achieved very high profitability in certain sites and seasons, but much lower profitability in other instances. Although the profitability varied significantly, it was nearly always positive for any given site and season. The only exceptions in 32 cases (four seasons in each of eight sites), for reasons explained earlier, were the second and fourth seasons at Suphan Buri and the first season in both Sukamandi and Hanoi.

16.6 Differential performance of SSNM among farms

Although SSNM is profitable on average for most farmers at all sites, important questions relate to differential performance across farmers. Does SSNM raise yields more for farmers with below-average yields? Although SSNM does not increase profits in every on-farm trial every year, does it increase profits on average for all farmers after several crops have been grown? Or does it fail to achieve profitability on average for some farmers, even after several crops have been harvested? If the latter, how easy is it to identify these farmers, so that the technology can be effectively targeted without harming their profitability?

A priori, it might be reasonable to suppose that SSNM would increase yields more for farmers whose yields are usually lower than those of other members of the group, the implicit assumption being that these lower yields are due at least in part to

Table 16.2. Regression results. Dependent variable is average difference in grain yield between site-specific nutrient management (SSNM) and farmers' fertilizer practice (FFP) treatments for a given farmer. Independent variable is FFP yield for a given farmer.

Site	Coefficient	P-value	r^2
Central Luzon, Philippines	–0.24	0.18	0.08
Central Plain, Thailand	–0.01	0.97	0.00
Mekong Delta, Vietnam	–0.04	0.80	0.00
West Java, Indonesia	–0.04	0.66	0.01
Tamil Nadu, India	–0.39	0.02	0.26
Red River Delta, Vietnam	–0.03	0.45	0.03
Zheijang, China	–0.02	0.88	0.00

poor nutrient management. To some extent, this is indeed true. A regression of the average increase in grain yield because of SSNM ($\Delta GY = GY_{SSNM} - GY_{FFP}$) for a given farmer against that farmer's average yield consistently gave a negative coefficient on the independent variable at all sites. However, the magnitude of that coefficient was typically quite small, with the P-value not allowing for a rejection of the null hypothesis of no effect (Table 16.2). The main exception was Aduthurai, where the coefficient was statistically significant at the 5% level and an average yield that was lower by 1 t ha^{-1} was associated with a larger yield increase because of SSNM of 0.39 t ha^{-1}. At most of the other sites, however, an average yield that was lower by 1 t ha^{-1} was associated with a larger yield increase because of SSNM of less than 0.05 t ha^{-1}.

To better understand the consistency of increased profitability among different farmers, two histograms were compared. One is a histogram of the absolute increase in profits using one crop for a specific farmer as the unit of observation (this will be termed the crop-specific histogram). The other is a histogram of the absolute increase in profits using the average level of profits over four crops for a specific farmer as the unit of observation (this will be termed the farmer-specific histogram). This second histogram gives a sense of how often SSNM is not profitable even when averaged over four crops. Figure 16.3 shows that SSNM was not profitable 28% of the time when considering specific crops. This figure is by necessity lower for the farmer-specific histogram (since extreme negative events do not always happen to the same farmer), but it remains relatively high at 22%, even when averaged over four crops. A little more than one-third of the farms where SSNM was not profitable were at Suphan Buri, but the rest of the cases were relatively evenly distributed across sites.

These data may suggest that SSNM does not work for certain small groups of farmers, and this is an important avenue for future research. One possibility is that a certain minimum level of crop care is required for SSNM to be profitable. This hypothesis deserves further investigation. The other implication of these data is that, for the farms where SSNM increases profitability on average, it appears to do so

Frequency (%)

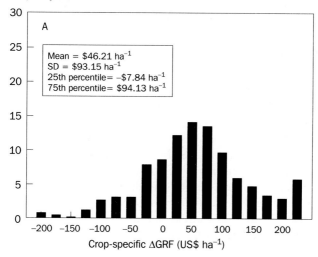

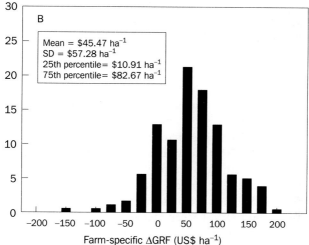

Fig. 16.3. Histograms of (A) crop-specific and (B) farm-specific differences in financial profitability (ΔGRF, gross returns above fertilizer cost) between site-specific nutrient management (SSNM) and farmers' fertilizer practice (FFP) treatments in US$ ha^{-1}. The crop-specific histogram is derived from data on 675 crops (179 different farms at eight different sites). The farm-specific histogram is derived from averages of three or four crops for 179 different farms at eight different sites. SD = standard deviation.

consistently from crop to crop. This consistency in performance may positively influence farmers' perception of the technology's usefulness.

16.7 Observability

Although SSNM is on average profitable for most farmers, a question remains of how visible (observable) the increased profitability will be to farmers. This is especially important if extension to a larger domain is anticipated because not all of these farmers will have the opportunity to interact closely with researchers or have access to records of controlled experiments that compare SSNM with FFP. Under these circumstances, the advantages of SSNM will need to be obvious even for casual observation. Large improvements in yield might constitute one such signal to farmers. For present purposes, large shall be considered as a yield improvement in excess of 0.5 t ha^{-1}. On average, over all four crops, SSNM generated a yield gain of at least this large in approximately 35% of the on-farm trials (see Fig 15.2 for a frequency distribution of these yield gains). It is not clear what standard is relevant for judging 35% to be large or small, but it seems large enough to make a substantial share of farmers notice the change. In any event, farmers' perceptions of the effectiveness of SSNM will be an important factor affecting adoption, and the visibility of yield gains will be an important factor influencing those perceptions.

16.8 Need for credit

Many investments, both inside and outside agriculture, are profitable but cannot be adopted because of the large costs of investment that must be incurred up front. This is particularly a constraint for poor farmers who have less working capital at their disposal. Because the SSNM strategy depends to some extent on recapitalization of soil K (and P) at sites where the soil supply of these nutrients is low, it does require some commitment of capital by farmers. These costs are measured by the increase in fertilizer costs in SSNM relative to those in FFP.

For the first year of SSNM, when a large part of soil recapitalization was done, the average increase in fertilizer costs was $15.10 ha^{-1} crop^{-1} across all sites. This declined to just $1.50 ha^{-1} crop^{-1} in the second year (Table 16.1). Although soil recapitalization still occurred in the second year, it was less than in the first year. Furthermore, savings in N use were much larger in the second year, thus offsetting some of the P and K recapitalization costs. The first-year costs of $15.10 ha^{-1} crop^{-1} would constitute about a 5% increase in total paid-out costs for farmers. This is not especially large and suggests that credit constraints should not be a major constraint to adoption in the intensive irrigated areas. Furthermore, it should also be noted that a survey of farmers conducted earlier in the project showed that only 7% of RTDP farmers mentioned that the availability of cash or credit influenced their fertilizer management decisions (Moya 1998).

16.9 Risk and yield variability

Risk is a fact of life for farmers and it affects technology adoption decisions. One of the most important components of risk is the variability of yields. If yield variability is substantially larger under SSNM than under current fertilizer management practices, this might constrain adoption. The evidence from the on-farm trials conducted so far strongly suggests that this is not an issue, as yield variability is approximately the same under SSNM as it is under FFP. The standard deviation of grain yield for all four crops at all sites under SSNM is 1,354 kg ha^{-1} versus 1,288 kg ha^{-1} for FFP. The coefficient of variation (CV) is lower for SSNM (24.4%) than for FFP (24.9%), however, because SSNM yields are higher than FFP yields by 7% on average. Regardless of which measure is used, the difference in variability is not substantial in terms of its effect on farmers.

16.10 The role of society: collective action and externalities

Many technologies require collective action to be most effective. For example, new water management strategies may require that all members of an irrigation group adopt the strategy, or at least that action be coordinated. Trapping rats usually requires collective action because attracting the rats to a particular field may result in losses for that particular farmer. In the case of SSNM, however, there does not appear to be any reason why the effectiveness of the technology will depend on the nutrient management practices of a neighboring farmer. This consideration should make it easier for SSNM to be adopted more widely. Some nutrient-pest interaction effects may occur where pests are selectively attracted to farms using SSNM, but this does not appear to be a major factor that will significantly inhibit technology adoption (especially since this problem was reduced once improved N management schemes were used in the SSNM treatments). Although collective action is thus not necessary for the adoption of SSNM, it would probably facilitate widespread adoption as one farmer's perception of a technology is influenced by the perceptions of neighboring farmers.

In terms of externalities, the effects of improved nutrient management on third parties and the environment are likely to be positive. Every increase in grain production that comes from higher yields per unit area reduces the pressure to bring other areas into cultivation. This allows the land to be used for other purposes and has the potential to reduce soil erosion and other forms of degradation.

There are other environmental benefits as well. First, because SSNM reduces N use to some extent and improves N recovery efficiency, there should be less N runoff and gaseous N losses into the environment. Bouman et al (2000) suggest that nitrate contamination of water supplies is not a major issue when rice is grown in anaerobic flooded systems (e.g., double-cropped rice), but the effects may be more important in rice-wheat systems, which occupy large areas in China and South Asia. Second, because SSNM increases the partial factor productivity of N, the amount of fertilizer needed to produce a unit of grain declines, which will reduce CO_2 emis-

Table 16.3. Share (%) of rice income in household income

Place	1987-88	1995-97
Bihar, India	–	13
Suphan Buri, Thailand	56	21
Khon Kaen, Thailand	46	8
Philippines	42	29
Vietnam	–	37
Myanmar	–	14
Bangladesh	38	–
Hunan, China	43	–
Tamil Nadu, India	50	–
Lampung, Indonesia	51	–
Nepal	46	–

Sources: Sombilla and Hossain (2000), Hossain et al (2000), Isvilanonda et al (2000).

sions resulting from the use of fossil energy to produce urea. Because urea production is responsible for only a small fraction of total fossil fuel consumption, this benefit will not significantly reduce CO_2 emissions on a global scale. Nevertheless, it is still a benefit. There may also be opportunities to sequester more C from the atmosphere into the soil because of increased biomass production, provided the straw is not burned (Dobermann and Witt 2000).

16.11 The opportunity cost of farmers' time

The traditional picture of rural life in Asia is one of small farmers who have little or nothing to do with their time other than tend painstakingly to farm tasks. Although this may have been the case many years ago, this viewpoint is no longer true in many parts of Asia. Economic development has increased wages and the diversity of income-earning opportunities and it has also increased the range of entertainment options available during leisure time. Rice is not the only source of household income for rice farmers; it often accounts for less than half of total household income even in key rice-growing areas (Sombilla and Hossain 2000; Table 16.3). Furthermore, the importance of rice in household income has declined over time (Marciano et al 2000, Isvilanonda et al 2000). Farmers want time to watch TV, see movies, listen to the radio, and participate in more traditional leisure activities. Thus, although farmers are interested in new technologies, they are also interested in other aspects of life and these other aspects may often be more important. The bottom line is that one of the most valuable assets that farmers possess is their own time.

Because of the opportunity cost of farmers' time, the widespread adoption of new technologies will be greatly facilitated if they have very large financial advantages for farmers, are relatively simple to implement at the farm level, or both. The Green Revolution package of seeds and fertilizers was widely adopted in many countries in a short time because it combined both of these features. Not only did it increase profits substantially compared with the planting of traditional varieties, it was also relatively simple for farmers to implement.

SSNM will not have the same impact as the Green Revolution, but it does increase yields and farmer profits. Those are important selling points. Yet, probably the most important factor that will determine whether it is widely adopted in the future will be how easy it is for farmers to implement without a major commitment of time. This problem has two aspects. First, how complex is the technology in terms of farmers being able to understand it and easily implement it? As pointed out by Pingali et al (1998), "Farmer adoption of knowledge-intensive management systems will be less likely where the cost of knowledge acquisition and decision making is high." Second, how does the quality of crop management practices (which are a function of the opportunity cost of farmers' time) affect the profitability of SSNM?

The complexity of any technology being disseminated in rural Asia is important because of at least two key differences between farmers in wealthy economies and those in Asia. First, farmers in wealthy economies have much higher levels of education, so they are more likely to be able to absorb and use more complicated technologies. The average education of farmers at RTDP sites generally ranges from four to eight years (see Table 3.3). RTDP farmers in India had substantially more education (nearly 12 years), but these farmers own very large farms and are not representative of the typical farmers in either Tamil Nadu or India in general. Second, farmers in wealthy economies typically have much larger farms than in Asia. Among RTDP sites, median farm sizes ranged from about 0.3 ha in northern Vietnam to about 4 ha in Thailand, and this range encompasses the situation in most Asian countries. This compares with an average farm size of nearly 200 ha in the United States. Larger farm sizes mean larger increases in the absolute level of profits when a new profitable technology is adopted, which in turn means that farmers in wealthy nations can afford to invest more time and energy in learning the technology. High farm output prices and subsidies in the wealthy countries tend to reinforce this effect, although a higher opportunity cost of time in those countries works in the opposite direction. All things considered, new technologies for farmers in Asia will probably need to be relatively simple if they are to be widely adopted.

In an effort to simplify SSNM for wider dissemination to both extension workers and interested farmers, researchers are developing a pocket fertilizer chart to aid in developing better fertilizer recommendations that implicitly or explicitly take into account season and soil quality (see Chapter 18 for more details). Another possible tool is the leaf color chart (LCC), which may prove to be effective in helping farmers to manage the timing of N applications more effectively (Balasubramanian et al 1999). At other sites, SSNM may take the form of improved region-specific recommendations that, for example, simply stress the importance of recapitalizing soil K. All of these would be steps in the right direction, but it is an open empirical question as to which strategy will be the most effective in communicating information to farmers. Research on the effectiveness of these efforts needs to be conducted because understanding farmers' knowledge and constraints is critical for the development of effective recommendations. The use of site-specific approaches to disseminating knowledge may be the most effective means of improving the nutrient management practices of farmers.

At the same time, it is also critical to understand whether simplified versions of SSNM can achieve increases in profitability similar to those achieved on the basis of field-by-field data collection. According to Pingali et al (1998), "The question we must ask is whether [nutrient management] decisions need to be made specifically for each farm, or even for particular parcels within a farm, or can they be generalized across farms within a particular area.... If nutrient management decisions are highly farm- and/or parcel-specific, then the farm-level costs of acquiring decision tools and using them can be a major deterrent to adoption." Thus, a crucial research task is to understand how much profitability is lost in moving from a complex model using field-specific data to a simple message that fits within the knowledge systems of farmers. This is an unresolved issue at present.

If the use of simple rules causes a large loss in profitability, then it may be possible to implement field-specific management by using the services of crop consultants whom farmers trust to provide input to the decision-making process. Although such crop consultants could in theory be public-sector extension agents, the general experience with extension services has been less than satisfactory in many instances in Asia. Furthermore, financially strapped central governments are continually privatizing such services. In other cases, responsibility for service provision is being devolved to local governments, as in the Philippines and Indonesia, but these local governments may be even more cash-strapped than the central government. An alternative may be for private-sector crop consultants to supply this service, provided that farmers feel the service is valuable enough to warrant paying for it. There is very little experience with such service provision, although the salespeople of large agrochemical companies provide something similar regarding pesticide use. Unfortunately, it is not always true that the advice offered by the private sector maximizes benefits to farmers and society at large.

The second key issue related to the opportunity cost of farmers' time is how the general quality of crop management affects the profitability of SSNM. This is yet another unresolved issue. Figure 15.8 shows that grain yields were the highest for a given level of plant N accumulation at the three sites (Zhejiang, China; Red River Delta, Vietnam; and Tamil Nadu, India) with good crop care and transplanting as the method of crop establishment. These results were achieved with relatively high labor use, which in turn was facilitated by a low opportunity cost of labor. In China and northern Vietnam, the low opportunity cost of labor was at least partially caused by the predominance of family labor at those sites (Fig. 3.1) as well as the small parcel sizes (and in northern Vietnam also by the low level of economic development). In India, the low opportunity cost of labor is primarily due to the availability of a large pool of landless laborers at low wages. At other sites where the opportunity cost of labor is higher, grain yield for a given level of N accumulation was lower.

One interpretation of this pattern is rather dismal. On the assumption that economic development will continue in Asia, wages and the opportunity costs of labor will continue to rise and good crop care will become more difficult to achieve. For example, farmers at the RTDP site in China began direct seeding of rice in 2000 (no data shown) in response to the rising opportunity costs of labor. Rozelle et al (1999) show that maize farmers in China tend to achieve lower yields when some family members have migrated to the cities. Their data are consistent with the hypothesis that migration of a family member provides an additional source of income and makes the family less reliant on farm income (and presumably less interested in crop management). Such trends will continue in China and are likely to occur at the other sites as well. By the standards of reaching high grain yields for a given level of N accumulation (Fig. 15.8) and achieving the yield target (Fig. 15.7), the future of SSNM may not be bright.

Yet, there are other standards by which to judge SSNM. Instead of judging SSNM by how closely it achieves "perfection" as measured by a yield target, it is perhaps more appropriate to judge it on its ability to improve on current practices. This is clearly a different standard and SSNM may be quite successful at increasing profits even in situations where the yield target is not achieved. For example, profitability increased quite substantially at Maligaya (see Fig. 16.2) despite achieving a relatively low percentage of the yield target (Fig. 15.7) and having relatively low grain yield given the level of N accumulation (Fig. 15.8). Much of the improvement in profitability at the sites in Zhejiang and Tamil Nadu was probably due to reduced doses of basal N and improved timing of N during the remainder of the growing season. Although some of this improvement may be due to good crop management at these sites, much of it was also probably caused by the relatively poor N management of existing farmers' practices. Thus, what is ultimately most important is not to reach high N-use efficiencies or achieve yield targets, but to improve profitability for farmers.

16.12 Conclusions

There is potential for improving farmers' nutrient management practices in Asia and SSNM has the potential to improve profits and increase yields for farmers. In tests in farmers' fields at different sites around Asia, it was shown to be financially profitable, to require little in the way of credit for financing, and to remain profitable even if rice prices were somewhat lower than current levels. It does not require complex coordination among farmers and it may also generate some positive environmental effects by reducing N use.

In general, minimal changes in government policies are probably needed to facilitate improved nutrient management. Past subsidies for nitrogen use had a rationale in the early years of the Green Revolution, but they may have caused farmers to focus excessively on N and ignore the role of P and K. N subsidies have already been phased out in many countries and the first step is to phase them out in the rest of Asia. After that, it will be important to conduct information campaigns to explain to farmers the importance of balanced nutrition. Large-scale demonstrations that are well publicized and encourage active participation by farmers will be an important part of such information dissemination.

The major challenge for SSNM will be to retain the success of the approach while reducing the complexity of the technology as it is disseminated to farmers. Reliance on nutrient omission plots and N management with SPAD meters would constrain technology adoption. It may be possible, however, to achieve good results without these techniques now that improved models of nutrient dynamics have been developed (Dobermann and White 1999, Witt et al 1999). A simplified approach for farmers may give good results in the absence of strong government extension systems, whereas less simplification may be needed where extension systems are strong. In the future, private-sector crop consultants might also have a role to play in dissemination, although such consultants are exceedingly rare in Asia today. The nature of the approach will need to be tailored to specific circumstances in different countries. In some areas, SSNM may be field- or farm-specific, but in many areas it is likely to be just region- and season-specific. What is certain is that SSNM will need to be flexible to adapt to the needs of different farmers and extension systems.

References

Ali M. 1999. Evaluation of green manure technology in tropical lowland rice systems. Field Crops Res. 61:61-78.

Balasubramanian V, Morales AC, Cruz RT, Abdulrachman S. 1999. On-farm adaptation of knowledge-intensive nitrogen management technologies for rice systems. Nutr. Cycl. Agroecosyst. 53:59-69.

Barker R, Herdt RW, Rose B. 1985. The rice economy of Asia. Washington, D.C. (USA) and Manila (Philippines): Resources for the Future and International Rice Research Institute.

Bouman BAM, Castañeda AR, Bhuiyan SI. 2002. Nitrate and pesticide contamination of groundwater under rice-based cropping systems: past and current evidence from the Philippines. Agric. Ecosyst. Environ. 92:185-199.

Cassman KG, De Datta SK, Amarante ST, Liboon SP, Samson M, Dizon M. 1996. Long-term comparison of the agronomic efficiency and residual benefits of organic and inorganic nitrogen sources for tropical lowland rice. Exp. Agric. 32:427-444.

Cassman KG, Dobermann A. 2001. Evolving rice production systems to meet global demand. In: Rice research and production in the 21st century: Proceedings of a symposium honoring Robert F. Chandler, Jr. Los Baños (Philippines): International Rice Research Institute. p 79-100.

Cohen M. 1997. Barren business: fertilizer monopoly goes to seed. Far Eastern Econ. Rev.:60-61.

Dawe D. 2000. The potential role of biological nitrogen fixation in meeting future demand for rice and fertilizer. In: Ladha JK, Reddy PM, editors. The quest for nitrogen fixation in rice. Los Baños (Philippines): International Rice Research Institute. p 1-9.

Dobermann A, White PF. 1999. Strategies for nutrient management in irrigated and rainfed lowland rice systems. Nutr. Cycl. Agroecosyst. 53:1-18.

Dobermann A , Witt C. 2000. The potential impact of crop intensification on carbon and nitrogen cycling in intensive rice systems. In: Kirk GJD, Olk DC, editors. Carbon and nitrogen dynamics in flooded soils. Makati City (Philippines): The International Rice Research Institute. p 1-25.

FAO. 1998. Guide to efficient plant nutrition management. Rome (Italy): FAO.

Hossain M, Gascon F, Marciano E. 2000. Income distribution and poverty: insights from a repeat village study. Econ. Polit. Weekly 35(52 and 53):4650-4656.

Isvilanonda S, Ahmed A, Hossain M. 2000. Recent changes in Thailand's rural economy: evidence from six villages. Econ. Polit. Weekly 35(52 and 53):4644-4649.

Kanazawa N. 1984. Trends and economic factors affecting organic manures in Japan. In: Organic matter and rice. Los Baños (Philippines): International Rice Research Institute. p 557-567.

Kendall HW, Beachy R, Eisner T, Gould F, Herdt RW, Raven PH, Schell JS, Swaminathan MS. 1997. Bioengineering of crops: Report of the World Bank Panel on transgenic crops. Environmentally and Socially Sustainable Development Studies and Monographs Series No. 23. 40 p.

Ladha JK, Reddy PM. 2000. Steps toward nitrogen fixation in rice. In: Ladha JK, Reddy PM, editors. The quest for nitrogen fixation in rice. Los Baños (Philippines): International Rice Research Institute. p 33-46.

Moya P. 1998. Fertilizer management practices at the farm level. Mega Project Progress Report 1997—Phase II:101.

Pasandaran E, Gultom B, Sri Adiningsih J, Apsari H, Rochayati S. 1999. Government policy support for technology promotion and adoption: a case study of urea tablet technology in Indonesia. Nutr. Cycl. Agroecosyst. 53:113-119.

Peng S, Garcia FV, Gines HC, Laza RC, Samson MI, Sanico AL, Visperas RM, Cassman KG. 1996. Nitrogen use efficiency of irrigated tropical rice established by broadcast wet-seeding and transplanting. Fert. Res. 45:123-134.

Pingali PL, Hossain M, Pandey S, Price LL. 1998. Economics of nutrient management in Asian rice systems: towards increasing knowledge intensity. Field Crops Res. 56:157-176.

Rozelle S, Taylor JE, de Brauw A. 1999. Migration, remittances, and agricultural productivity in China. Am. Econ. Rev. 89:287-291.

Savant NK, Stangel PJ. 1995. Recent developments in urea briquette use for transplanted rice. Fert. News. 40:27-33.

Singh U, Cassman KG, Ladha JK, Bronson KF. 1995. Innovative nitrogen management strategies for lowland rice systems. In: Fragile lives in fragile ecosystems. Proceedings of the International Rice Research Conference, 13-17 February. Manila (Philippines): International Rice Research Institute. p 229-254.

Sombilla MA, Hossain M. 2000. Rice and food security in Asia: a long-term outlook. In: Food Security in Asia. p 35-59.

Witt C, Cassman KG, Olk DC, Biker U, Liboon SP, Samson MI, Ottow JCG. 2000. Crop rotation and residue management effects on carbon sequestration, nitrogen cycling, and productivity of irrigated rice systems. Plant Soil 225:263-278.

Witt C, Dobermann A, Abdulrachman S, Gines HC, Guanghuo W, Nagarajan R, Satawathananont S, Son TT, Tan PS, Tiem LV, Simbahan GC, Olk DC. 1999. Internal nutrient efficiencies of irrigated lowland rice in tropical and subtropical Asia. Field Crops Res. 63:113-138.

Notes

Authors' addresses: D. Dawe, A. Dobermann, C. Witt, International Rice Research Institute, Los Baños, Philippines, email: d.dawe@cgiar.org; A. Dobermann, University of Nebraska, Lincoln, Nebraska, USA; S. Abdulrachman, Research Institute for Rice, Sukamandi, Indonesia; H.C. Gines, Philippine Rice Research Institute, Maligaya, Nueva Ecija, Philippines; R. Nagarajan, Soil and Water Management Research Institute, Thanjavur, India; S. Satawathananont, Pathum Thani Rice Research Center, Pathum Thani, Thailand; T.T. Son, National Institute for Soils and Fertilizers, Hanoi, Vietnam; P.S. Tan, Cuu Long Delta Rice Research Institute, Omon, Cantho, Vietnam; G.H. Wang, Zhejiang University, Hangzhou, China.

Acknowledgments: The authors gratefully acknowledge the assistance of Marites Tiongco in performing the total factor productivity calculations using Reversing Trends of Declining Productivity project data. The authors would like to acknowledge Greg Simbahan for his help in the ANOVA analysis and all of the farmers across Asia who participated in this project.

Citation: Dobermann A, Witt C, Dawe D, editors. 2004. Increasing productivity of intensive rice systems through site-specific nutrient management. Enfield, N.H. (USA) and Los Baños (Philippines): Science Publishers, Inc., and International Rice Research Institute (IRRI). 410 p.

7 | Toward a decision support system for site-specific nutrient management

C. Witt and A. Dobermann

Developing a suitable nutrient management strategy for delivery is a complex process. Different interest groups are involved in the decision making, including farmers, extension workers, researchers, fertilizer companies and retailers, campaign planners, and policymakers, all of which have different interests, knowledge, and expertise that need to be integrated in the decision-making process. Decision support systems (DSS), a specific class of computerized information systems that support decision-making activities (Power 1999), could greatly enhance the ability to structure and solve such complex problems. However, given the limited computer access of extension workers in Asia, a broader definition of DSS must be used, which recognizes the need for simple guidelines and decision trees that can be used in developing fertilizer recommendations.

This chapter describes further improvements of the site-specific nutrient management (SSNM) approach developed for irrigated rice systems, focusing on developing generic guidelines, decision trees, and computer software for providing decision support. The framework of a nutrient decision support system (NuDSS) for irrigated rice is described, where computers can be used in addition to printed material to provide assistance in complex mathematical calculations that would be difficult to perform otherwise (Witt et al 2001).

Specific objectives of this chapter were to

- Develop simplified equations for estimating nutrient and fertilizer requirements of N, P, and K based on the SSNM approach described in Chapter 5,
- Include a nutrient balance model in the estimation of fertilizer P and K maintenance rates, particularly for situations in which a short-term direct yield response to fertilizer application is uncertain, but maintenance of soil fertility is a requirement for sustaining the productivity of the cropping system, and
- Integrate refined principles for SSNM in a general framework with a set of software tools for decision support.

17.1 Nutrient and fertilizer requirements

Quantifying fertilizer requirements based on the expected yield gain

The plant nutrient requirements for balanced N, P, and K nutrition of irrigated rice in Asia have been quantified by Witt et al (1999) using the model QUEFTS (Janssen et al 1990). The QUEFTS model divides the relationship between yield and nutrient supply into several steps by taking interactions in supply, acquisition, and use of N, P, and K into account. Yield is predicted as a function of (1) yield potential, (2) the relationship between grain yield and plant nutrient accumulation, (3) estimated recovery efficiencies of fertilizer N, P, and K, and (4) field-specific estimates of the indigenous nutrient supplies of N, P, and K. In the approach described in Chapter 5, an optimization routine was used to estimate fertilizer requirements for a specified yield goal, taking into account the indigenous supply of N, P, and K that was directly measured in nutrient omission plots. The optimization routine accounted for a decrease in internal nutrient efficiencies (yield per unit nutrient uptake) as target yields approached the potential yield, and optimized the yield-producing uptake efficiencies of N, P, and K (Janssen et al 1992, Witt et al 1999). The QUEFTS principles can be simplified into a straightforward equation:

$$FX \ (kg \ ha^{-1}) = \frac{UX - UX_{0X}}{REX} \qquad (1)$$

where X is one of the three macronutrients N, P, or K, FX is the fertilizer nutrient requirement to achieve a specified yield target, UX is the predicted optimal plant nutrient uptake requirement for the specified yield target (kg ha^{-1}, Fig. 17.1), UX_{0X} is the indigenous nutrient supply measured as plant nutrient uptake in an omission plot (kg ha^{-1}), and REX is the expected first-season recovery efficiency of the applied fertilizer nutrient (kg kg^{-1}).

Because it is not feasible for extension staff in developing countries to measure plant nutrient uptake, Dobermann et al (2003b) suggested estimating the indigenous nutrient supply of N, P, and K from grain yields in the respective omission plots. Estimates had a precision of ±5–10 kg N ha^{-1}, ±2–3 kg P ha^{-1}, and ±10–20 kg K ha^{-1}, which was considered sufficiently robust to develop meaningful fertilizer recommendations. Equation 1 can therefore be simplified by using grain yield as an indicator of indigenous nutrient supply (= yield gain approach):

$$FX \ (kg \ ha^{-1}) = \frac{(GY - GY_{0X}) \times UX'}{REX} \qquad (2)$$

where GY is the grain yield target (t ha^{-1}), GY_{0X} is the grain yield (t ha^{-1}) in the respective nutrient omission plot (nutrient-limited yield as an indicator of indigenous nutrient supply), and UX' is the assumed optimal plant nutrient uptake requirement of 14.7 kg N, 2.6 kg P, and 14.5 kg K to produce 1 t of grain yield (Witt et al 1999).

Using equation 2, location-specific fertilizer requirements can be calculated for most irrigated rice areas based on the expected yield increase over the respective omission plot and using certain assumptions on plant nutrient requirements and re-covery efficiencies of applied fertilizer nutrients. The QUEFTS model predicts a linear increase in grain yield if nutrients are taken up in balanced amounts of 14.7 kg N, 2.6 kg P, and 14.5 kg K (UX', equation 2) per 1 t of grain yield produced, until the yield reaches about 70–80% of the climate-adjusted potential yield (Y_{max}). The model predicts a decrease in internal nutrient-use efficiencies when yield targets approach Y_{max}, which results in an increased plant nutrient requirement per unit yield (Fig. 17.1). The nutrient requirement (UX' in equation 2) is therefore only a constant if yield goals are chosen that are equal to or lower than 70–80% of the potential yield. This is the case in most areas with irrigated rice in Asia, where current rice yields in farmers' fields are mostly below 60% to 70% of Y_{max} (Dobermann et al, Chapter 15, this volume). A realistic yield goal for a particular cropping season should therefore target not more than a 10–20% increase over the average farmers' yield achieved in the last 3–5 years. Below, further considerations are given separately for each of the three macronutrients.

Nitrogen. With SSNM, internal N efficiency in farmers' fields averaged 58 kg grain kg^{-1} plant N, with an interquartile range from 52 to 63 kg grain kg^{-1} plant N (Table 17.1). This is equivalent to a plant N requirement (UN' in equation 2) of 17.2 kg to produce 1 t of grain yield or 15.9–19.2 kg kg^{-1} for the above given interquartile range of internal nitrogen efficiency. Measured internal N efficiencies under field conditions were lower than the 67 kg grain kg^{-1} plant N predicted by QUEFTS, suggesting that actual plant nutrient requirements tend to be larger than the model-predicted 14.7 kg plant N per ton of grain yield. This difference occurs because the predicted optimal nutrient use assumes optimal growth conditions in the field, with no or few abiotic or biotic stresses, which is difficult to ensure in practice. To avoid N supply limiting yield under current conditions, we therefore propose to use a de-fault value of 17 kg N t^{-1} grain yield for UN' in equation 2.

With SSNM, the recovery efficiency of applied fertilizer N averaged 0.43 kg plant N kg^{-1} fertilizer N (Table 17.1), but there was a large variation among sites (Table 17.2). With good management, 25% of the farmers were able to achieve fertil-izer recovery efficiencies of more than 0.50 kg kg^{-1} at all sites, except at Thanjavur, India (0.48 kg kg^{-1}), and Jinhua, China (0.35 kg kg^{-1}). Based on the above given interquartile ranges of internal nitrogen efficiency and assuming a recovery efficiency of 0.43 kg kg^{-1}, fertilizer requirements range from 37 to 45 kg N per ton of grain yield increase over an unfertilized control. For optimal growing conditions with re-covery efficiencies of about 0.50 kg kg^{-1}, fertilizer requirements would be only 32–38 kg N ha^{-1} t^{-1} grain yield increase. As a rule of thumb, we therefore assume that about 40 kg N ha^{-1} is required to raise grain yield by 1 ton over the yield in an N omission plot. Local adjustment may be necessary, for example, in southeast China, where 50 kg N ha^{-1} t^{-1} grain yield appears to be a more realistic estimate because water management practices make it difficult to achieve higher recovery efficiency

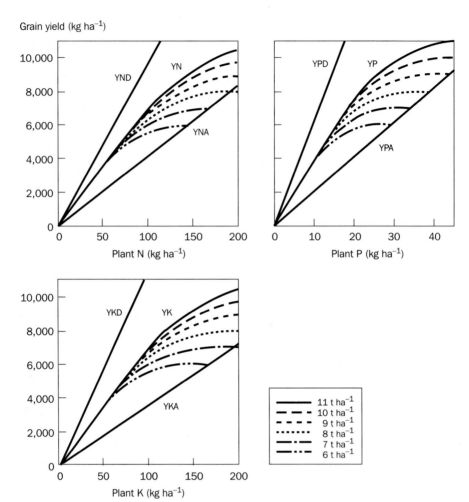

Fig. 17.1. The balanced N, P, and K uptake requirements for different rice yield goals depending on the location- and season-specific climatic yield potential as calculated by QUEFTS (adapted from Witt et al 1999). The borderlines in each graph describe situations of maximum nutrient dilution (YND, YPD, and YKD) and accumulation (YNA, YPA, YKA) in the plant, while YN, YP, and YK describe the optimal nutrient requirements for a given yield potential (Y_{max}).

of fertilizer N. For comparison, a fertilizer requirement of 40–50 kg N ha^{-1} t^{-1} grain yield is equivalent to agronomic efficiencies of 20–25 kg grain per kg fertilizer N applied, which can be achieved with good crop and nutrient management (Balasubramanian et al 1999, Dobermann et al 2002). Note that the suggested N fertilizer requirements assume proper timing and splitting patterns of N applications through the use of a location-specific N splitting scheme or tools such as a chlorophyll meter or leaf color chart (see also Witt et al, Chapter 18, this volume).

Table 17.1. Grain yield, total aboveground plant dry matter (DM), harvest index (HI), nutrient accumulation in plant DM, internal nutrient efficiencies (IEN, IEP, IEK; kg grain per kg nutrient in aboveground plant dry matter), and recovery efficiencies (REN, REP, REK, kg kg⁻¹) of fertilizer N, P, or K applied to irrigated lowland rice using the site-specific nutrient management (SSNM) approach in farmers' fields across five tropical Asian countries, 1997-98. Data with an HI of less than 0.40 kg kg⁻¹ were excluded ($n = 314$).

	Unit	Mean	SD[a]	Min.	25% quartile	Median	75% quartile	Max.
Grain yield	kg ha⁻¹	5,940	1,199	2,644	5,130	5,918	6,813	9,186
Plant DM	kg ha⁻¹	10,792	1,950	5,155	9,442	10,998	11,973	16,183
Harvest index	kg kg⁻¹	0.49	0.04	0.40	0.46	0.49	0.52	0.64
Plant N	kg ha⁻¹	104	21	46	91	102	118	167
Plant P	kg ha⁻¹	20	6	8	16	20	24	47
Plant K	kg ha⁻¹	109	30	45	88	108	125	208
IEN	kg kg⁻¹	58	9	38	52	59	63	93
IEP	kg kg⁻¹	304	60	185	268	293	335	514
IEK	kg kg⁻¹	57	13	32	47	57	64	99
Fertilizer N	kg ha⁻¹	116	31	40	90	119	140	185
Fertilizer P	kg ha⁻¹	22	11	9	12	19	30	43
Fertilizer K	kg ha⁻¹	62	30	15	30	56	85	163
REN	kg kg⁻¹	0.43	0.18	0.00	0.31	0.40	0.54	0.96
REP	kg kg⁻¹	0.25	0.27	0.00	0.11	0.22	0.35	1.00
REK	kg kg⁻¹	0.44	0.52	0.00	0.16	0.35	0.66	1.00

[a]SD = standard deviation.

Table 17.2. Nutrient accumulation in total aboveground plant dry matter at physiological maturity of irrigated lowland rice grown in farmers' fields in +PK (N omission), +NK (P omission), and NP (K omission) plots (i.e., indigenous nutrient supplies, INS, IPS, and IKS) and in plots with site-specific nutrient management (UN, UP, and UK) receiving balanced amounts of fertilizer nutrients (FN, FP, FK) at seven sites in five Asian countries, 1997-99. Recovery efficiencies of fertilizer nutrients (REN, REP, and REK) were calculated using the difference method. Medians with interquartile ranges are in parentheses.

	Unit	Aduthurai, India	Thanjavur, India	Sukamandi, Indonesia	Maligaya, Philippines	Omon, Vietnam	Hanoi, Vietnam	Jinhua, China
INS	kg ha^{-1}	51	41	65	53	51	59	69
	kg ha^{-1}	(44-59)	(32-52)	(39-80)	(44-63)	(36-55)	(51-68)	(62-80)
UN	kg ha^{-1}	106	94	106	99	96	101	109
FN	kg ha^{-1}	126	123	100	110	90	90	152
	kg ha^{-1}	(110-140)	(114-146)	(80-156)	(80-130)	(78-107)	(80-105)	(139-163)
REN	kg kg^{-1}	0.43	0.39	0.43	0.42	0.50	0.39	0.24
	kg kg^{-1}	(0.34-0.53)	(0.32-0.48)	(0.29-0.62)	(0.33-0.66)	(0.41-0.59)	(0.32-0.54)	(0.19-0.35)
IPS	kg ha^{-1}	17	18	11	13	13	13	20
	kg ha^{-1}	(14-19)	(15-20)	(10-12)	(10-17)	(11-15)	(11-17)	(18-24)
UP	kg ha^{-1}	23	25	13	20	20	16	24
	kg ha^{-1}	(21-26)	(22-27)	(12-16)	(16-24)	(15-23)	(14-19)	(22-27)
FP	kg ha^{-1}	26	15	20	18	22	10	15
	kg ha^{-1}	(15-37)	(12-35)	(10-28)	(14-24)	(14-34)	(10-40)	(10-23)
REP	kg kg^{-1}	0.23	0.30	0.15	0.28	0.26	0.14	0.19
	kg kg^{-1}	(0.15-0.37)	(0.19-0.41)	(0.06-0.24)	(0.08-0.43)	(0.17-0.37)	(0.08-0.27)	(0.03-0.43)
IKS	kg ha^{-1}	85	60	102	77	72	77	106
	kg ha^{-1}	(75-97)	(50-69)	(74-135)	(66-97)	(60-81)	(55-84)	(93-139)
UK	kg ha^{-1}	113	75	112	98	110	102	136
	kg ha^{-1}	(101-124)	(64-87)	(63-146)	(84-114)	(96-123)	(92-114)	(113-162)
FK	kg ha^{-1}	73	75	50	49	62	45	77
	kg ha^{-1}	(44-92)	(48-96)	(30-84)	(30-62)	(42-80)	(30-60)	(50-103)
REK	kg kg^{-1}	0.33	0.21	0.08	0.36	0.56	0.68	0.34
	kg kg^{-1}	(0.19-0.53)	(0.18-0.33)	(0-0.44)	(0.10-0.71)	(0.34-0.76)	(0.27-1)	(0.14-0.55)
	n^a	86	15	36	60	30	45	42

$^a n$ = number of data sets (farm × crops).

Phosphorus. The internal P efficiency in farmers' fields with SSNM averaged 304 kg grain kg^{-1} plant P, ranging from 268 to 335 kg grain kg^{-1} plant P (interquartile ranges, Table 17.1). The equivalent plant P uptake was 3.3 kg plant P t^{-1} grain yield, ranging from 3.0 to 3.7 kg kg^{-1}. This compares to a plant P requirement of 385 kg grain kg^{-1} plant P or 2.6 kg plant P t^{-1} grain yield as predicted by QUEFTS, indicating that the P supply from soil plus fertilizer was more than sufficient in SSNM treatments to support the yields achieved. Even in 0-P plots, the IEP averaged only 345 kg grain kg^{-1} P (2.9 kg plant P t^{-1} grain yield), with interquartile ranges of 287 to 379 kg kg^{-1} (Witt et al 1999), so that optimal internal P efficiencies were reached in only 25% of all crops. Based on QUEFTS, we propose a conservative default value of 2.6 kg P t^{-1} grain yield for UP' in equation 2 to avoid excessive P uptake.

With SSNM, the recovery efficiency of applied fertilizer P averaged 0.25 kg kg^{-1}, ranging from 0.11 to 0.35 kg kg^{-1} (interquartile ranges, Table 17.1), and there was also a large variation among sites, ranging from an average of 0.14 kg kg^{-1} at Hanoi, Vietnam, to 0.30 kg kg^{-1} on coarse-textured soils at Thanjavur, India (Table 17.2). Recovery efficiencies of P and K are influenced by several factors, including the amount of fertilizer applied, the difference between yield in fertilized and unfertilized plots, the application method, splitting and timing of applications, and soil properties. In general, recovery efficiencies of P and K, as measured by the difference method, decline with an increasing amount of fertilizer applied. Fertilizer P and K rates used in the initial SSNM field testing (Table 17.2) were generally on the high side to ensure that these nutrients were not limiting yield and to increase indigenous nutrient supplies at sites where soil nutrient levels had been depleted in the past. In subsequent years, rates were reduced, also because of changes in the model parameterization (Dobermann and Witt, Chapter 5, this volume). To calculate fertilizer P and K requirements, we therefore propose to use default values of recovery efficiencies for P and K that are between the median and the 75th percentiles given in Table 17.1. For a default P requirement of 2.6 kg P t^{-1} grain yield and with P recovery efficiencies ranging from 0.22 to 0.35 kg kg^{-1}, fertilizer requirements to produce 1 ton of grain yield increase may range from about 7.4 to 11.8 kg P. As a conservative rule of thumb, on lowland rice soils with little potential for P fixation, about 9 kg fertilizer P or 20 kg fertilizer P_2O_5 ha^{-1} may therefore be required to raise grain yield by 1 ton over the yield in a P omission plot.

Potassium. The internal K efficiency in farmers' fields with SSNM averaged 57 kg grain kg^{-1} plant K, ranging from 47 to 64 kg grain kg^{-1} plant K (interquartile ranges, Table 17.1). The equivalent plant K uptake was 17.5 kg plant K t^{-1} grain yield, ranging from 15.6 to 21.3 kg kg^{-1}. This compares to a plant K requirement of 69 kg grain kg^{-1} plant K or 14.5 kg plant K t^{-1} grain yield as predicted by QUEFTS. Thus, plant K uptake in SSNM was slightly greater than the optimal values predicted by QUEFTS. In 0-K plots, the IEK averaged 71 kg grain kg^{-1} K (14.1 kg plant K t^{-1} grain yield), with interquartile ranges of 52 to 83 kg kg^{-1} (Witt et al 1999), so that optimal internal K efficiencies were reached in more than 50% of all crops. We propose a conservative default value of 14.5 kg K t^{-1} grain yield for UK' in equation 2. Conservative default values for P and K requirements are proposed because nutrient uptake observed under field conditions is most likely to be sufficient to support higher yields. A less conservative nutrient requirement value is proposed for N, the most limiting nutrient, to ensure that N supply is not limiting yield.

With SSNM, the recovery efficiency of applied fertilizer K averaged 0.44 kg kg^{-1}, with a wide range from 0.16 to 0.66 kg kg^{-1} (interquartile ranges, Table 17.1). There was a large variation among sites, with averages ranging from 0.08 kg kg^{-1} at Sukamandi, Indonesia, to 0.68 kg kg^{-1} at Hanoi, Vietnam (Table 17.2). Using the default value of 14.5 kg K t^{-1} grain yield and moderate to high K recovery efficiencies of 0.35 and 0.66 kg kg^{-1} (median and 75th percentile, Table 17.1), fertilizer requirements range from 22 to 41 kg K to produce 1 ton of grain yield. Given the uncertainties associated with K recovery efficiencies and the greater plasticity of internal K efficiencies compared to other nutrients, we propose as a conservative rule of thumb applying about 25 kg fertilizer K ha^{-1} or 30 kg K_2O ha^{-1} to raise grain yield by 1 ton over the yield in a K omission plot.

Quantifying fertilizer P and K requirements based on a nutrient input-output balance

The yield gain approach described above suggests applying fertilizer in the amount of 40–50 kg N, 7–12 kg P, or 22–41 kg K ha^{-1} to raise yield by 1 t ha^{-1} over the yield measured in the respective nutrient omission plots. This concept provides a simple framework for both the evaluation of an existing nutrient management practice and for the development of new recommendations. However, both the original QUEFTS approach to calculate optimal rates of P and K fertilizer (Dobermann and Witt, Chapter 5, this volume) and the QUEFTS-based yield gain approach neglect long-term residual effects of fertilizer application, and they suggest applying fertilizer P and K only if a yield response is expected. This may be a sensible recommendation where farmers face short-term constraints in the availability of funds to purchase fertilizer, but would likely lead to a depletion of soil nutrient reserves when practiced for several seasons. This is of less concern for N because the residual effect of fertilizer N application is small, and, under constant management, the indigenous N supply changes little over medium- to long-term time periods that are relevant for making fertilizer decisions (Cassman et al 1996, Dobermann et al 2003a). In contrast, fertilizer inputs and crop residue management have a long-term impact on input-output balances of P and K and the long-term supply of these nutrients (Greenland 1997). Recent reports indicate, for example, that negative K balances are widespread and K deficiency has become a constraint to increasing yields (Dobermann et al 1996, 1998). Long-term management strategies must therefore focus on maintaining adequate nutrient balances in the topsoil layer, particularly in high-yielding areas such as Zhejiang Province, China, where indigenous soil nutrient supplies would decline rapidly if nutrients removed with grain and straw were not replenished (Witt et al 2003).

In the original SSNM approach tested during 1997 to 2000 (Chapters 5–12 and 14–16, this volume), lower limits of 10 kg P ha^{-1} and 30 kg K ha^{-1} were set as the minimum fertilizer rates to maintain soil nutrient supplies. In addition, a simple P and K balance model was used to predict changes in IPS and IKS resulting from the previous crop cycle, and to empirically adjust fertilizer rates accordingly in the subsequently grown rice crop. However, the season-to-season adjustment of indigenous nutrient supplies is not only impractical, it is also associated with great uncertainties because of the variability of the much larger soil indigenous nutrient pool. Instead of seasonally adjusting soil indigenous nutrient supplies, we therefore propose integrating a nutrient balance model in the development of fertilizer rates to arrive at P and K fertilization strategies that are cost-efficient and sustainable over the long term.

A simple nutrient balance model was constructed using standard parameters for a typical rice-growing area in Asia, taking into account (1) nutrient inputs from crop residues and other organic amendments, (2) nutrient inputs from irrigation and rainwater, (3) percolation losses, and (4) nutrient removal with grain and straw (Table 17.3). Maintenance fertilizer rates were then calculated according to

$$FP \; (kg \; ha^{-1}) = GY \times UP' + (GY - GY_{0P}) \times UP' - P_{ST} - P_{OM} - P_W + P_L \quad (3)$$

$$FK \; (kg \; ha^{-1}) = GY \times UK' + (GY - GY_{0K}) \times UK' - K_{ST} - K_{OM} - K_W + K_L \; (4)$$

where FP and FK are the fertilizer rates (kg ha^{-1}), GY is the grain yield target (t ha^{-1}), GY_{0P} and GY_{0K} are the grain yields (t ha^{-1}) in P and K omission plots, UP' and UK' are the empirically estimated plant nutrient requirements of 2.6 kg P and 14.5 kg K to produce 1 ton of grain yield (see above), P_{ST} and K_{ST} are the net nutrient inputs with crop residues, P_{OM} and K_{OM} are the nutrient inputs with organic amendments, P_W and K_W are the nutrient inputs with irrigation and rainwater, and P_L and K_L are the percolation losses (all in kg ha^{-1} per crop).

The fertilizer rates given in equations 3 and 4 aim to fertilize according to the deficit of the nutrient input-output balance to replenish nutrients removed with grain and straw and nutrient losses caused by percolation. If the nutrient demand for the targeted yield is greater than the indigenous nutrient supply, fertilizer rates are increased by the amount of the uptake deficit to slowly build up soil indigenous nutrient supplies. This approach considers the soil indigenous nutrient supply as the status quo and takes more information (e.g., on straw management) into account when developing fertilizer P and K requirements. As with the yield gain approach, fertilizer P and K application is not recommended and soil nutrients may be mined for a short period of time if the indigenous nutrient supply exceeds the nutrient demand of the yield target. However, regular reevaluation of IPS and IKS is required every 5–10 years to avoid depletion of soil nutrients to levels that may cause a yield reduction (Dobermann et al 2003b).

Sensitivity analysis
A sensitivity analysis was conducted to evaluate the two approaches for calculating fertilizer P and K requirements by changing the most relevant input parameters. According to equations 3–4 and Table 17.3, input parameters with the expected greatest effects on fertilizer use and nutrient balances were indigenous nutrient supplies, recovery efficiencies of applied fertilizer, plant nutrient requirements to produce 1 ton

Table 17.3. Standard parameters and example for a P and K balance without fertilizer input in the nutrient balance model.

No.	Parameters for P and K balance	Unit	Default value	Ranges	Reference[a]
1	Grain yield target	t ha^{-1}	7.0	4–10	User-defined
2	Straw yield	t ha^{-1}	7.0	4–10	User-defined
3	Harvest index	kg kg^{-1}	0.50	0.44–0.51	Witt et al (1999)[b]
4	Straw returned	kg kg^{-1}	0.35	0.05–1.00	User-defined
5	Organic nutrient sources	t ha^{-1}	1.0	0–15	User-defined
6	P in grain at harvest	g kg^{-1}	2.0	1.8–2.6	Witt et al (1999)[b]
7	K in grain at harvest	g kg^{-1}	2.5	2.3–3.3	Witt et al (1999)[b]
8	P in straw at harvest	g kg^{-1}	1.0	0.7–1.3	Witt et al (1999)[b]
9	K in straw at harvest	g kg^{-1}	12.5	11.7–17.3	Witt et al (1999)[b]
10	P in incorporated straw	g kg^{-1}	1.0		From line 8
11	K in incorporated straw	g kg^{-1}	12.5		From line 9
12	P organic sources	g kg^{-1}	2.0		RTDP, unpublished[c]
13	K organic sources	g kg^{-1}	5.0		RTDP, unpublished[c]
14	Irrigation water requirement	L m^{-2} crop^{-1}	1,250	700–2,100	B. Bouman, IRRI[d]
15	Rainfall	L m^{-2} crop^{-1}	1,000		User-defined
16	Percolation per day	L m^{-2} d^{-1}	5.0	1.0–15.0	B. Bouman, IRRI[d]
17	Duration of submergence	d	100		User-defined
18	P in irrigation water	mg L^{-1}	0.03	0.0–0.12	RTDP, unpublished[c]
19	K in irrigation water	mg L^{-1}	1.75	1.1–3.2	RTDP, unpublished[bc]
20	P in rainwater	mg L^{-1}	0.025	0.01–0.04	Abedin Mian et al (1991)
21	K in rainwater	mg L^{-1}	0.5	0.24–0.5	Greenland (1997)
22	P in soil solution	mg L^{-1}	0.1	0.07–0.47	Greenland (1997)
23	K in soil solution	mg L^{-1}	2.1	2.1–14.3	Greenland (1997)
24	P input with straw	kg ha^{-1}	2.5		L2×L4×L10
25	P input with organic sources	kg ha^{-1}	2.0		L5×L12
26	P input with irrigation water	kg ha^{-1}	0.4		L14×L18:100
27	P input with rainwater	kg ha^{-1}	0.3		L15×L20:100
28	P losses with percolation	kg ha^{-1}	0.5		L16×L17×L22:100
29	P removal with grain and straw	kg ha^{-1}	18.5		L1×L6+L2×L8–L24
30	Phosphorus balance	kg ha^{-1}	–16.3		Σ(L25–L27) – Σ(L28–L29)
31	K input with straw	kg ha^{-1}	31.3		L2×L4×L11
32	K input with organic sources	kg ha^{-1}	5.0		L5×L13
33	K input with irrigation water	kg ha^{-1}	21.9		L14×L19:100
34	K input with rainwater	kg ha^{-1}	5.0		L15×L21:100
35	K losses with percolation	kg ha^{-1}	11.0		L16×L17×L23:100
36	K removal with grain and straw	kg ha^{-1}	73.8		L1×L7+L2×L9–L31
37	Potassium balance	kg ha^{-1}	–52.4		Σ(L32–L34) – Σ(L35–L36)

[a]References refer to ranges on the left; numbers referring to lines are preceded by the letter L. [b]Interquartile ranges. [c]Data collected at sites of the Reversing Trends of Declining Productivity (RTDP) project, 1997-2000. [d]Personal comment.

Table 17.4. Parameters used in the sensitivity analysis of the yield gain and nutrient balance approach for calculating fertilizer requirements. Note that P- and K-limited yield are used as an indicator of indigenous P and K supplies, respectively. See text for further details.

Parameter	Unit	−	Reference value	+
P-limited yield in 0-P plots	t ha^{-1}	5.0	6.0	7.0
K-limited yield in 0-K plots	t ha^{-1}	5.0	6.0	7.0
Fertilizer P recovery efficiency[a]	kg kg^{-1}	0.22	0.28	0.35
Fertilizer K recovery efficiency[a]	kg kg^{-1}	0.35	0.50	0.66
Plant P requirement[b]	kg t^{-1}	2.2	2.6	3.0
Plant K requirement[b]	kg t^{-1}	12.3	14.5	16.7
P input with water[c]	kg ha^{-1}	0	0.1	0.2
K input with water[c]	kg ha^{-1}	0	16	32
Amount of incorporated straw[c]	t ha^{-1}	0.5	2.5	4.5

[a]Median (−) and 75th percentiles (+) were taken from Table 17.1. [b]Reference values (±15%) as suggested by the QUEFTS model (Witt et al 1999). [c]Reference values were taken from Table 17.3.

of grain yield, nutrient inputs with water (input from irrigation and rainwater minus percolation losses), and nutrient inputs from incorporated crop residues (straw). Reference values of these input parameters were reduced (−) or increased (+) to cover a meaningful range of values for the sensitivity analysis (Table 17.4). A yield target of 7 t ha^{-1} was chosen for all scenarios, and the standard setting assumed an indigenous nutrient supply sufficient to support a yield of 6 t ha^{-1} for both P and K. Indigenous nutrient supplies given as nutrient-limited yield then varied from severe nutrient deficiency (yield in 0-P and 0-K plots = 5 t ha^{-1}, Table 17.4) to a situation where a yield response to fertilizer P and K was not expected (yield in 0-P and 0-K plots = 7 t ha^{-1}). Lower and upper recovery efficiencies covered only the range between the median and 75th percentiles given in Table 17.1, whereas reference values for plant P and K requirements ±15% were based on the QUEFTS model (Witt et al 1999). Other nutrient inputs ranged from 0% to 200% (water balance) or 20% to 180% (applied straw) of the reference value taken from Table 17.3. Only one parameter was changed at a time, while the standard value was kept for all other parameters. The results of the sensitivity analysis are given in Figures 17.2 and 17.3.

In the yield gain approach, fertilizer P rates were expectedly most sensitive to changes in IPS (±100%), while effects of changes in fertilizer recovery efficiency (about ±25%) and plant P requirements (±15%) were smaller (Fig. 17.2). However, fertilizer P requirements covered a substantial range (7.4 to 23.6 kg P ha^{-1}) when the interquartile ranges for recovery efficiencies given in Table 17.1 were used in the sensitivity analysis. The P balance was generally negative, with −4.5 kg P ha^{-1} crop^{-1} for the standard yield deficit of 1 t ha^{-1} (yield goal − P-limited yield or IPS). Removal of P was greatest at high IPS since fertilizer P is not applied when a yield

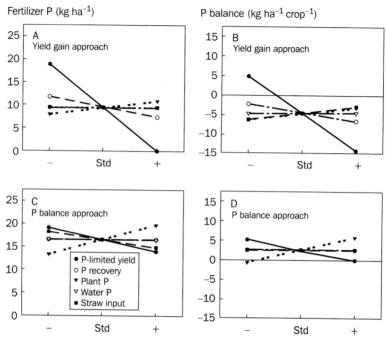

Fertilizer P (kg ha^{-1})

P balance (kg ha^{-1} crop^{-1})

Fig. 17.2. Sensitivity analysis of parameters used for calculating fertilizer P requirements (A, C) following two different approaches (yield gain and P balance) and associated effects on nutrient balances (B, D) at a yield level of 7 t ha^{-1}. Only one parameter was reduced (–) or increased (+) at a time, while all other parameters were kept at the standard value (Std). Parameter values are given in Table 17.4. Note that P-limited yield is used as an indicator of indigenous P supply (IPS).

response is not expected. The P balance became positive when soil P supply was severely limiting.

The fertilizer requirements in the P balance approach largely depend on the yield goal (equation 3), so that changing the standard parameters of IPS, plant P requirements, or straw P inputs had relatively small effects on fertilizer requirements (±10–19%, Fig. 17.2). The recovery efficiency has no effect on the calculation of fertilizer requirements (equation 3) and P rates increased by 16% when the yield deficit increased by 1 t ha^{-1}. The P balance was generally positive (mean 2.6 kg ha^{-1} crop^{-1}) as this approach aims to build up soil P if the P-limited yield is less than the yield goal. The P balance is zero when no yield response is expected and the maintenance rate is applied. Results suggested that P supply and losses with water and straw can be neglected in this approach. However, nutrient inputs from other sources such as farmyard manure (FYM) were not considered in the sensitivity analysis, but should be taken into account as required. Based on the example given in Table 17.3, a nutrient load of 7.5 t FYM ha^{-1} would be sufficient to replenish the P completely removed with grain and straw of a crop yielding 5–6 t ha^{-1}.

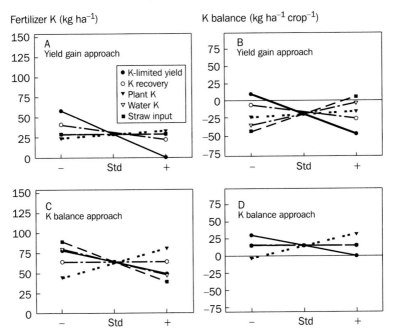

Fig. 17.3. Sensitivity analysis of parameters used for calculating fertilizer K requirements (A, C) following two different approaches (yield gain and K balance) and associated effects on nutrient balances (B, D) at a yield level of 7 t ha⁻¹. Parameter values are given in Table 17.4. Note that P-limited yield is used as an indicator of indigenous K supply (IKS).

Similar results were obtained when evaluating the sensitivity of the two approaches for calculating fertilizer K requirements to changes in standard parameters (Fig. 17.3). Fertilizer K requirements in the yield gain approach depended largely on yield deficit (±100%), recovery efficiencies (–24% to +41% for the range given in Table 17.4, but –46% to +122% for the interquartile ranges given in Table 17.1), and plant K requirement (about ±15%). For the K balance approach, fertilizer rates were most affected by changes in straw input (about ±40%), plant K requirements (about ±30%), IKS (about ±25%), and K water balance (±25%). In contrast to P, K inputs with irrigation water and straw should be considered when calculating fertilizer K maintenance rates. Also, K inputs with FYM help maintain IKS, and 6 t fresh FYM ha⁻¹ contributed 100% of the K load contained in straw dry weight based on the example given in Table 17.3. Note that the nutrient-supplying capacity of external nutrient sources such as FYM should be determined on-farm to properly account for this capacity in the nutrient balance. We recommend applying FYM to omission plots to estimate the combined nutrient supply of FYM and indigenous sources. We further propose to consider the nutrient inputs of FYM in the nutrient balance by taking into account 75% of the P load and 50% of the K load of FYM based on the assumed first-season recovery efficiencies of 25% for P and 50% for K supplied with FYM (see

Chapter 11, this volume). These assumptions should be validated for other organic nutrient sources as and where required.

Each of the two approaches has its strengths and weaknesses. The yield gain approach offers a meaningful adjustment of fertilizer rates considering yield goal and indigenous nutrient supply, which are both important criteria for efficient and economical site-specific nutrient management. Weaknesses include uncertainties related to the estimation of recovery efficiencies and the danger of nutrient depletion. The nutrient balance approach offers little opportunity to adjust fertilizer rates to differences in soil nutrient supply and is therefore less site-specific in that regard, but it maintains or, if necessary, builds up soil nutrient supplies required to support site-specific yield levels. Combining the two approaches should eliminate the weaknesses while keeping the strength of the two approaches. The following assumptions and adjustments were made to develop a combined yield gain + nutrient balance approach in order to avoid both overfertilization and severe mining of soil P and K resources:

- Fertilizer rates provided by the yield gain approach (equation 2) are considered as minimum rates if they are above the maintenance recommendations (equations 3 and 4).
- Where appropriate, location-specific parameters should be used to fine-tune the recommendations, provided that good measurements of key components of the nutrient input-output balance or other parameters used in Table 17.3 are available. For example, irrigation water amounts and concentrations of K can vary widely among different areas, depending on the water source, climate, cropping season, soil type, and predominant water management practices.
- To calculate maintenance rates, nutrient inputs with organic sources such as farmyard manure should be considered for both P and K, while nutrient inputs with straw and a water-related nutrient balance need to be considered only for K.

Note that the fertilizer P and K requirements calculated with equations 3–4 represent moderate application rates because conservative estimates of nutrient requirements, recovery efficiencies, and input parameters for the nutrient balances were chosen as default values and percolation losses are probably smaller in many irrigated rice areas (Table 17.3). Where a clear yield response is expected, we suggest applying the minimum of 20 kg P_2O_5 and 30 kg K_2O ha^{-1} per ton of targeted yield increase according to the yield gain approach to ensure that farmers fully exploit the attainable yield gap and achieve a visible yield increase in regular years without major problems. However, maintenance rates may be further reduced considering that the desired yield goal will not be reached every season because of constraints other than nutrient management (e.g., climate, pests, etc.). Based on the accumulated experience with SSNM in more than 200 farmers' fields with irrigated rice in six Asian countries, achievements of the yield goal varied from site to site, ranging from about 75% to 95% (Dobermann et al 2002). For maintenance, the true crop removal

Straw input 2.5 t ha^{-1}

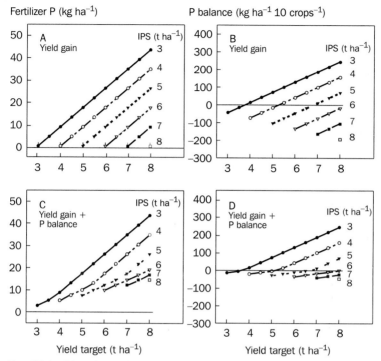

Fig. 17.4. Phosphorus fertilizer rates (A, C) and balances (B, D) for two different calculation approaches (yield gain and yield gain + P balance) as affected by yield target and indigenous P supply measured as grain yield in a 0-P omission plot (P-limited yield) at a straw input of 2.5 t ha^{-1}. The P balances were calculated for 10 cropping cycles using the standard parameters given in Table 17.3. See text for further details.

that needs to be replenished in the long run is thus rarely more than 85% of the yield-goal-based removal.

Potential impact of different fertilizer P and K strategies on input-output balances

Fertilizer P and K requirements were calculated using the two different approaches (yield gain and yield gain + nutrient balance) evaluating a wider range of yield goals and indigenous nutrient supplies measured as grain yield in 0-P and 0-K omission plots (P- and K-limited yield, respectively). The model settings included a fixed straw return of 2.5 t straw ha^{-1} for the calculation of fertilizer P rates and P balances (Fig. 17.4). Considering the higher K than P concentration in straw, fertilizer K rates and K balances were calculated for low straw inputs of 0.5 t ha^{-1} (Fig. 17.5) as found in many parts of India, Nepal, Bangladesh, and North Vietnam, and higher straw inputs

Straw input 0.5 t ha^{-1}

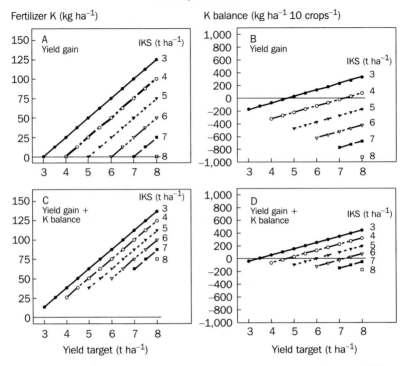

Fig. 17.5. Potassium fertilizer rates (A, C) and balances (B, D) for two different calculation approaches (yield gain and yield gain + K balance) as affected by straw input (0.5 t ha^{-1}), yield target, and indigenous K supply measured as grain yield in a 0-K omission plot (K-limited yield). The K balances were calculated for 10 cropping cycles using the standard parameters given in Table 17.3. See text for further details.

of 4.5 t ha^{-1} (Fig. 17.6), which can already be achieved when leaving 20–25 cm of rice stubble in the field after harvesting a crop of 6–7 t grain yield ha^{-1} (Witt et al 2000). Such a straw management practice is common in the Philippines (Gines et al, Chapter 8, this volume), and comparable straw inputs are also achieved for lower-yielding crops in countries where longer stubble remain in the field (Philippines, Indonesia) or combine harvest is common (Thailand, South Vietnam, northern India). All other parameters were based on the standard values given in Table 17.3. Maintenance rates estimated with equations 3 and 4 were reduced by 15%, taking into account that the yield goal will not be reached in every season. However, the nutrient balances were calculated for 10 cropping cycles, assuming that the yield goal was reached every season to depict the situation of greatest crop removal. This scenario would represent a 5-year period of a typical double-rice cropping system in Asia with good crop management and few constraints other than nutrients.

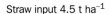

Straw input 4.5 t ha^{-1}

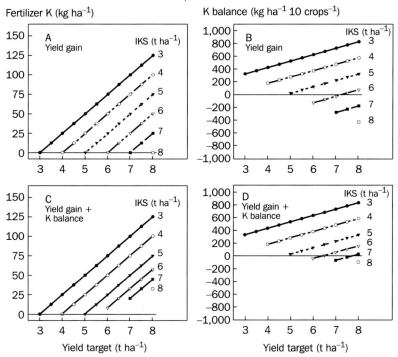

Fig. 17.6. Potassium fertilizer rates (A, C) and balances (B, D) for two different calculation approaches (yield gain and yield gain + K balance) as affected by straw input (4.5 t ha^{-1}), yield target, and indigenous K supply measured as grain yield in a 0-K omission plot (K-limited yield). The K balances were calculated for 10 cropping cycles using the standard parameters given in Table 17.3. See text for further details.

In the yield gain approach, fertilizer P rates were increased at a constant rate of 9 kg P ha^{-1} per ton of expected yield deficit (Fig. 17.4A). Depending on the yield level, this strategy resulted in a net P removal of about –50 to –200 kg P ha^{-1} 10 crops^{-1} where a yield response was not expected (Fig. 17.4B). Balances were more negative when yield goals increased and expected yield gains were small. In such situations, the combined yield gain + P balance approach suggested maintenance rates of 3 to 15 kg P ha^{-1} to replenish the increased nutrient removal with grain and straw (Fig. 17.4C), and P balances were only slightly negative, ranging from about –15 to –50 kg P ha^{-1} 10 crops^{-1} (Fig. 17.4D). Where IPS levels were low and yield gains large, fertilizer rates followed the yield gain principle in both approaches, resulting in positive P balances and a buildup of IPS (Fig. 17.4B and D).

Similar trends were obtained when evaluating the two approaches for K, but straw management had a pronounced effect on K fertilizer requirements and balances of the two approaches. Where straw input is small and crop responses are not expected, the yield gain approach would lead to a strong net removal of K with grain and straw in the range of –175 to –925 kg K ha^{-1} 10 crops^{-1} (Fig. 17.5A and B). Regardless of the yield target and amount of fertilizer applied with the yield gain approach, K balances would be negative for all IKS above 4 t ha^{-1}. It is certainly advisable to evaluate the long-term effects of such a K management strategy on IKS at low straw inputs before recommending it on a larger scale. According to the combined yield gain + K balance approach, higher fertilizer K rates are required at low straw input to reverse severe negative K balances and avoid rapid soil K depletion (Fig. 17.5C and D). To account for the greater K removal with grain and straw at elevated yield levels, fertilizer rates in the combined approach mostly followed the K balance strategy (equation 4) and rates were more closely related to the yield goal than the yield deficit. Fertilizer maintenance rates ranged from 15 to 75 kg K ha^{-1} for yields ranging from 3 to 8 t ha^{-1}. If sufficient amounts of straw are returned, maintenance rates are essentially not required for yields below 7 t ha^{-1}, and only a small dose of 20–30 kg K ha^{-1} would be required at yields of 7–8 t ha^{-1} (Fig. 17.6C). Both approaches therefore largely followed the yield gain principle, and K balances with the yield gain approach would be positive in most cases because of the high K concentration in straw (Fig. 17.6C).

In conclusion, the combined yield gain + nutrient balance approach provided a useful framework for preparing meaningful long-term fertilizer P and K strategies to overcome apparent nutrient limitations and avoid nutrient depletion as and where required. Fertilizer rates and nutrient balances were mostly influenced by yield goal and indigenous nutrient supply, but other factors should be considered where appropriate, such as yield stability, straw management (K), nutrient inputs with farmyard manure (P and K) and irrigation water (K), and nutrient losses with percolation (K). If uncertainties exist whether local conditions greatly divert from the proposed standard nutrient balance given in Table 17.3, research should aim to improve estimates of relevant components of the input-output balance and also assess changes in IPS and IKS using long-term experiments. The latter could provide useful information on whether P- and K-limited yields are expected to decline without the application of maintenance rates. This would be particularly important where the soil reservoir for supplying P and K is expected to be small because of cropping history or limitations related to soil mineralogy or soil volume. Long-term experiments are also useful where input-output balances are difficult to estimate, as in the Mekong River Delta of Vietnam, where annual flooding provides additional nutrient inputs through sedimentation.

Validation of fertilizer calculation approaches

The data set from 179 farmers' fields with irrigated rice in Asia described in Chapters 6–12 plus two additional cropping seasons was used to evaluate these simplified and refined fertilizer calculation approaches. Grain yield and fertilizer use in treatments with SSNM and the farmers' fertilizer practice (FFP) for high- and low-yielding seasons (HYS, LYS) are given in Tables 17.5 to 17.7. Fertilizer rates were estimated following the yield gain principle (N, P, and K) and yield gain + nutrient balance approach (P and K) using site-specific, realistic yield goals and indigenous

Table 17.5. Fertilizer N requirements calculated with the yield gain approach for adjusted yield goals in high- and low-yielding seasons compared with yield and fertilizer N use with site-specific nutrient management (SSNM) and the farmers' fertilizer practice (FFP) in each three to four cropping seasons in 179 farmers' fields in 1997-2000. The indigenous N supply was measured as grain yield in N omission (0-N) or minus fertilizer (–F) plots each four seasons (Dobermann et al 2003b).

Site[a]	Grain yield (t ha^{-1})					Fertilizer N (kg ha^{-1})			
	FFP	SSNM	Goal		0-N or –F plots	FFP	SSNM[b]	Yield gain	
			SSNM	Adjusted				AEN$_{25}$[c]	AEN$_{20}$[d]
High-yielding season									
JI	6.6	7.0	8.0	7.2	5.3	177	126	95	125
HA	6.2	6.4	7.2	7.0	4.9	105	99	85[e]	105[e]
AD	6.5	7.0	8.0	7.0	4.5	121	127	100	125
TH	5.6	6.2	7.2	6.5	4.2	103	127	90	115
SU	4.9	5.1	6.5	6.0	3.7	131	96	90	115
MA	5.1	5.8	7.8	7.0	4.0	139	143	120	150
OM	5.3	5.7	7.0	6.0	3.6	106	106	95	120
Mean	5.7	6.1	7.4	6.7	4.4	126	118	96	122
Low-yielding season									
JI	5.5	5.9	7.2	6.5	4.7	165	126	90	120
HA	6.1	6.3	6.5	6.5	4.3	105	86	90[e]	110[e]
AD	5.7	6.1	7.0	6.5	4.1	111	125	110	135
TH	4.7	5.4	6.5	6.0	3.1	105	135	90	110
SU	4.2	4.3	5.0	5.0	2.4	131	96	105	130
MA	3.9	4.2	6.0	5.0	3.1	86	82	75	95
OM	3.4	3.5	5.3	4.5	2.0	114	94	55	70
Mean	4.8	5.1	6.2	5.7	3.5	117	106	88	110

[a]JI = Jinhua, Zhejiang (China); HA = Hanoi, Red River Delta (Vietnam); AD = Aduthurai, Old Cauvery Delta, Tamil Nadu (India); TH = Thanjavur, New Cauvery Delta, Tamil Nadu (India); SU = Sukamandi, West Java (Indonesia); MA = Maligaya, Central Luzon (Philippines); OM = Omon, Mekong Delta (Vietnam). [b]Fertilizer N rates were based on the SSNM yield goals. [c]Fertilizer N rates for the adjusted yield goals were calculated using an agronomic efficiency of 25 kg grain kg^{-1} fertilizer N except for JI (20 kg kg^{-1}). [d]Fertilizer N rates for the adjusted yield goals were calculated using an agronomic efficiency of 20 kg grain kg^{-1} fertilizer N except for JI (15 kg kg^{-1}). [e]Note that an application of 8.5 t farmyard manure ha^{-1} typical for this site is not considered in this estimate.

nutrient supplies measured separately in HYS and LYS. Indigenous nutrient supplies were calculated for two (IPS, IKS) to four (INS) cropping seasons (Dobermann et al 2003b). The yield goals used in 1997-2000 were adjusted to more realistic levels considering the actual yields achieved with SSNM (Table 17.5). In both seasons, adjusted yield goals were up to 0.6 t ha^{-1} greater than yields achieved with SSNM except for sites where higher yields can be expected once other constraints are removed (see particularly Chapters 8 and 9, this volume, for production constraints in Maligaya, Philippines, and Sukamandi, Indonesia). Note that adjusted yield goals were about 0.9 t ha^{-1} greater than yields achieved with the FFP in both HYS and LYS, which compares to +17% in the HYS and +21% in the LYS.

Table 17.6. Fertilizer P (FP) requirements in high- and low-yielding seasons (HYS and LYS) calculated with two different approaches (YG = yield gain, YG + P = yield gain + P balance) compared with average fertilizer P use in treatments with site-specific nutrient management (SSNM) and the farmers' fertilizer practice (FFP) in each three to four cropping seasons in 179 farmers' fields in 1997-2000. The indigenous P supply (IPS) was measured as grain yield in P omission (0-P) plots each two HYS and LYS (Dobermann et al 2003b). The yield gain (ΔGY) is the difference between the adjusted yield goal given in Table 17.5 and yield in 0-P plots.

Site	IPS (t ha^{-1})	ΔGY (t ha^{-1})	FP$_{YG}$ (kg ha^{-1})	FP$_{YG + P}$ (kg ha^{-1})	FP$_{SSNM}$ (kg ha^{-1})	FP$_{FFP}$ (kg ha^{-1})
High-yielding season						
JI	6.4	0.8	7	15	15	20
HA	6.1	0.9	8	8[a]	20	23
AD	6.5	0.5	4	16	21	24
TH	5.8	0.7	6	15	16	18
SU	4.9	1.1	10	14	19	9
MA	6.3	0.7	6	15	16	16
OM	4.7	1.3	11	13[b]	20	19
Mean	5.9	0.9	7	14	20	19
Low-yielding season						
JI	6.4	0.1	1	12	13	19
HA	4.9	1.6	14	14[a]	10	19
AD	5.6	0.9	8	15	18	20
TH	4.1	1.9	17	17	16	18
SU	3.5	1.5	13	13	13	6
MA	4.3	0.2	6	11	13	14
OM	2.8	1.2	15	15	19	21
Mean	4.6	1.1	11	14	15	17

[a]Assuming an input of 17 kg P ha^{-1} with an application of 8.5 t farmyard manure ha^{-1}. [b]Note that additional P inputs from sedimentation prior to the dry season are not considered.

Newly calculated fertilizer rates were evaluated and compared with rates used in SSNM and FFP considering the (1) agronomic efficiencies of N (AEN, kg grain kg^{-1} fertilizer N), (2) internal efficiencies of P and K (IEP, IEK, kg grain kg^{-1} plant nutrient), and (3) P and K input-output balances.

Nitrogen. Two different fertilizer N rates were calculated following the yield gain principle given in equation 2 using fertilizer requirements of 40 and 50 kg N ha^{-1} t^{-1} grain yield, which is equivalent to agronomic efficiencies (AEN) of 25 and 20 kg grain per kg fertilizer N applied, respectively. For the site in Zhejiang, China, a lower AEN of 20 and 15 kg kg^{-1} was used considering the low N-use efficiencies that are achieved at this site (see Chapter 12, this volume). Note that a targeted AEN of only 20 kg kg^{-1} would be 4.6 kg kg^{-1} or 31% greater than the actual AEN achieved in farmers' fields with SSNM, excluding the sites in China and Thailand (see Table 15.7, Chapter 15, this volume).

Table 17.7. Fertilizer K (FK) requirements in high-yielding seasons (HYS) calculated with two different approaches (YG = yield gain, YG + K = yield gain + K balance) compared with average fertilizer K use in treatments with site-specific nutrient management (SSNM) and the farmers' fertilizer practice (FFP) in 3–4 HYS in 179 farmers' fields in 1997-2000. The indigenous K supply (IKS) was measured as grain yield in K omission (0-K) plots in each two HYS and low-yielding seasons (LYS) (Dobermann et al 2003b). The yield gain (ΔGY) is the difference between the adjusted yield goal given in Table 17.5 and yield in 0-K plots.

Site	Straw input (t ha^{-1})	IKS (t ha^{-1})	ΔDGY (t ha^{-1})	FK_{YG} (kg ha^{-1})	FK_{YG+K} (kg ha^{-1})	FK_{SSNM} (kg ha^{-1})	FK_{FFP} (kg ha^{-1})
High-yielding season							
JI	3.3	6.3	0.9	22	51	57	66
HA	0.5	5.8	1.2	30	46[a]	63	67
AD	1.1	6.6	0.4	11	66	68	37
TH	0.9	5.6	0.9	23	68	75	40
SU	2.1	4.9	1.1	27	51	30	4
MA	2.7	6.3	0.7	17	52	61	29
OM	3.0	4.9	1.1	29	42[b]	37	25
Mean	1.9	5.8	0.9	23	54	56	38
Low-yielding season							
JI	3.3	6.4	0.1	3	32	47	51
HA	0.5	4.4	2.1	51	53[a]	47	63
AD	1.1	5.6	0.9	23	66	55	35
TH	0.9	4.1	1.9	48	74	73	37
SU	2.1	3.5	1.5	38	44	42	4
MA	2.7	4.3	0.2	18	28	35	25
OM	3.0	3.3	0.7	33	33	49	28
Mean	1.9	4.6	1.1	31	47	50	35

[a]Assuming an input of 42.5 kg K ha^{-1} with an application of 8.5 t farmyard manure ha^{-1}. [b]Note that additional P inputs from sedimentation prior to the dry season are not considered.

Using measured INS and realistic yield goals that were comparable with actual yields achieved with SSNM in high-yielding seasons, the yield gain approach produced a range of fertilizer N rates that were comparable with actual fertilizer N use in SSNM. Depending on the AEN chosen, average fertilizer requirements ranged from 96 to 122 kg N ha^{-1}, vis-à-vis an average fertilizer use of 118 kg N ha^{-1} in SSNM and 126 kg N ha^{-1} in FFP. At most high-yielding sites (6–7 t ha^{-1}), fertilizer N rates used with SSNM were close to fertilizer N requirements calculated with an AEN of 20 kg kg^{-1} (15 kg kg^{-1} at Jinhua, China). In the LYS, average fertilizer N requirements ranged from 88 to 110 kg N ha^{-1}, vis-à-vis 106 kg N ha^{-1} in SSNM and 117 kg N ha^{-1} in FFP, but yield goals were higher than yields achieved with SSNM and FFP. Thus, fertilizer N rates calculated with the yield gain approach were in good congruence with the actual rates used in SSNM, although there was a substantial variation among sites and seasons. In the HYS, differences in fertilizer N between SSNM and the yield gain approach ranged from +6 to +37 kg N ha^{-1} for the higher AEN and from +12 to –19 kg N ha^{-1} for the lower AEN. Variation was even greater in the LYS, and the corresponding differences ranged from +45 to –9 kg N ha^{-1} for the higher AEN and from +25 to –34 kg N ha^{-1} for the lower AEN. Efficient fertilizer N use will in most cases depend on the opportunities for real-time N management, and further improvement of fertilizer N strategies practiced with SSNM in 1997-2000 should be possible through a more fine-tuned use of on-farm tools such as the chlorophyll meter or leaf color chart (LCC). Clearly, the major objective of SSNM is to increase farmers' profit through the efficient use of fertilizer, to increase the agronomic N efficiency, and to avoid environmental risks caused by inefficient fertilizer N use. The share of N fertilizer cost in production value was only about 4.7% for the average yield, fertilizer N use, and N fertilizer cost reported across all sites (Dawe et al, Chapter 16, this volume). A savings of 30 kg fertilizer N ha^{-1} would reduce the amount of fertilizer N released into the environment and lead to a relatively small profit increase of US$10.50, which can also be achieved with a yield increase of only 70 kg ha^{-1}. Profit increases with SSNM will therefore mainly have to come from yield increases through efficient fertilizer N use rather than savings in costs for fertilizer N. Optimizing agronomic efficiency through fertilizer N reduction without increasing yields would therefore contribute little to increasing farmers' profit, unless pesticide inputs could be reduced where plants are less susceptible to pests and diseases at lower N levels. This may also have external benefits in terms of health improvements. Greater efforts are probably required to remove non-nutrient-related constraints to increase yields and N efficiencies.

The most suitable N management strategy will have to be selected from a comparison of several available N management options, including location-specific split schedules for preventive N management (preseason calculation of fertilizer N), corrective N management using an LCC (real-time N management based on plant N status), or a combination of both (see Witt et al, Chapter 18, this volume). Using participatory approaches to meet farmers' needs, N management strategies will then have to be refined at new locations, which may require a few cropping seasons. Where a full real-time N management strategy provides the greatest opportunities to increase farmers' yield and profit, INS measurements will be needed only to determine the necessity for basal N applications. For other strategies, for example, where fertilizer N is applied at critical growth stages following a more fixed N splitting pattern, the yield gain approach offers a preseason calculation of fertilizer N requirements. In this case, it may be sufficient to provide a meaningful, season-specific range of fertilizer N rates given the inefficient fertilizer N use in certain rice domains in Asia (e.g., in China and Indonesia, Table 17.5) and the substantial year-to-year variation in yield and yield responses to fertilizer N application (Dobermann et al, Chapter 15, this volume). Individual split N applications should then be fine-tuned using tools such as the chlorophyll meter or LCC to provide additional options in the more static approaches (Witt et al, Chapter 18, this volume).

Phosphorus. Two different fertilizer P rates were calculated using the yield gain and yield gain + P balance approach and compared with fertilizer P rates in

SSNM and FFP in 1997-2000 (Table 17.6). Average fertilizer P rates were similar in FFP and SSNM in both HYS (19 and 20 kg P ha^{-1}, respectively) and LYS (17 and 15 kg P ha^{-1}, respectively), but variation in rates was greater among sites in the FFP treatment. Average fertilizer P rates in SSNM were more than sufficient to support the higher yields achieved with SSNM as plant P uptake was sufficient to support even higher yields. The average IEP in SSNM was about 300 kg grain kg^{-1} plant P, and more than 75% of the cases had an IEP of below 335 kg kg^{-1} (Table 17.1). This compares to an optimal IEP of 385 kg kg^{-1} suggested by the QUEFTS model (Witt et al 1999). Thus, fertilizer P rates could probably be further reduced with SSNM. The expected yield gain or P-related yield deficit in the HYS averaged 0.9 t ha^{-1}, ranging from 0.5 to 1.3 t ha^{-1}. Following the yield gain + P balance approach, fertilizer P requirements would average 14 kg P ha^{-1}, ranging from 8 to 16 kg P ha^{-1} depending on the site (Table 17.6). This would be a reduction of 30% vis-à-vis fertilizer P rates used in SSNM in 1997-2000. Fertilizer P could be reduced by 60% at the Hanoi site, where farmers apply substantial amounts of P through FYM and P input-output balances are generally positive (Fig. 17.7A). Suggested fertilizer P reductions at other sites would be smaller and result from setting more realistic yield goals. However, caution is required at the four sites in Jinhua (China), Aduthurai and Thanjavur (both India), and Maligaya (Philippines), where P input-output balances with SSNM were negative for at least 50% of the cases. There was little difference in fertilizer P rates in the LYS, and an average of 14–15 kg P ha^{-1} should be sufficient to support the adjusted yield goals. At Hanoi, suggested fertilizer P rates were higher in the LYS than in the HYS because the P-related yield deficit was estimated to be 1.6 t ha^{-1} in the LYS compared with only 0.9 t ha^{-1} in the HYS. Additional measurements would be required to verify observed P limitations in the LYS, taking into account that seasonal differences in indigenous P supplies are also influenced by temperature.

Following the yield gain approach (equation 2), average fertilizer P requirements in the HYS would amount to only 7 kg P ha^{-1}, ranging from 4 to 11 kg P ha^{-1} (Table 17.6). This is about 35% of the fertilizer P applied with SSNM or FFP in 1997-2000 and 50% of the fertilizer P rates calculated with the yield gain + P balance approach. In the LYS, fertilizer P rates based on yield gain would amount to 11 kg P ha^{-1}, which is 65%, 73%, and 79% of the rates in FFP, SSNM, and yield gain + P balance, respectively. Fertilizer rates based on the yield gain approach would be insufficient to maintain soil P supplies in the long term considering the relatively even input-output balances with SSNM and FFP at the seven major irrigated rice domains (Fig. 17.7A). In most cases, P input-output balances with the yield gain approach would be negative and fertilizer P rates calculated with the yield gain + P balance approach would mostly follow the P balance (Table 17.6).

Potassium. Although there was little difference in the average fertilizer P use among SSNM and FFP in 1997-2000, average fertilizer K rates were about 45% greater with SSNM than with FFP (Table 17.7). However, a comparison of internal efficiencies suggested that plant K uptake was more efficiently translated into grain yield than plant P (Table 17.1). Although IEP was generally below optimal levels

P balance (kg P ha⁻¹ crop⁻¹)   K balance (kg K ha⁻¹ crop⁻¹)

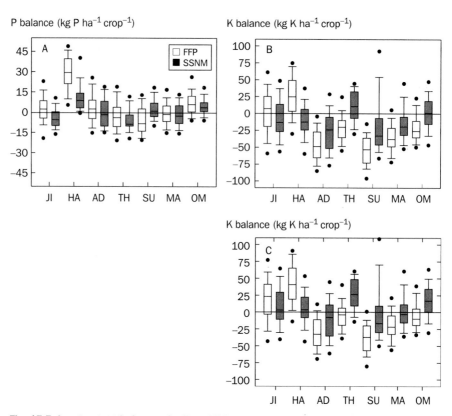

K balance (kg K ha⁻¹ crop⁻¹)

Fig. 17.7. Input-output balances for P and K in treatments with site-specific nutrient management (SSNM) and the farmers' fertilizer practice (FFP) during six cropping seasons in 179 farmers' fields in 1997-2000. Balances in Figures A and B were calculated according to Dobermann et al (Chapter 2, this volume), while additional net K inputs of 16 kg K ha⁻¹ were assumed in Figure C. The box plots show medians (horizontal lines), 25th to 75th percentiles (boxes), 10th and 90th percentiles (error bars), and minimum and maximum values (bullets). See text for further information.

(see above), average IEK was 59 kg grain kg⁻¹ plant K, and more than 25% of the cases were close to or above the optimal IEK of 69 kg kg⁻¹ calculated with the QUEFTS model (Witt et al 1999). This would suggest that there was generally little excess plant K uptake to support the yields achieved, and that efforts should concentrate on removing non-nutrient-related constraints to increasing productivity. As shown in Table 17.7, fertilizer K use with SSNM increased significantly in Aduthurai and Thanjavur (both Tamil Nadu, India), Maligaya (Philippines), Sukamandi (Indonesia), and Omon (Vietnam).

Potassium input-output balances were calculated for two scenarios following the approach described by Dobermann et al (Chapter 2, this volume), assuming (1) K inputs from atmosphere or water equivalent to K losses by leaching in a standard scenario (Fig. 17.7B) and (2) net K inputs of 16 kg K ha⁻¹ in a conservative scenario (Fig. 17.7C) based on estimated K inputs with irrigation water and rain (27 kg K ha^{-1}, Table 17.3) and K losses through percolation (11 kg K ha^{-1}). The conservative scenario represents a more optimistic estimate of the K balance, but better estimates of net K inputs with water are needed considering the sensitivity of the K balance to changes in this component.

The standard K input-output balances suggested that, even with increased fertilizer K rates, K balances with SSNM were negative in more than 50% of the cases at three sites and in more than 75% of the cases at another three sites (Fig. 17.7B). A clearly positive K balance with SSNM was observed only at Thanjavur. Where K rates in SSNM were similar to those in FFP as in Jinhua (China) and Hanoi (Vietnam), positive K balances in FFP became neutral in SSNM because of higher yields and greater K removal with grain and straw in SSNM. Except for these two sites, K balances would still be largely negative with FFP in the more optimistic scenario, with assumed greater natural K inputs given in Fig. 17.7C, whereas negative K balances were reversed with SSNM at most sites.

Using the newly adjusted yield goals, the average K-related yield gain or deficit averaged 0.9 t ha^{-1} in the HYS, ranging from 0.4 t ha^{-1} in Aduthurai to 1.2 t ha^{-1} in Hanoi. In the LYS, average yield gains were 1.1 t ha^{-1} and covered a wider range, from 0.1 t ha^{-1} in Jinhua to 2.1 t ha^{-1} in Hanoi (Table 17.7). Average fertilizer K rates calculated with the yield gain + K balance approach were comparable with rates used with SSNM in 1997-2000, but there were differences in previous and revised K rates among sites. Fertilizer K rates calculated with the yield gain approach are likely to be insufficient to support targeted yields in the long term, and rates would even be 39% and 11% lower than current fertilizer K use in FFP in high- and low-yielding seasons, respectively. In addition to the detailed discussions in the individual Chapters 6–12 (this volume), the following conclusions can be drawn for individual sites from the comparisons provided with Table 17.7:

- At the sites near Jinhua (China) and Hanoi (Vietnam), fertilizer K use by farmers is adequate or could be reduced slightly based on estimated K input-output balances (Fig. 17.7B and C) and expected yield gains (Table 17.7). Estimated yield gains in the LYS should be confirmed at Hanoi, as they were higher than in the HYS. Long-term experiments at Jinhua showed rapid decreases in IKS (and IPS) when maintenance K (and P) rates were suspended for a few seasons (Wang et al 2001).
- The two sites in Aduthurai and Thanjavur (India) are characterized by insufficient fertilizer K use by farmers, and recent research confirmed yield responses in the range of 0.2–0.4 t ha^{-1} to increased fertilizer K rates calculated with the yield gain + K balance approach in both HYS and LYS (RTOP[1], unpublished data). Thus, fertilizer K rates suggested by the yield gain approach were insufficient to fully exploit an existing yield gap in both HYS and LYS.

---

[1]Reaching Toward Optimal Productivity (RTOP), 2001-2004, is the continuation of the project on Reversing Trends in Declining Productivity (RTDP), 1997-2000.

- Farmers' fertilizer K rates are insufficient in the HYS in Maligaya (Philippines) but adequate in the LYS. The yield gain + K balance approach provided adequate fertilizer K recommendations, whereas fertilizer K rates calculated with the yield gain approach were low, particularly in the HYS, in which K deficiencies have been observed under the current farmers' practice (see Chapter 8).

- Farmers' fertilizer K use in Sukamandi (Indonesia) is only 4 kg K ha^{-1}, vis-à-vis recommended fertilizer K rates of 27–50 kg K ha^{-1} in the HYS and 38–44 kg K ha^{-1} in the LYS with SSNM, yield gain, and yield gain + K balance approaches. Yield responses to increased fertilizer K rates should be validated and compared with current fertilizer K use in farmers' fields in participatory on-farm trials that aim to improve other crop management practices as well (see Chapter 9).

- Farmers' fertilizer K rates at Omon (Vietnam) are probably adequate in both HYS and LYS. Note that the fertilizer K rates in the HYS suggested with the yield gain + K balance approach would be lower if nutrient inputs with sedimentation caused by flooding prior to the HYS (see Chapter 10) were known and considered in the calculation. Fertilizer K rates calculated with the yield gain approach would be adequate at this site because of the substantial annual nutrient inputs with returned crop residues and sedimentation.

Seasonal differences in the estimation of indigenous nutrient supplies. Grain yield and nutrient uptake in omission plots were consistently greater in HYS crops than in LYS crops (Dobermann et al 2003a). The average difference between HYS and LYS was 0.9 t ha^{-1} or 20% for N and 1.3 t ha^{-1} or 28% for both P and K (Tables 17.5–17.7). To obtain an estimate of the potential or effective estimate of indigenous nutrient supplies, it has been proposed to conduct crop-based measurements mainly in HYS crops because crop growth in LYS would be more affected by abiotic and biotic constraints to yield (Dobermann et al 2003b). Measuring indigenous nutrient supplies in HYS crops and using such crop-based estimates for LYS crops, however, may considerably underestimate fertilizer requirements, particularly for N. For example, the average actual yield in 0-N plots in the HYS was 4.4 t ha^{-1} vis-à-vis 3.5 t ha^{-1} in the LYS (Table 17.5). Differences in nutrient uptake between HYS and LYS crops are driven not only by seasonal differences in plant growth but also by differences in other factors governing nutrient availability such as root mass, soil temperature, nutrient inputs with irrigation water, fallow period management, crop establishment, and many more.

Using IPS and IKS measured at the individual sites, we further compared the effect of indigenous nutrient supplies estimated in HYS and LYS on fertilizer P and K requirements in LYS (Table 17.8). At most sites, the nutrient-limited yield measured in the HYS (potential IPS and IKS) was slightly lower than or equal to the yield goal in the LYS, so that little or no fertilizer P and K would be required according to the yield gain approach. On average, fertilizer rates were only a fraction (9–16%) of rates based on the IPS and IKS measured in the LYS. Differences were smaller with

Table 17.8. Fertilizer P and K (FP, FK) requirements in low-yielding seasons based on estimates of indigenous P and K supplies (see Tables 17.6 and 17.7, respectively, Chapter 17, this volume) measured in high- and low-yielding seasons (HYS, LYS). Fertilizer rates were calculated with two different approaches (YG = yield gain; YG + P = yield gain + P balance or YG + K = yield gain + K balance).

	Origin of IPS estimate				Origin of IKS estimate			
Site	HYS		LYS		HYS		LYS	
	FP_{YG}	$FP_{YG + P}$ $(kg\ ha^{-1})$	FP_{YG}	$FP_{YG + P}$	FK_{YG}	$FK_{YG + K}$ $(kg\ ha^{-1})$	FK_{YG}	$FK_{YG + K}$
JI	1	12	1	12	5	34	3	32
HA	3	0	14	14	18	33	51	53
AD	0	13	8	15	0	55	23	66
TH	2	13	17	17	10	55	48	74
SU	0	9	13	13	0	25	38	44
MA	0	0	6	11	0	0	18	28
OM	0	0	15	15	0	0	33	33
Mean	1	7	11	14	5	29	31	47

the combined yield gain + nutrient balance approach, but average fertilizer rates suggested for the LYS were 34–50% lower when based on HYS than on LYS estimates of indigenous nutrient supplies. Differences were smallest at sites where fertilizer rates of the combined approach largely followed rates calculated following the nutrient balance (e.g., Jinhua, China). At Maligaya (Philippines) and Omon (Vietnam), both calculation approaches suggested not to apply fertilizer P and K when rates were based on HYS estimates of IPS and IKS. While evidence is good that the potential IPS and IKS can be more accurately measured in the HYS, the calculation of fertilizer rates should be based on the actual yield deficit observed in a particular season. Fertilizer rates that were based on the potential IPS and IKS measured in the HYS would not consider actual nutrient limitations that can occur in the LYS measured as the difference between actual yield and nutrient-limited yield. We therefore suggest developing fertilizer N, P, and K rates based on estimates of indigenous nutrient supplies obtained in the same season.

In conclusion, the equations for estimating nutrient and fertilizer requirements of N, P, and K suggested in this section provide not only a simplification but also an improvement of the SSNM approach described in Chapter 5. The yield gain approach offers simple but robust principles for both the evaluation of current fertilizer strategies in farmers' fields and the development of improved recommendations. Integrating a nutrient balance in the yield gain approach appears essential for developing meaningful short- and long-term fertilizer P and K strategies. Few additional input parameters would be needed, which should at least include the average amount of straw returned and the use of organic nutrient sources such as FYM. The combined yield gain + nutrient balance strategy provides scientific principles for the development of fertilizer P and K maintenance strategies, which were lacking in the original SSNM approach.

17.2 General framework for technical decision support

A major advantage of the SSNM approach is that extension staff or farmers can develop fertilizer requirements on-farm using readily available information and tools. Fertilizer N strategies can be developed and fine-tuned using such inexpensive on-farm tools as the LCC, and expensive chemical analysis in laboratories is not required for the development of fertilizer P and K rates where on-farm estimates of soil indigenous nutrient supplies are based on yield measurements in omission plots. However, the development of fertilizer recommendations and the installation of omission plots may be challenging for individual farmers more since excellent management of omission plots is crucial to obtaining reliable crop-based estimates of indigenous nutrient supplies. Furthermore, field-specific management is probably only useful for very dynamic nutrients such as nitrogen to achieve optimal congruence between N supply and crop demand and therefore high fertilizer N-use efficiency. In contrast, management of P and K mainly requires decisions about rates to apply (Dobermann and White 1999) and recommendations can probably be given for larger domains rather than being field-specific. Where farmers' fertilizer use is inadequate, it may be most effective and economical to develop, evaluate, and locally adapt improved fertilizer recommendations through farmer participation and then promote new guidelines in suitably large areas, including guidelines for further adjustments.

On the basis of the general framework for decision support depicted in Figure 17.8, the development of improved fertilizer recommendations may include six major steps with the following outputs:

1. *Recommendation domains and indigenous nutrient supplies.* Larger areas are divided into smaller agroecological recommendation domains. Domain sizes determine the required number of nutrient omission plots that are used to obtain average N-, P-, and K-limited yields (estimates of indigenous nutrient supplies) valid for the domain.
2. *Yield target.* Season-specific yield targets are based on a yield gap analysis of yield potential and current yield levels in farmers' fields.
3. *Fertilizer nutrient requirements.* Suitable fertilizer N strategies are developed on the basis of the options summarized in Chapter 18 (this volume). Domain-specific P and K fertilizer requirements are calculated based on yield target and indigenous nutrient supply using the yield gain + nutrient balance approach.
4. *Least-costly fertilizer types.* Fertilizer rates of elemental nutrients (ka ha⁻¹) are expressed in nutrient sources per local area unit to facilitate wider-scale promotion.
5. *Profit estimate.* The existing practice is compared with the newly developed alternative nutrient management strategy to obtain an estimate of the ex-

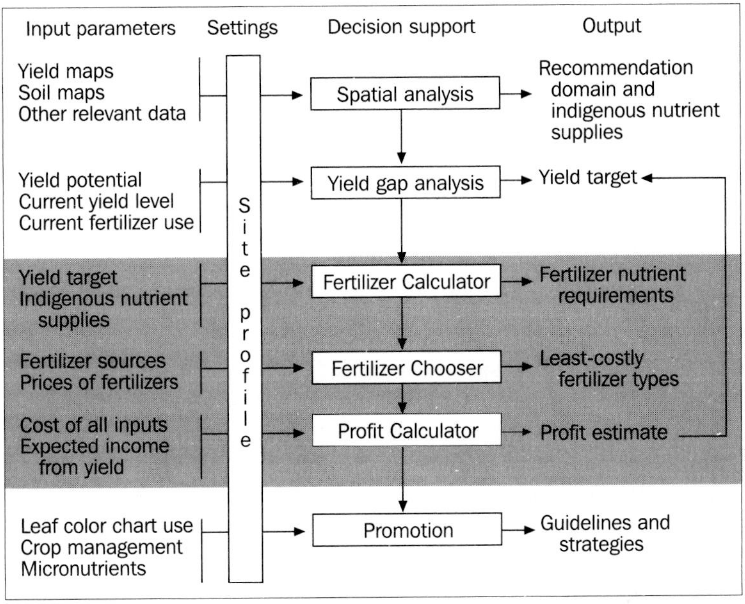
Input parameters
Settings
Decision support
Output
Yield maps
Soil maps
Other relevant data
Spatial analysis
Recommendation domain and indigenous nutrient supplies
Yield potential
Current yield level
Current fertilizer use
Yield gap analysis
Yield target
Site profile
Yield target
Indigenous nutrient supplies
Fertilizer Calculator
Fertilizer nutrient requirements
Fertilizer sources
Prices of fertilizers
Fertilizer Chooser
Least-costly fertilizer types
Cost of all inputs
Expected income from yield
Profit Calculator
Profit estimate
Leaf color chart use
Crop management
Micronutrients
Promotion
Guidelines and strategies

Fig. 17.8. Flow chart of the nutrient decision support system for irrigated rice (modified after Witt et al 2001). The gray area indicates the availability of three MS Excel spreadsheet models—Fertilizer Calculator (Witt 2002), Fertilizer Chooser (Fairhurst and Witt 2001a), and Profit Calculator (Fairhurst and Witt 2001b).

pected profit increase (ex ante analysis). Fertilizer strategies are adjusted depending on the outcome of the economic analysis.

6. *Guidelines and strategies for promotion.*

The new recommendation should be developed in close interaction with relevant stakeholders, integrating crucial information such as farmers' preferences for certain management practices or fertilizer types. Issues related to the promotion and wider-scale delivery of improved nutrient management strategies are discussed in Chapter 18 (this volume), so that the following sections will focus on the first five issues, which require technical decision support.

1. Recommendation domains and indigenous nutrient supplies

The SSNM approach described in this book (see Chapter 5, this volume) is field-specific, that is, fertilizer recommendations were worked out specifically for each crop grown in a single rice field. However, detailed analysis has shown that nutrient omission plots are not required in each individual farmer's field (Dobermann et al 2003b) and technical expertise is required to decide on the number of required omission plots and the area for which recommendations are valid (recommendation domains). At issue is then (1) how spatial recommendation domains can be derived from biophysical and socioeconomic information that determines yield potential, indigenous nutrient supplies, and response to fertilizer, and (2) how many crops, fields within a domain, or plots within a field need to be sampled to obtain a representative estimate of the mean IPS and IKS for each fertilizer recommendation domain.

The domains sampled in our on-farm studies were mostly of about 200 to 300 km^2 in size. The following selected conclusions on sampling requirements at this spatial scale were based on a detailed analysis reported earlier:

- Grain yield can be measured in a single omission plot per field for each nutrient because differences between replicate plots embedded in small rice fields are typically less than 5% (Dobermann et al 2003b).
- The precision of the indigenous nutrient supply estimates based on nutrient-limited yield in omission plots within a domain was mainly a function of the number of crops (years) and farms sampled (Dobermann et al 2003b).
- Domain- and season-specific means of grain yield in nutrient omission plots can be estimated using a small sample of farms during a period of only one or two crops (Dobermann et al 2003b). For example, assuming measurements in one HYS crop only and the use of just one omission plot per nutrient and field, the precision of the domain mean ranged from ±6% to 20% or ±4% to 19% if five or ten farms were sampled within a domain, respectively. Sampling for two years increased the attainable precision to ±4% to 14% with five fields sampled or ±3% to 14% with 10 fields sampled per domain. Increasing the number of farms beyond 10 per domain or the number of years beyond two did not increase the achievable precision enough to justify the extra cost associated with this.
- Grain yield and nutrient uptake in omission plots were consistently higher in HYS crops than in LYS crops because of more abiotic and biotic constraints to plant growth in the latter (Dobermann et al 2003a). At some sites, this also caused greater differences between yield and nutrient uptake in omission plots in the LYS than in the HYS. Yield measurements in nutrient omission plots should therefore be adjusted at these sites to avoid an underestimation of indigenous supplies. Fertilizer requirements should be estimated on the basis of season-specific estimates of indigenous supplies as discussed in section 17.1.

In summary, sampling of just one typical HYS and one LYS crop on about 10 farms would allow us to estimate the season-specific domain mean of the effective indigenous supplies with about ±10% precision at many sites, which may be sufficient for generating an initial domain-specific fertilizer recommendation for an area of about 200 to 300 km^2 in size. There were differences in sampling requirements depending on the uniformity of the domains sampled in our study (Dobermann et al 2003b), but these estimates may provide sufficiently robust guidance when estimating sampling requirements for larger areas.

Sampling requirements for estimating domain-specific values of INS, IPS, or IKS depend on the size and the homogeneity of the domain of interest. Boundaries of fertilizer recommendation domains must therefore be defined on the basis of a minimum set of available biophysical and socioeconomic characteristics that determine the uniformity of yield potential, indigenous nutrient supplies, response to fertilizer, or validity of guidelines within a spatial domain. We propose to divide a larger area into smaller agroecological domains on the basis of maps of climate, yield potential, yield history, soil texture groups, soil-testing data, soil depth, soil drainage, cropping systems, and major cropping practices such as crop establishment method, water management, and crop residue management. Borderlines can be delineated either manually or through the use of geo-statistical methods and geographic information systems (GIS). Manual delineation of borderlines can be done in a village, where extension staff and farmers would draw a map on the basis of information available to them, including local expertise. Participatory studies also offer the possibility to include a new layer of information based on indigenous knowledge within GIS-based analysis (Zurayk et al 2001). Many of the thematic maps mentioned above should be readily available for the major rice-growing areas in Asia. However, there is merit in disintegrating the existing information displayed on maps (e.g., maps that already have borderlines on soil series), applying geo-statistical tools such as a fuzzy-k-means cluster analysis to reclassify the data, and developing a meaningful number of new classes or recommendation domains taking local expertise into account. The domains provide the borderlines of activities in which fertilizer recommendations and guidelines are developed together with farmers in participatory approaches.

This process should be repeated in intervals of about 10–20 years to account for major changes in indigenous nutrient supplies, varieties, and cropping technologies with time. The development of recommendation domains offers an entry point for discussions among the various stakeholders and should be seen as an essential component of the omission plot technology and the refined approach for calculating fertilizer P and K requirements.

2. Yield target

As a rule of thumb, we propose realistic yield targets based on the average yield of the last 3–5 crops (same season) obtained in farmers' fields, plus not more than 10–20%, to achieve a visible yield increase, unless the primary goal is to avoid fertilizer overuse at current yield levels. The target yield should be not more than 75–80% of the yield potential to ensure efficient internal use of nutrients taken up by the plant (Witt et al 1999). More specific guidelines are provided elsewhere (Fairhurst and Witt 2002). Note that recommendation domains may have to be subdivided into smaller areas if yield targets differed in a large domain.

3. Fertilizer nutrient requirements

Fertilizer requirements can be looked up on simple charts where rates are based on a broad classification of yield goals and indigenous nutrient supplies (Chapter 18, this volume). Several options are provided for developing fertilizer N strategies, and fertilizer P and K rates were calculated following the yield gain + nutrient balance approach described in section 17.1. The underlying model for calculating fertilizer requirements is available on request in MS Excel spreadsheet format to facilitate the evaluation and local adaptation of model settings (Witt 2002). The refined and simplified SSNM approach has recently been summarized in a pocket-sized guide for extension personnel (Fairhurst and Witt 2002).

4. Least-costly fertilizer types

Elemental fertilizer rates need to be translated into locally available and cost-effective fertilizer types for wider-scale delivery of improved recommendations. The evaluation of different types of fertilizer can be challenging, particularly where farmers use a wide range of combined fertilizers. A simple spreadsheet model was therefore developed to facilitate the mathematical calculations by employing the optimization routine of a standard software. The MS Excel spreadsheet model *Fertilizer Chooser* (Fairhurst and Witt 2001a) can be used to (1) translate nutrient recommendations into the correct amount of available fertilizer types and (2) select the least-costly nutrient sources on the basis of locally available fertilizer types, their nutrient concentrations, and their farm-gate prices. A built-in optimization routine in MS Excel called Solver is employed to select the least-costly combination of fertilizer types on the basis of user-defined settings such as types to consider in a particular model run or constraints such as the minimum amount of a nutrient given with a particular fertilizer type. The major advantage of the model is the effortless integration of combined fertilizer types in the calculation process when translating elemental fertilizer rates into amounts of recommended fertilizer types. For example, if farmers prefer a certain combined fertilizer for the basal application of N, P, and K, a setting could be introduced in the model run to apply at least 50% of the fertilizer K with the respective fertilizer type. This would leave the program the choice of selecting muriate of potash (MOP) as the potentially least-costly K source for topdressing at panicle initiation. The ideal combination of fertilizer types is found through an iterative process evaluating different options (settings). Fertilizers should finally be expressed in units that are familiar to the farmer (e.g., bags of nutrient source ha^{-1}).

5. Profit estimate

The expected profit of alternative nutrient management strategies should be evaluated against the current farmers' practice. In the following, we discuss an example for such a profit analysis, introducing another computer-based decision aid, the *Profit Calculator* (Fairhurst and Witt 2001b). The MS Excel spreadsheet model was developed to evaluate the profitability of alternative management strategies considering the revenue from grain yield × paddy farm-gate price and the most important costs for inputs such as materials and labor (Table 17.9). The simple spreadsheet model compares and evaluates cost centers (e.g., costs for fertilizers, pesticides, labor, etc.) and provides other relevant key ratios. The most important parameters to consider when evaluating alternative fertilizer management strategies are probably expected changes in net benefit (marginal benefit), fertilizer cost (marginal cost), and the marginal benefit-cost ratio (marginal benefit divided by marginal cost). As long as the latter is greater than one, it is sensible to adopt the strategy that provides the greatest net benefit, although the marginal benefit-cost ratio may have to be larger than two

Table 17.9. Gross margin analysis, distribution of cost centers, and key ratios for evaluating alternative management strategies in an irrigated rice season as calculated in the spreadsheet model Profit Calculator (Fairhurst and Witt 2001b). The given example was based on data from Central Luzon, Philippines, as modified from Table 3.13 (Chapter 3, this volume).

No.	Parameters	Unit	Example	Calculation[a]	Labor[b] (8-h days ha^{-1}) Hired	Family
1	Seeds	US$ ha^{-1}	31.50			
2	Inorganic fertilizer	US$ ha^{-1}	69.50			
3	Organic fertilizer	US$ ha^{-1}	0.00			
4	Pesticides	US$ ha^{-1}	23.50			
5	Machine rent and fuel	US$ ha^{-1}	54.50			
6	Nursery	US$ ha^{-1}	0.00			
7	Land preparation	US$ ha^{-1}	30.00		3.2	1.0
8	Planting	US$ ha^{-1}	48.50		5.1	1.5
9	Manual weeding	US$ ha^{-1}	13.80		1.5	0.5
10	Pesticide application	US$ ha^{-1}	0.00			
11	Inorganic fertilizer application	US$ ha^{-1}	0.00			
12	Organic fertilizer application	US$ ha^{-1}	0.00			
13	Irrigation	US$ ha^{-1}	0.00			
14	Harvesting	US$ ha^{-1}	115.20			
15	Threshing	US$ ha^{-1}	0.00		12.1	3.5
16	Cleaning	US$ ha^{-1}	0.00			
17	Drying	US$ ha^{-1}	0.00			
18	Land rent	US$ ha^{-1}	0.00			
19	Irrigation	US$ ha^{-1}	0.00			
20	Harvest (share)	US$ ha^{-1}	0.00			
21	Tax	US$ ha^{-1}	0.00			
	Gross margin analysis					
22	Grain yield	kg ha^{-1}	5,000			
23	Paddy price	US$ kg^{-1}	0.21			
24	Gross benefit or revenue	US$ ha^{-1}	1,050.00	L1 × L2		
25	Total costs	US$ ha^{-1}	386.50	Σ L1 to L21		
26	Net benefit	US$ ha^{-1}	663.50	L24 to L25		
	Distribution of cost centers					
27	Inorganic fertilizer	%	18.0	L2 ÷ L25		
28	Materials	%	28.3	Σ(L1 to L5) ÷ L25		
29	Labor	%	53.7	Σ(L6 to L17) ÷ L25		
	Key ratios					
30	Breakeven yield	t ha^{-1}	1.84	L25 ÷ L23 ÷ 1,000		
31	Production cost kg^{-1} paddy	US$ kg^{-1}	0.077	L25 ÷ L22		
32	Net benefit kg^{-1} paddy	US$ kg^{-1}	0.133	L26 ÷ L22		
33	Family labor input	8-h day	6.5	Σ right column		
34	Return to family labor 8-h day^{-1}	US$ 8-h d^{-1}	102.08	L26 ÷ L33		

[a]Numbers referring to lines are preceded by the letter L. [b]No labor cost is assumed for family labor.

for farmers to adopt new technologies. In addition to fertilizer cost, the share of fertilizer in total cost provides further useful information on the investment requirements of alternative fertilizer strategies. In most cases, additional fertilizer costs associated with SSNM were marginal (Dawe et al, Chapter 15, this volume), which, however, would be important to calculate and communicate to farmers.

17.3 Conclusions

The original SSNM approach was simplified and refined for wider-scale distribution while maintaining the scientific principles of the underlying QUEFTS model for rice (Janssen et al 1990, Witt et al 1999). Major steps in the calculation of fertilizer N, P, and K requirements in the refined SSNM strategy include (1) selection of a suitable yield goal of not more than 70–80% of the potential yield, (2) the use of omission plots to estimate soil nutrient supplies based on yield, (3) the estimation of fertilizer requirements based on the expected yield deficit between yield goal and the respective nutrient-limited yield estimated in omission plots (yield gain approach), (4) the use of empirically derived standard values for fertilizer requirements per unit yield deficit, and (5) integration of a simple nutrient balance for calculating fertilizer P and K maintenance rates (yield gain + nutrient balance approach).

Fertilizer requirements in the yield gain approach were estimated following the principles of the QUEFTS model. On the basis of plant nutrient requirements and expected fertilizer recovery efficiencies, about 40–50 kg fertilizer N ha^{-1}, 7–12 kg fertilizer P ha^{-1}, or 22–41 kg fertilizer K ha^{-1} would be needed to raise yield by 1 t over the nutrient-limited yield in the respective omission plot. In a second approach, fertilizer P and K rates were calculated using a newly developed nutrient balance approach to replenish nutrient removal with grain and straw. The two strategies were compared on the basis of (1) a sensitivity analysis modifying the most relevant input parameters, (2) the potential impact on long-term fertilizer use and nutrient input-output balances, and (3) site-specific scenarios using actual on-farm data from seven sites with irrigated rice. It was concluded that integrating a nutrient balance in the yield gain approach appears essential for developing meaningful short- and long-term fertilizer P and K strategies to avoid nutrient depletion as and where required. Few additional input parameters would be needed, which should at least include the average amount of straw returned and the use of organic nutrient sources such as FYM. Future research should aim to (1) properly model recovery efficiencies of applied fertilizer nutrients, (2) obtain better estimates of relevant components of the nutrient balance model at major rice-growing sites in Asia, and (3) validate the long-term performance of the yield gain + nutrient balance approach in further field studies, including economic analysis of alternative fertilizer management strategies.

Field-specific strategies are probably required at most sites to efficiently manage fertilizer N using decision tools such as the leaf color chart, but fertilizer P and K recommendations are likely to be valid for larger areas. The estimation of indigenous nutrient supplies is particularly recommended for P and K because of limited opportunities for within-season adjustments of fertilizer rates. Summarizing earlier findings on the variability of indigenous nutrient supplies, a limited number of about 10 omission plots would be needed to obtain a sufficiently robust season-specific estimate of indigenous nutrient supplies for an area of about 200 to 300 km^2 in size. Omission plots should be installed during each major cropping season to obtain a realistic estimate of the expected season-specific yield deficit. We further propose to develop fertilizer P and K rates for so-called recommendation domains, that is, larger areas that are characterized by relatively uniform biophysical and socioeconomic conditions, resource management, and therefore production characteristics. Existing information such as maps on soil properties, water availability, yield history, and other readily available parameters can be used to delineate borderlines of recommendation domains. This can be done manually for smaller areas, for example, in a village-based extension approach, or for larger areas through the use of geo-statistical methods using GIS.

A general framework for decision support was proposed, including six major steps: (1) recommendation domain development and estimation of indigenous nutrient supplies, (2) yield gap analysis and adjustment of domain borders, (3) calculation of fertilizer requirements, (4) conversion of elemental fertilizer rates into adequate amounts of nutrient sources, (5) economic evaluation of alternative management strategies, and (6) promotion of recommendations and guidelines. In addition to printed promotional materials, Excel spreadsheet models were developed to provide assistance in certain complex mathematical calculations that would be difficult to perform otherwise, such as applying optimization routines in the selection of the least-costly fertilizer types.

References

Abedin Mian MJ, Blume HP, Bhuiya ZH, Eaqub M. 1991. Water and nutrient dynamics of a paddy soil of Bangladesh. Z. Pflanzenernähr. Bodenk. 154:93-99.

Balasubramanian V, Morales AC, Cruz RT, Abdulrachman S. 1999. On-farm adaptation of knowledge-intensive nitrogen management technologies for rice systems. Nutr. Cycl. Agroecosyst. 53:59-69.

Cassman KG, De Datta SK, Amarante S, Liboon SP, Samson MI, Dizon MA. 1996. Long-term comparison of the agronomic efficiency and residual benefits of organic and inorganic nitrogen sources for tropical lowland rice. Exp. Agric. 32:427-444.

Dobermann A, Cassman KG, Mamaril CP, Sheehy SE. 1998. Management of phosphorus, potassium, and sulfur in intensive, irrigated lowland rice. Field Crops Res. 56:113-138.

Dobermann A, Sta.Cruz PC, Cassman KG. 1996. Fertilizer inputs, nutrient balance, and soil nutrient-supplying power in intensive, irrigated rice systems. I. Potassium uptake and K balance. Nutr. Cycl. Agroecosyst. 46:1-10.

Dobermann A, White PF. 1999. Strategies for nutrient management in irrigated and rainfed lowland rice systems. Nutr. Cycl. Agroecosyst. 53:1-18.

Dobermann A, Witt C, Abdulrachman S, Gines HC, Nagarajan R, Son TT, Tan PS, Wang GH, Chien NV, Thoa VTK, Phung CV, Stalin P, Muthukrishnan P, Ravi V, Babu M, Simbahan GC, Adviento MA. 2003a. Soil fertility and indigenous nutrient supply in irrigated rice domains of Asia. Agron. J. 95:913-923.

Dobermann A, Witt C, Abdulrachman S, Gines HC, Nagarajan R, Son TT, Tan PS, Wang GH, Chien NV, Thoa VTK, Phung CV, Stalin P, Muthukrishnan P, Ravi V, Babu M, Simbahan GC, Adviento MA, Bartolome V. 2003b. Estimating indigenous nutrient supplies for site-specific nutrient management in irrigated rice. Agron. J. 95:924-935.

Dobermann A, Witt C, Dawe D, Abdulrachman S, Gines HC, Nagarajan R, Satawathananont S, Son TT, Tan CS, Wang GH, Chien NV, Thoa VTK, Phung CV, Stalin P, Muthukrishnan P, Ravi V, Babu M, Chatuporn S, Sookthongsa J, Sun Q, Fu R, Simbahan GC, Adviento MAA. 2002. Site-specific nutrient management for intensive rice cropping systems in Asia. Field Crops Res. 74:37-66.

Fairhurst T, Witt C. 2001a. Fertilizer Chooser. Software. Prerelease version as MS Excel spreadsheet model (online). Available at www.irri.org/irrc/nutrients (last update 11 Oct. 2001; accessed 16 May 2003). Singapore and Los Baños: Potash and Phosphate Institute & Potash and Phosphate Institute Canada, and International Rice Research Institute.

Fairhurst T, Witt C. 2001b. Profit Calculator (gross margin analysis). Software. Prerelease version as MS Excel spreadsheet model. Version 1.0. Singapore and Los Baños, Philippines: Potash and Phosphate Institute and Potash and Phosphate Institute of Canada, and International Rice Research Institute.

Fairhurst T, Witt C. editors. 2002. Rice: a practical guide to nutrient management. Singapore and Makati City (Philippines): Potash & Phosphate Institute, Potash & Phosphate Institute of Canada (PPIC), and International Rice Research Institute (IRRI). p 1-89.

Greenland DJ. 1997. The sustainability of rice farming. Wallingford (UK) and Manila (Philippines): CAB International and International Rice Research Institute. 273 p.

Janssen BH, Guiking FCT, Braakhekke WG, Dohme PAE. 1992. Quantitative evaluation of soil fertility and the response to fertilizers. Wageningen (Netherlands): Department of Soil Science and Plant Nutrition, Wageningen Agricultural University. p 1-92.

Janssen BH, Guiking FCT, van der Eijk D, Smaling EMA, Wolf J, van Reuler H. 1990. A system for quantitative evaluation of the fertility of tropical soils (QUEFTS). Geoderma 46:299-318.

Power DJ. 1999. Decision support systems glossary (online). Available at www.dssresources.com; accessed 30 April 2003.

Wang GH, Dobermann A, Witt C, Sun QZ, Fu RX. 2001. Analysis of the indigenous nutrient supply capacity of rice soils in Jinhua, Zhejiang Province. Chinese J. Rice Sci. 15(3):201-205.

Witt C. 2002. Fertilizer Calculator. Software. Prerelease version as MS Excel spreadsheet model. Los Baños (Philippines): International Rice Research Institute.

Witt C, Cassman KG, Olk DC, Biker U, Liboon SP, Samson MI, Ottow JCG. 2000. Crop rotation and residue management effects on carbon sequestration, nitrogen cycling, and productivity of irrigated rice systems. Plant Soil 225:263-278.

Witt C, Dobermann A, Abdulrachman S, Gines HC, Wang GH, Nagarajan R, Satawathananont S, Son TT, Tan PS, Le Van Tiem, Simbahan GC, Olk DC. 1999. Internal nutrient efficiencies in irrigated lowland rice of tropical and subtropical Asia. Field Crops Res. 63:113-138.

Witt C, Dobermann A, Arah JRM, Pamplona RR. 2001. Nutrient decision support system (NuDSS) for irrigated rice. Int. Rice Res. Notes. 26(2):14-15.

Witt C, Wang GH, Dobermann A, Sun Q, Fu R. 2003. Balanced nutrition and nutrient balances in irrigated rice: a case study in Zhejiang Province, PR China. Proceedings of the 9th IPI/ISSAS Regional Workshop on Nutrient Cycling and Management in Cropping Systems of Different Agro-ecoregions in China, Haikou, Hainan, China, 6-8 December 1999. International Potash Institute (IPI) and Institute of Soil Science, Academia Sinica (ISSAS): Basel, Switzerland, and Nanjing, China. p 1-12.

Zurayk R, el-Awar F, Hamadeh S, Talhouk S, Sayegh C, Chehab AG, al Shab K. 2001. Using indigenous knowledge in land use investigations: a participatory study in a semi-arid mountainous region of Lebanon. Agric. Ecosyst. Environ. 86:247-262.

Notes

Authors' addresses: C. Witt, A. Dobermann, International Rice Research Institute, Los Baños, Philippines; A. Dobermann, University of Nebraska, Lincoln, Nebraska, USA.

Acknowledgments: We are grateful for the many suggestions made by NARES collaborators and IRRI colleagues on the simplification of the SSNM approach, and we particularly thank Drs. Stephan Haefele, David Dawe, Roland Buresh, and V. Balasubramanian for their helpful comments.

Citation: Dobermann A, Witt C, Dawe D, editors. 2004. Increasing productivity of intensive rice systems through site-specific nutrient management. Enfield, N.H. (USA) and Los Baños (Philippines): Science Publishers, Inc., and International Rice Research Institute (IRRI). 410 p.

Principles and promotion of site-specific nutrient management

C. Witt, R.J. Buresh, V. Balasubramanian, D. Dawe, and A. Dobermann

Innovative fertilizer management has to integrate both preventive and corrective strategies to manage nutrients efficiently, sustain the soil resource base, and increase the profitability of irrigated rice farming in Asia. Successful strategies will need to provide principles that can be developed into a range of management options based on location-specific needs to address (1) the seasonal and year-to-year variation in climate (particularly solar radiation) and (2) the spatial and temporal variation of indigenous soil nutrient supplies. Both factors lead to a large variation in the optimal rates of fertilizer inputs, and in crop performance among sites, seasons, and years. Current fertilizer recommendations in Asia, however, typically consist of "blanket" recommendations with fixed rates and timings for large rice-growing areas. Much progress has been made in recent years in developing field- and season-specific nutrient management approaches as alternatives to such blanket recommendations for N, P, and K fertilizers. The approaches have been widely evaluated in farmers' fields in Asia, and they are now positioned for wide-scale evaluation and farmer adaptation.

The site-specific nutrient management (SSNM) strategy described in this book has demonstrated promising agronomic and economic potential, and provided a strong conceptual and scientific basis for the simplified and refined SSNM approach developed in the previous chapter. Major progress was also made in the development, on-farm evaluation, and promotion of leaf color charts for real-time N management (Balasubramanian et al 1999, 2000, Bijay-Singh et al 2002, Buresh et al 2002, Yang et al 2003). In this chapter, we provide an update on earlier efforts to integrate and summarize the major principles of these plant-based nutrient management approaches into tools and guidelines for the delivery of improved nutrient management in Asia's irrigated rice systems (Buresh et al 2001, 2003, Dobermann et al 2002, Witt et al 2002b). We further propose strategies that will facilitate the relatively knowledge-intensive development and thus challenging delivery of SSNM recommendations in larger areas, and conclude with a summary of policy recommendations and future research needs.

18.1 The principles of SSNM

Principle 1: Balanced fertilization based on crop requirements

The principles described here offer a basic plan for a preseason calculation of balanced fertilizer rates considering the deficit between plant nutrient requirement and soil nutrient supply. This deficit largely depends on the expected yield gain, which we define as the required yield increase over the nutrient-limited yield to reach a season-specific yield goal. To consider differences in indigenous supply among nutrients, yield gains have to be estimated for N, P, and K separately. For example, if the nutrient-limited yield was 5 t ha^{-1} for P and 6 t ha^{-1} for K as measured in omission plots (see Principle 2), the required yield increases to achieve a yield goal of 6 t ha^{-1} would be 1 t ha^{-1} for P and 0 t ha^{-1} for K. Thus, fertilizer P rates would have to be sufficiently high to support the required yield increase, whereas preventive fertilizer strategies would focus on replenishing most of the crop nutrient removed to maintain soil K supplies (see Principle 4). As a rule of thumb, we estimate that 40 kg fertilizer N, 20 kg P$_2$O$_5$, or 30 kg K$_2$O are required to raise the respective nutrient-limited yield by 1 t ha^{-1} (for details, see Chapter 17, this volume).

Principle 2: Plant-based estimation of soil nutrient supplies

The opportunities to improve current fertilizer recommendations through the use of conventional soil tests are limited in irrigated rice in Asia. Soil properties and rapid chemical extractions of soil samples showed few correlations with indigenous N, P, and K supply measured as plant nutrient uptake in nutrient omission plots across a wide range of on-farm environments in South and Southeast Asia (Dobermann et al 2003b). As an attractive alternative to soil testing, soil nutrient supplies can be indirectly estimated as plant nutrient uptake in nutrient omission plots. Plant-based estimates of soil nutrient supply integrate the supply of all indigenous sources estimated under field conditions and also offer the possibility for estimating the nutrient-supplying power of organic manures, irrigation, and biological N$_2$ fixation (Dobermann et al 2003a).

More suitable for extension purposes, however, is the estimation of soil nutrient supply expressed as nutrient-limited yield in the respective omission plot (Dobermann et al 2003b). A major advantage of this approach is that the soil supply is expressed in a unit that can be directly used in the calculation of fertilizer requirements (see Principle 1). Furthermore, soil nutrient supply becomes "visible" to farmers, making omission plots a simple and effective demonstration tool for nutrient limitations in extension. A limited number of nutrient omission plots placed in areas with different cropping systems, soil types, and topographies may help extension workers in developing an improved understanding of the local distribution of soil fertility in partnership with farmers (see section 18.2). It takes only one season to obtain a practical estimate of the indigenous nutrient supply, and this season also provides the opportunity to evaluate need-based fertilizer N management strategies together with collaborating farmers.

Principle 3: Need-based fertilizer N management

Asian farmers generally apply fertilizer N in several split applications, but the number of splits, amount of N applied per split, and the time of application vary substantially. The apparent flexibility of rice farmers in adjusting the time and amounts of fertilizer application offers potential to synchronize N application with the real-time demand of the rice crop. In the following, we briefly summarize three main forms of N management recommendations: (1) location-specific split schedules for preventive N management, (2) corrective N management using a leaf color chart (LCC), and (3) a combination of both in which the LCC is used at certain growth stages to identify the need for fertilizer N ("split N + LCC").

a. Location-specific split schedules for preventive N management involve preset fertilizer N applications at key growth stages of rice. General recommendations for N application regimes are widespread, and were often developed through N fertilizer response experiments. Limitations of fertilizer response experiments include (1) the costly and time-consuming identification of the best splitting pattern and corresponding N rates and (2) a limited extrapolation potential because of wide variation in both soil N supply within the often large recommendation domains and crop response caused by climatic factors. Although corrective N management strategies offer greater potential for efficient fertilizer N management (see below), recommendations for location-specific N regimes may be sufficiently accurate under stable climatic conditions with low pest pressure (Haefele et al 2001), or where large benefits can be expected from fundamental adjustments in fertilizer N management, for example, at sites with highly excessive fertilizer N use such as in Zhejiang, China (see Chapter 12, this volume).

Location-specific split schedules can be developed following Principles 1 and 2 given above, where fertilizer N requirements are calculated on the basis of crop requirements and soil indigenous N supply. An estimate of the latter can be obtained by analyzing current farm yields and farmers' N management strategies in combination with local knowledge on soil fertility. Thus, N omission plots may not always be required to obtain a sufficiently accurate estimate of indigenous N supply. Locally refined splitting patterns have to take into account specific needs for differences in climatic seasons, varieties, crop establishment, basal N application, and water management (Witt et al 2002a, Dobermann and Fairhurst 2000).

b. LCC-based corrective N management is a true real-time N management approach, in which the plant N status is periodically assessed and application of fertilizer N is delayed until (almost when) N-deficiency symptoms appear. This "need-based N management" does not require the estimation of soil N supply or the calculation of a preseason fertilizer rate. The scientific basis for need-based N management was developed with the introduction of the chlorophyll (SPAD) meter (Peng et al 1996, Balasubramanian et al 1999). Recognizing the limitations of the costly SPAD meter as an on-farm tool, a leaf color chart (LCC) modified from prototypes developed in Japan (Furuya 1987) and China (by Prof. Tao Qinnan, Zhejiang University,

Zhejiang, China) was developed through collaboration between the International Rice Research Institute (IRRI) and the Philippine Rice Research Institute (IRRI 1999, Balasubramanian 1999b, Balasubramanian et al 2000). Leaf color is a visual and subjective indicator of plant N deficiency, and the LCC with its six color panels of different shades of green is used as a reference tool. Numerous LCC units have been fabricated and distributed to farmers through collaboration with national agricultural research and extension systems (NARES) in a number of Asian countries. Several versions of LCCs currently exist (IRRI, Japan, China, and University of California Cooperative Extension), and research efforts at IRRI are under way to compare LCCs and explore options for refining and standardizing the colors of LCCs (Cabrera-Pasuquin and Witt 2003, Witt et al 2003, Yang et al 2003). Need-based N management requires the identification of an optimal leaf color that needs to be maintained throughout the season to obtain high yields. The optimal leaf color (or critical LCC value) varies depending on cultivar and crop establishment method. Guidelines for the use of the LCC include reading of leaf color at 7–10-day intervals from early tillering until flowering. When the average leaf color falls below the critical value, a predetermined rate of N fertilizer is applied immediately to prevent N deficiency. Considering differences in yield potential, season-specific standard rates were developed, with recommended N rates per application not exceeding 40 kg N ha^{-1} to ensure efficient fertilizer N use (Balasubramanian et al 2000).

c. Location-specific split schedules, including the LCC, combine preventive and corrective N management strategies. Total fertilizer N requirements are calculated as described for location-specific split schedules (see above), including guidelines for the need of basal N application. At advanced growth stages, the LCC is then used to adjust predetermined N doses upward or downward depending on the plant requirement for fertilizer N (Witt et al 2002a). This dual strategy is similar to the SPAD meter approach described by Dobermann et al (Chapter 5, this volume). Using the LCC in combination with application schedules may address farmers' preferences and needs at certain sites to reduce reliance on frequent visits to the field. This strategy also reduces the risk of temporal N deficiency caused by the inaccurate use of the pure LCC approach.

The abovementioned strategies offer a portfolio of N management options. The most promising strategy may vary from location to location, and will have to be identified through farmer participatory evaluation and validation, including an economic analysis of the strategies tested. Common to all approaches aiming at efficient use of fertilizer N is the need for good calibration of tools and training to assess plant N status with sufficient frequency.

Principle 4: Sustainable P and K management
The estimation of P and K requirements is challenging for individual farmers because of small landholdings and substantial variation in soil P and K supplies within small domains (Dobermann et al 2003a). Information on soil nutrient supply is of particular importance for the commonly less limiting macronutrients P and K be-

Table 18.1. Fertilizer P$_2$O$_5$ requirements depending on yield goal and P-limited yield measured in 0-P omission plots following the yield gain + P balance approach (Chapter 17, this volume).

Yield in 0-P plots (t ha^{-1})	Yield goal (t ha^{-1})				
	4	5	6	7	8
Fertilizer P$_2$O$_5$ requirement (kg ha^{-1})					
3	20	40	60	[a]	[a]
4	15	25	40	60	[a]
5	0	20	30	40	60
6	0	0	25	35	45
7	0	0	0	30	40
8	0	0	0	0	35

[a]A lower yield goal is recommended when the required yield increase exceeds 3 t ha^{-1}.

cause of (1) uncertainties in short- and long-term crop responses to P and K application and (2) limited options to correct for deficiencies of these nutrients within a season as compared to N. In general, P and most K should be applied early in the season for greatest efficiency and to avoid nutrient deficiencies at early growth stages. This requires a conceptual framework to assist farmers in the estimation of total fertilizer P and K requirements to maintain indigenous P and K supplies or increase them when necessary.

The refined SSNM approach described in Chapter 17 (this volume) was used to develop simple charts with fertilizer P and K rates that were based on a broad classification of yield goals, indigenous nutrient supplies, and nutrient inputs with crop residues. Rice straw contains relatively little P, so that only a simplified chart is presented (Table 18.1), assuming an incorporation of moderate amounts of 2–3 t straw ha^{-1}. The fertilizer P requirements largely depend on the deficit between yield goal and soil nutrient supply, and the suggested maintenance fertilizer P rates for conditions where a direct crop response is not expected (yield goal equals yield in 0-P plot) would increase slightly with an increase in the targeted yield level. Three different levels of crop residue inputs were considered for estimating fertilizer K rates (Table 18.2), since straw management had a pronounced effect on the maintenance of soil K supply (Chapter 17, this volume). Simple decision trees facilitate the on-farm estimation of straw inputs depending on yield (biomass production) and straw management practice (Witt et al 2002a). For example, bulk straw incorporation or widespread burning have similar positive effects on P and K recycling (Dobermann and Fairhurst 2000). Substantial amounts of fertilizer K would be needed, especially at elevated yield levels, to balance K removal where little straw is retained after harvest (Table 18.2), as practiced in Bangladesh, India, Nepal, and northern Vietnam. Where 4–5 t straw ha^{-1} are retained in the field after harvest, fertilizer K application would in most cases be required only if a crop response is expected, and

Table 18.2. Fertilizer K_2O requirements depending on straw return, yield goal, and K-limited yield measured in 0-K omission plots following the yield gain + K balance approach (Chapter 17, this volume).

Yield in 0-K plots (t ha^{-1})	Yield goal (t ha^{-1})				
	4	5	6	7	8
Fertilizer K_2O requirement (kg ha^{-1}) at rice straw inputs of 0–1 t ha^{-1}					
3	45	75	105	[a]	[a]
4	30	60	90	120	[a]
5	0	45	75	105	135
6	0	0	60	90	120
7	0	0	0	75	105
8	0	0	0	0	90
Fertilizer K_2O requirement (kg ha^{-1}) at rice straw inputs of 2–3 t ha^{-1}					
3	30	60	90	[a]	[a]
4	0	35	65	95	[a]
5	0	20	50	80	110
6	0	0	35	65	95
7	0	0	0	50	80
8	0	0	0	0	65
Fertilizer K_2O requirement (kg ha^{-1}) at rice straw inputs of 4–5 t ha^{-1}					
3	30	60	90	[a]	[a]
4	0	30	60	90	[a]
5	0	0	30	60	90
6	0	0	10	35	70
7	0	0	0	25	55
8	0	0	0	0	40

[a]A lower yield goal is recommended when the required yield increase exceeds 3 t ha^{-1}.

minimum rates of 30 kg K_2O ha^{-1} per ton required yield increase would be sufficient as outlined under Principle 1. The fertilizer charts presented take the most relevant input and output parameters into account and aim to examine such important issues as mining and replenishment of soil P and K reserves. Local adaptation and refinement of these generic principles may be required, integrating research findings, for example, on the soil P- and K-supplying capacity as determined in long-term experiments.

Principle 5: Increasing profitability

The major benefit for farmers from improved nutrient management strategies can be expected from an increase in the profitability of rice cropping (plausible promise). SSNM principles can accommodate a wide range of socioeconomic conditions, including situations of labor shortage. Small amounts of additional labor may be required, but labor costs for nutrient management are relatively small compared with those for land preparation, transplanting, or harvesting. Efficient N management may also result in off-farm environmental benefits through a reduction in fertilizer N use without sacrificing yield, especially in situations in which N input is very high (e.g., China, Java island in Indonesia). This may increase the profitability to some extent, especially in cases of very high fertilizer N inputs (China, Indonesia). Large reductions in N use in such locations may also increase farm profits, but the cost of fertilizer N across seven sites with irrigated rice was typically less than 5% of the gross revenue from paddy (Dawe et al, Chapter 16, this volume). The major potential for increasing farm profitability through innovative nutrient management therefore lies in increasing yield through efficient N management and balanced nutrition, unless a reduction in other inputs such as pesticides offers substantial additional savings. To select the most profitable strategies, farmer participatory evaluation of innovative nutrient management should be accompanied by an evaluation of fertilizer costs comparing combined and straight fertilizers and a gross margin analysis. A small software package has been developed to facilitate this analysis (Fairhurst and Witt 2001a,b).

18.2 Promotion of SSNM technologies

The SSNM approach developed into a set of principles in recent years, where individual components can be selected and adapted based on local conditions and farmers' needs. The development of recommendations and guidelines for wider-scale promotion requires technical expertise, while the promoted messages can be simple to fit into the farmers' knowledge systems (Haefele et al 2002). According to Pingali et al (1998), a successful transfer of knowledge-intensive technologies to farmers will largely depend on the ability to develop cost-effective methods for wider-scale dissemination. A major challenge in the development of training strategies and promotional materials is to maintain the scientific principles summarized in section 18.1. In the following, we will examine the complexity of transferring SSNM principles to farmers assuming that N, P, and K management-related constraints to increasing productivity have been identified through adequate participatory assessments and surveys (Adhikarya 1994, FAO 2000, Balasubramanian et al 2001).

Knowledge-based training programs

The SSNM strategy requires large changes in the way fertilizer recommendations are now formulated. Extension campaigners or trainers will need to acquire new knowledge and become themselves responsible for developing recommendations by adapting SSNM principles to local conditions in contrast to the past, when they were asked to promote fixed fertilizer rates developed by others for large areas. Training strategies will depend on the complexity of the nutrient management-related constraints to increasing productivity.

For larger-scale extension campaigns, a coordination unit of technical experts from various disciplines will be needed to (1) coordinate regional initiatives as and where required and (2) conduct training of trainers for local adaptation of SSNM principles, including the development of promotional material at the local level. Regional strategies would cut across administrative boundaries to divide larger areas into smaller recommendation domains with common agroclimatic characteristics (see Chapter 17, this volume). Few experts are needed to develop broad categories of domains on a large scale because borders are not likely to change for many years and further adjustments could be implemented at the local level. The broader domain classification would need to be developed by researchers in consultation with the relevant stakeholders at the regional, provincial, and district level. Subsequent strategies for individual recommendation domains will then depend on the prevalent constraints. The main tasks of the coordination unit would be to (1) provide training of trainers in the principles of SSNM, including the use of tools such as the LCC or the omission plot technique (see also www.knowledgebank.irri.org/SSNM), (2) establish a centralized database to store relevant information collected in the recommendation domains (yield, fertilizer use, estimates of indigenous nutrient supplies, etc.) to refine borderlines of recommendation domains, (3) provide decision makers with decision aids such as maps with hot spots of nutrient management-related constraints to help focus extension efforts, (4) assist in the development of concepts for on-farm evaluation and adaptation of nutrient management options, (5) develop or assist in the development of promotional print materials or mass media campaigns to ensure that the scientific principles of SSNM are maintained, and (6) provide a platform to integrate and facilitate extension efforts of different initiatives (e.g., by combining integrated pest management and SSNM initiatives).

Local initiatives and activities by extension staff and farmers at the community or village level could include (1) local training of field staff and farmers in the principles of SSNM, including the use of tools such as the LCC or the omission plot technique, (2) the manual delineation of recommendation domains in the village based on a regional classification and/or locally available biophysical and socioeconomic data and information (local expertise), (3) the implementation of the omission plot technique in selected farmers' fields in the village, (4) the development of improved nutrient management strategies in consultation with other stakeholders and technical experts of the coordination unit as required, (5) the on-farm evaluation and adaptation of improved nutrient management strategies by farmers, and (6) the wider-scale promotion of tested nutrient management strategies through demonstrations, training, and distribution of promotional material during farmers' meetings and field days in the village. If a local extension initiative operates in an area that is larger than a single community or village, knowledge generated at the village level could be disseminated in a larger recommendation domain cutting across several villages (see example given in Figure 18.1). The domain concept is particularly relevant for farmer participatory development of P and K recommendations.

Simple SSNM recommendations for farmers

The adoption potential of any technology will largely depend on the expected financial advantages for farmers and the simplicity of the technology for implementation

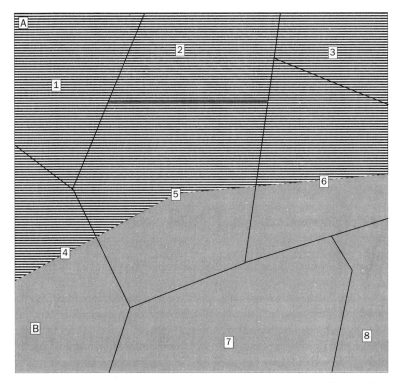

Fig. 18.1. Example for boundaries of recommendation domains (A, B, broken line) in comparison to administrative boundaries for extension activities in villages (1–8, solid lines).

(Pingali et al 1998, Pandey 1999). Where knowledge-intensive technologies are required, the authors argued that blanket fertilizer recommendations should be replaced by knowledge transfer that empowers farmers to make appropriate decisions. A recommendation domain would then be a domain in which the decision-making process is uniform rather than the actual recommendation being uniform. Certainly, a major issue is the extent to which farmers would need to acquire more complex knowledge of modern nutrient management strategies compared to conditions where simple SSNM recommendations may well fit into their existing knowledge systems.

There are limited options and also little need for farmers and extension staff to assess field- or farm-specific soil-nutrient status. The costs for implementing such programs would be too high and little additional information would be gained compared to approaches that would aim at developing guidelines for villages or larger recommendation domains (Dobermann et al 2003b). Only a few farmers would need to participate in a village- or domain-wide initiative to develop novel fertilizer P and K recommendations by estimating indigenous nutrient supplies in omission plots embedded in their fields. Domain-specific recommendations may then provide several season-specific alternatives for basal or early-season application of types and rates of N, P, and K fertilizer, and guidelines for farmers.

Farmers would need to acquire more complex knowledge of field-specific N management strategies to achieve optimal congruence between N supply and plant demand. Efficient N management would require some knowledge of initial soil fertility to decide on the need for basal N application (see above), and a more detailed knowledge of plant growth stages and plant symptoms indicating the N status of the rice crop, including N deficiency and surplus. Guidelines and decision tools such as the LCC can assist in the decision making on appropriate timing of nutrient application and the quantities to apply, but would not replace farmers' knowledge (Pingali et al 1998). Suitable N, P, and K management strategies would need to be developed in participatory on-farm trials possibly evaluating several management options prior to wider-scale promotion. A comprehensive summary of participatory research methods for technology evaluation is provided by Bellon (2001). Such strategies will initiate a learning selection process, in which the farmers as first adopters, extension staff, NGOs, or representatives from cooperatives learn how to adopt and improve the tools and strategies (Deugd et al 1998, Douthwaite et al 2002).

Domain-specific recommendations could be divided into two categories depending on the requirement for local adjustment:
1. Blanket recommendations and general guidelines
   - Season-specific alternatives for basal or early-season application of N, P, and K using different fertilizer types
   - General guidelines for N management, including LCC use for real-time N management
   - Recommendation for midseason topdressing of fertilizer K at panicle initiation
   - General crop, weed, water, and pest management-related guidelines
2. Flexible recommendations for field-specific adjustments by farmers
   - Opportunities for farmer experimentation based on if-then type of rules

The promotion of improved fertilizer recommendations in a domain may require demonstration trials (e.g., fertilizer addition plots) and other means of direct interaction between farmers and extension staff. Where suitable, fixed recommendations should be avoided and replaced by guidelines that encourage farmers to apply newly acquired knowledge by providing guidelines of the if-then type. For example, the following message could be promoted in areas and seasons with high-yielding hybrid rice: "If crop stand is good and pest pressure low, apply a late N dose at flowering to enhance grain filling." Many if-then rules have been developed to be adapted and incorporated into local strategies, including rules for the identification of inefficient fertilizer N use and unbalanced nutrition, selection of suitable yield targets, fine-tuning of fertilizer N applications, and adjustment of fertilizer N, P, and K rates to various crop management practices (Balasubramanian 1999a,b, Dobermann and Fairhurst 2000, Fairhurst and Witt 2002).

18.3 Conclusions

The principles of site-specific nutrient management presented were developed in partnership with NARES on the basis of on-farm research and evaluation of SSNM in the workgroups Reaching Toward Optimum Productivity (RTOP) and Impact of the Irrigated Rice Research Consortium (IRRC). The strategies outlined can accommodate local adaptation across a wide range of cropping conditions in rice-based systems and integrated approaches involving the use of organic nutrient sources. Local adaptation mainly involves adjusting the nutrient management principles discussed to specific crop rotations, major germplasm differences, crop management practices, climatic seasons, and available nutrient sources.

These SSNM principles could not have been developed without several years of on-farm and on-station research collecting data that were used to further improve strategies and guidelines. We have captured this evolution and documented the development and evaluation of strategies in many chapters of this book. Research continues to evaluate the simplified and refined approaches summarized in this and the previous chapter. Major research issues are

1. The on-farm comparison of N management options,
2. The validation of fertilizer P and K requirements in major irrigated rice domains,
3. Differences in plant nutrition between inbred and hybrid rice, including genotypic variation of nutrient requirements, long-term changes in soil P and K, and evaluation of the yield gain + nutrient balance approach for P and K,
4. Improved algorithms to estimate fertilizer recovery efficiencies,
5. The development of recommendation domains for major irrigated rice areas as and where required,
6. Development of integrated crop and resource management strategies, including nutrient × pest interactions, and
7. The development of suitable delivery strategies. Even with all these researchable issues, the SSNM principles are mature and well positioned for widerscale validation and farmer evaluation.

The SSNM strategy can be disseminated at the community or village level as there are substantial opportunities to reduce the complexity of the SSNM approach by developing guidelines and recommendations that are valid for larger areas and allow farmers to learn through experimentation and make appropriate field-specific decisions. This will require investments in training, on-farm evaluation, and promotion of improved strategies at both the regional and local level, but it promises considerable increases in fertilizer-use efficiency, productivity, and profit at the farm and household level.

IRRI is involved in wider-scale farmer evaluation and adaptation of SSNM through partnership with NARES, as part of the IRRC (www.irri.org/irrc), and through the Rice-Wheat Consortium (RWC). Interdisciplinary NARES teams are involved in on-farm evaluation of innovative nutrient management strategies in Bangladesh, China, India, Indonesia, Myanmar, Nepal, Pakistan, Thailand, and Vietnam. The involvement of public- and private-sector partners is being strengthened to facilitate the dissemination of information and delivery of SSNM to rice farmers.

References

Adhikarya R. 1994. Strategic extension campaign: a participatory oriented method of agricultural extension. Rome (Italy): FAO. p 1-209.

Balasubramanian V. 1999a. Farmer adoption of improved nitrogen management technologies in rice farming: technical constraints and opportunities for improvement. In: Balasubramanian V, Ladha JK, Denning GL, editors. Resource management in rice systems: nutrients. Dordrecht (Netherlands): Kluwer Academic Publishers. p 153-165.

Balasubramanian V. 1999b. Use of leaf color chart (LCC) for N management in rice. CREMNET Technology Brief No. 2. Los Baños (Philippines): International Rice Research Institute. p 1-4.

Balasubramanian V, Bell MA, Marcotte P. 2001. Needs and opportunity assessment (NOA): a prerequisite for understanding farmers' production systems, constraints/problems and opportunities in a target area (unpublished training manual). Los Baños (Philippines): International Rice Research Institute. p 1-34.

Balasubramanian V, Morales AC, Cruz RT, Abdulrachman S. 1999. On-farm adaptation of knowledge-intensive nitrogen management technologies for rice systems. Nutr. Cycl. Agroecosyst. 53:59-69.

Balasubramanian V, Morales AC, Cruz RT, De NN, Tan PS, Zaini Z. 2000. Leaf color chart (LCC): a simple decision tool for nitrogen management in lowland rice. Poster presented at the American Society of Agronomy meeting, Minneapolis, Minnesota, 5-9 November 2000.

Bellon MR. 2001. Participatory research methods for technology evaluation: a manual for scientists working with farmers. El Batán (Mexico): CIMMYT. p 1-80.

Bijay-Singh, Yadvinder-Singh, Ladha JK, Bronson KF, Balasubramanian V, Jagdeep-Singh, Khind CS. 2002. Chlorophyll meter- and leaf color chart-based nitrogen management for rice and wheat in Northwestern India. Agron. J. 94:821-829.

Buresh RJ, Balasubramanian V, Witt C, Ladha JK, Peng S, Dawe D, Morin S. 2001. Development and delivery of nutrient management innovations for lowland rice farmers. Paper presented at the Workshop on Integrated Management for Sustainable Agriculture, Forestry, and Fisheries, 28-31 Aug. 2001. CIAT, Cali, Colombia.

Buresh RJ, Peng S, Witt C, Balasubramanian V, Laureles EV. 2002. Improving nitrogen fertilizer use efficiency for rice through site-specific nutrient management. In: International Rice Congress, Innovation, Impact, and Livelihood, 16-20 September 2002, Beijing, China. Abstracts, p 399.

Buresh RJ, Witt C, Balasubramanian V, Peng S, Dobermann A, Ladha JK. 2003. The principles of site-specific nutrient management for rice. Poster presented at the IFA/FAO Conference on Global Food Security and the Role of Sustainable Fertilization, Rome, Italy, 26-28 March 2003. (Abstract only.)

Cabrera-Pasuquin JMCA, Witt C. 2003. Leaf color and leaf color charts for efficient N management in rice: How do they compare? Poster presented at the 17th Annual Scientific Conference on Crop Science and Technology: Key to Global Competetiveness. Federation of Crop Science Societies in the Philippines (FCSSP), 22-25 April 2003, Aklan, Philippines.

Deugd M, Roling N, Smaling EMA. 1998. A new praxeology for integrated nutrient management facilitating innovation with and by farmers. Agric. Ecosyst. Environ. 71:269-
283.

Dobermann A, Fairhurst T. 2000. Rice: nutrient disorders and nutrient management. Singapore
and Los Baños (Philippines): Potash & Phosphate Institute (PPI), Potash & Phosphate
Institute of Canada (PPIC), and International Rice Research Institute (IRRI). 191 p.

Dobermann A, Witt C, Abdulrachman S, Gines HC, Nagarajan R, Son TT, Tan PS, Wang GH,
Chien NV, Thoa VTK, Phung CV, Stalin P, Muthukrishnan P, Ravi V, Babu M, Simbahan
GC, Adviento MA. 2003a. Soil fertility and indigenous nutrient supply in irrigated
rice domains of Asia. Agron. J. 95:913-923.

Dobermann A, Witt C, Abdulrachman S, Gines HC, Nagarajan R, Son TT, Tan PS, Wang GH,
Chien NV, Thoa VTK, Phung CV, Stalin P, Muthukrishnan P, Ravi V, Babu M, Simbahan
GC, Adviento MA, Bartolome V. 2003b. Estimating indigenous nutrient supplies for
site-specific nutrient management in irrigated rice. Agron. J. 95:924-935.

Dobermann A, Witt C, Dawe D, Abdulrachman S, Gines HC, Nagarajan R, Satawathananont
S, Son TT, Tan CS, Wang GH, Chien NV, Thoa VTK, Phung CV, Stalin P,
Muthukrishnan P, Ravi V, Babu M, Chatuporn S, Sookthongsa J, Sun Q, Fu R, Simbahan
GC, Adviento MAA. 2002. Site-specific nutrient management for intensive rice crop-
ping systems in Asia. Field Crops Res. 74:37-66.

Douthwaite B, Keatinge JDH, Park JR. 2002. Learning selection: an evolutionary model for
understanding, implementing and evaluating participatory technology development.
Agric. Syst. 72:109-131.

Fairhurst T, Witt C. 2001a. Fertilizer Chooser. Software. Pre-release version as MS Excel
spreadsheet model (online). Available at www.irri.org/irrc/nutrients (last update 11
Oct. 2001; accessed 16 May 2003). Singapore and Los Baños (Philippines): Potash
and Phosphate Institute & Potash and Phosphate Institute of Canada, and International
Rice Research Institute.

Fairhurst T, Witt C. 2001b. Profit Calculator (gross margin analysis). Software. Pre-release
version as MS Excel spreadsheet model. Version 1.0. Singapore and Los Baños (Philippines): Potash and Phosphate Institute and Potash and Phosphate Institute of Canada,
and International Rice Research Institute.

Fairhurst T, Witt C, editors. 2002. Rice: a practical guide to nutrient management. Singapore
and Makati City (Philippines): Potash & Phosphate Institute, Potash & Phosphate Institute of Canada (PPIC), and International Rice Research Institute (IRRI). 89 p.

FAO. 2000. Guidelines for participatory diagnosis of constraints and opportunities for soil
and plant nutrient management. Rome (Italy): FAO. 98 p.

Furuya S. 1987. Growth diagnosis of rice plants by means of leaf color. Jpn. Agric. Res. Q.
20:147-153.

Haefele SM, Wopereis MCS, Donovan C. 2002. Farmers' perceptions, practices and performance in a Sahelian irrigated rice scheme. J. Exp. Agric. 38:197-210.

Haefele SM, Wopereis MCS, Donovan C, Maubuisson J. 2001. Improving the productivity
and profitability of irrigated rice production in Mauritania. Eur. J. Agron. 14:181-196.

IRRI. 1999. Use of leaf color chart (LCC) for N management in rice. CREMNET Technology
Brief No. 2. Los Baños (Philippines): International Rice Research Institute (IRRI).
p 1-4.

Pandey S. 1999. Adoption of nutrient management technolgies for rice production: economic
and institutional constraints and opportunities. Nutr. Cycl. Agroecosyst. 53:103-111.

Peng S, Garcia FV, Laza RC, Sanico AL, Visperas RM, Cassman KG. 1996. Increased N use efficiency using a chlorophyll meter on high-yielding irrigated rice. Field Crops Res. 47:243-252.

Pingali PL, Hossain MZ, Pandey S, Price LL. 1998. Economics of nutrient management in Asian rice systems: towards increasing knowledge intensity. Field Crops Res. 56:157-176.

Witt C, Balasubramanian V, Dobermann A, Buresh RJ. 2002a. Nutrient management. In: Fairhurst T, Witt C, editors. Rice: a practical guide for nutrient management. Singapore and Los Baños (Philippines): Potash and Phosphate Institute & Potash and Phosphate Institute of Canada and International Rice Research Institute. p 1-45.

Witt C, Buresh RJ, Balasubramanian V, Dawe D, Dobermann A. 2002b. Improving nutrient management strategies for delivery in irrigated rice in Asia. Better Crops Int. 16(2):10-17.

Witt C, Cabrera-Pasuquin JMCA, Mutters R, Peng S. 2003. Nitrogen content of rice leaves as predicted by SPAD, NIR, spectral reflectance and leaf color charts. Paper presented at the ASA-CSSA-SSSA Annual Meeting on Changing Sciences for a Changing World: Building a Broader Vision, 2-6 November 2003, Denver, Colorado, USA. (Abstract only.)

Yang W-H, Peng S, Sanico AL, Buresh RJ, Witt C. 2003. Estimation of leaf nitrogen status using leaf color charts in rice. Agron. J. 95:212-217.

Notes

Authors' addresses: C. Witt, R.J. Buresh, V. Balasubramanian, D. Dawe, A. Dobermann, International Rice Research Institute, Los Baños, Philippines; A. Dobermann, University of Nebraska, Lincoln, USA.

Acknowledgments: We are grateful for the many suggestions made by NARES collaborators and IRRI colleagues on the principles and promotion of SSNM, and we particularly thank Dr. S. Haefele for his helpful comments.

Citation: Dobermann A, Witt C, Dawe D, editors. 2004. Increasing productivity of intensive rice systems through site-specific nutrient management. Enfield, N.H. (USA) and Los Baños (Philippines): Science Publishers, Inc., and International Rice Research Institute (IRRI). 410 p.